Submarine Telegraphy and the Hunt for Gutta Percha

Global Economic History Series

VOLUME 15

The titles published in this series are listed at *brill.com/gehs*

Submarine Telegraphy and the Hunt for Gutta Percha

Challenge and Opportunity in a Global Trade

By

Helen Godfrey

BRILL

LEIDEN | BOSTON

Cover illustration: Collecting gutta percha in the Baram district of Sarawak, Borneo.

Library of Congress Cataloging-in-Publication Data

Names: Godfrey, Helen L., author.
Title: Submarine telegraphy and the hunt for gutta percha : challenge and opportunity in a global trade / by Helen Godfrey.
Description: Leiden ; Boston : Brill, [2018] | Series: Global economic history series, ISSN 1872-5155 ; volume 15 | Includes bibliographical references and index.
Identifiers: LCCN 2017056318 (print) | LCCN 2018002929 (ebook) | ISBN 9789004357280 (Ebook) | ISBN 9789004344334 (hardback : alk. paper)
Subjects: LCSH: Cables, Submarine--History. | Gutta-percha--Economic aspects--Asia--History. | Telegraph--Economic aspects--History.
Classification: LCC HE7711 (ebook) | LCC HE7711 .G63 2018 (print) | DDC 382/.4567863--dc23
LC record available at https://lccn.loc.gov/2017056318

Typeface for the Latin, Greek, and Cyrillic scripts: "Brill". See and download: brill.com/brill-typeface.

ISSN 1872-5155
ISBN 978-90-04-34433-4 (hardback)
ISBN 978-90-04-35728-0 (e-book)

This book is printed on acid-free paper and produced in a sustainable manner.

Contents

List of Illustrations

Figures

Maps

Diagram

Plates

Tables

Appendix 2 Statistical Material

Abbreviations

BCL	Borneo Company Limited
BSP	British Sessional Papers (Parliamentary Papers)
CO	Colonial Office
Cwt	Hundredweight (see Weights below)
FO	Foreign Office
GP	Gutta Percha
IRGPETJ	India Rubber and Gutta Percha and Electrical Trades Journal
IRW	The India Rubber World
NIM	Nederlandsch-Indische Maatschappij van Nijverheid en Landbouw
SG	Sarawak Gazette
SMH	The Sydney Morning Herald
SDSS	Singapore Directory for the Straits Settlement
SFPMA	Singapore Free Press and Mercantile Advertiser
SSD	Straits Settlements Directory
ST	Straits Times
STSJC	Straits Times and Singapore Journal of Commerce
TCM	Telcon Archives
TJER	Telegraphic Journal and Electrical Review
USHO	U.S. Hydrographic Office

Currency Values, Weights and Measurements

Currency

Sarawak Dollars	Tied to the Straits Settlements Dollar. Sarawak trade data is given in Sarawak Dollars.

Weights

cwt	Hundredweight. Unit of Imperial weight: it equaled 112 pounds. In Imperial weight 20 cwt = 1 ton or 2240 pounds
lb	Pound. Unit of Imperial weight, approximately 0.45 kilograms
Kati	Unit of weight; 100 katis = 1 pikul
Pikul	Unit of Weight =133.3 Pounds (lbs.) or 60.5 kgs.

Measurements

N.M.	Nautical Mile, maritime measurement also used for submarine telegraph cables: 1 n.m. = 1.852 kilometers or approximately 1.15 statute miles.

Glossary

Adat	Customary law
Antu or Hantu	Spirit/Ghost
Bala	War party or fleet of boats on hostile expedition in Sarawak
Bandong	Chinese trading boat
Bejalai	Iban expedition to earn wealth and prestige
India rubber	Nineteenth century term for latex, a generic term sometimes covering para rubber but often used to refer simply to indigenous latex collected in Southeast Asia and other regions
N.M.	Nautical Mile (see Measurements)
Padi	Unhusked rice
Pangeran	Brunei aristocrat
Para rubber	*Hevea brasiliensis*
Penghulu	Village headman
Pikul	Unit of Weight (see Weights)
Prahu/Perahu	Sailing boat
Pusaka	Prestige or heirloom items, perhaps inhabited by spirits or *antu*
Straits Settlements	Combined colonies of Singapore, Malacca and Penang
Tampun	Iban collection of human trophy heads
Temenggong	Non-hereditary ruler or official
Towkay	Leading Chinese merchant

Introduction

> The Gutta Percha is probably a complicated substance[1]
>
> MICHAEL FARADAY 1848

Gutta percha is now largely forgotten and the history of its trade not widely explored. Yet for decades it was a product of such strategic and economic significance that in 1899 the Director of the Royal Botanic Gardens at Kew observed that the supply of gutta percha was 'a matter of interest to the whole civilized world'.[2]

In the mid-nineteenth century, a communication revolution came about with the development of the telegraph, in particular, the submarine telegraph. Telegraph cables transmitted messages by electrical impulses, so truly global telegraphic links could not be achieved until a way was found for cables to be insulated so that they could be laid across the ocean floor. Eventually an effective material was identified for this purpose – gutta percha, a latex collected from trees growing in parts of Southeast Asia. Contact around the world that once could have taken months now might take just minutes. International communication networks developed quickly after the first successful international cable was laid in 1851, such that by 1903 there were approximately 230,000 nautical miles of submarine telegraph cables connecting many countries across the world, with onward connections to overland networks ensuring links to even out of the way locations.[3] Almost every cable was insulated with gutta percha. This new era of global communications was in large measure dependent upon the supply of gutta percha and upon the people who collected and traded it. Gutta percha was the most effective insulator for submarine telegraph cables because it had better tensile strength and the ability to withstand cold, damp and saltwater when compared with other electrical insulators. This made gutta percha the ideal solution for the requirements of both insulating

1 M. Faraday, *Faraday's Diary, Volume 5, September 6, 1847 – October 17, 1851*, London, G Bell and Sons Ltd, Rep. 1934, 1848, p. 45.

2 W.T. Thiselton-Dyer, Correspondence to C.P. Lucas (Colonial Office), in Royal Botanic Gardens Kew, Archives, *Kew: Miscellaneous Reports, Gutta Percha*, 1899, 26 December, f.45.

3 See Appendix 2, Table 1 for details.

 | DOI 10.1163/9789004357280_002

and laying submarine telegraph cables. There was little likelihood of anything other than a continuing demand for this product that insulated undersea cables until an alternative product or technology was developed – but that took well over half a century to occur.

Submarine telegraphy was based on a new technology and on scientific study that was still in its developmental stages. Cable manufacture presented, then, both challenges and opportunities since it was a profitable, but expensive and somewhat risky undertaking. During the first few years after the 1851 cable was laid, the scientific study of electricity was at a relatively formative stage and the procedures and materials required to produce quality cables were not fully understood. As a result, there were some highly expensive cable failures (despite many successes) until 1859 when the British Board of Trade established a Joint Committee to investigate the entire question of submarine telegraphy. One of the Committee's conclusions was an endorsement of gutta percha as the best insulator for submarine cables.[4] The Committee's Report ensured a demand for considerable quantities of gutta percha, and that the quality of gutta percha would be addressed by cable manufacturers who would establish rigorous testing regimes for its use. After that time, the manufacturing processes used in the production of submarine cables saw relatively few changes; partly because many of the problems had been solved, but also because the industry and its customers took a conservative approach to change due to a fear of more expensive failures.[5]

Solving the scientific and production issues associated with the manufacture of submarine cables using gutta percha insulation was just part of the challenge facing the development of global telegraphic communication. Because gutta percha was found only in the forests of Southeast Asia, its supply was limited, yet the demand for it continued unabated for decades – a challenge for cable producers and, in the wider sense, for the international community because of its growing dependence upon telegraphic communication. The limited area of supply also presented an opportunity for those in the region supplying it because it concentrated the trade on a relatively small number of commercial networks. The development of submarine telegraph communications set in train events that were to alter the lives of many people living in Southeast Asia – around Singapore, from where most of the world's supply of gutta percha was exported – and in the areas from which it was

4 *Report of the Joint Committee Appointed by the Lords of the Committee of the Privy Council for Trade and the Atlantic Telegraph Company to Inquire into Construction of Submarine Telegraph Cables, Together with the Minutes of Evidence and Appendix*, 1861, Command Papers, [2744] LXII.591, HMSO, London, p. xxv.

5 See Chapter 2.

collected for export, such as Borneo, and Sarawak, where some of the most skilled collectors of gutta percha lived. The latex was not easy to locate and collect; it was mainly indigenous collectors who had the knowledge and skill required to bring it from the forest for trade to the outside world.

Continued supply was complicated by an additional challenge – the problem of collecting the gutta percha in a sustainable manner. The trees that supply gutta percha do not possess a cell structure of interconnecting 'tubes' carrying the latex. As a result, incisions in a tree supplying gutta percha provided latex only from very proximate areas of the tree – in effect 'the bleeding ... is less free'.[6] This made tapping of the gutta percha tree to collect the latex extremely difficult. The most practical way for forest collectors to work was to fell the tree and make multiple incisions into the bark. The common practice was simply to make the incisions on the upper side of the felled tree trunk as it lay on the ground, leaving the underside untouched, as can be seen in the photograph below at Plate 1 of Kayan collectors in Sarawak.

PLATE 1 *Kayan Gutta Percha collectors, Sarawak*
Four Kayan collecting Gutta Percha from a tree.
PHOTOGRAPH C. 1896, CHARLES HOSE, V0037406.
Source: Wellcome Library, London, With Permission.

6 I.H. Burkill (ed.), with contributions by W. Birtwistle, F.W. Foxworthy, and J.B. Scrivenor, A *Dictionary of the Economic Products of the Malay Peninsula*, 2 Vols, London, Crown Agents, 1935, II, p. 1624.

This meant not only that the collecting practice was not sustainable because the trees were cut down, but also that the yield obtained was far less than the potential yield because the gutta percha was being collected from only part of the tree. Even mature trees provided only small amounts: sometimes as little as 250 grams or about half a pound of gutta for twenty-year-old trees and some telegraph cables were manufactured with up to 400 pounds of gutta percha per nautical mile.[7] Almost every mile of cable was insulated with gutta percha collected in this way. It is possible that around twenty-seven million trees were felled to collect gutta percha in the half century after the first cable was laid.[8] The process was arduous. Expeditions into the forest to collect gutta percha could take months, even years. It was dangerous work: resource disputes were a feature of the trade. Some collectors in Borneo were headhunters; but, for those willing to take the risk, the trade was an important means of obtaining both subsistence and prestige goods. At the local level, the challenges and opportunities presented by the trade were many.

Notwithstanding the limitations of supply and unsustainable methods of collection, commercially successful plantations were not established for many decades. In fact, during the great era of the gutta percha trade considered in this book, *none* were in operation. The trees that supplied gutta percha were slow growing and it was thought they would take up to thirty years before they would be commercially productive. This fact and low yield were thought to make the trees unprofitable for cultivation by small holders or larger European operations. These factors contributed to a lack of investment in commercial plantations. This was despite the fact that the lack of a sustainable source of supply could have put at risk the long-term operation of worldwide telegraphic communication. The collectors and traders of gutta percha at the periphery of the global economy were in large measure left to manage the supply of this product.

A comparative analysis of evidence from commercial archives, trade data and colonial government reports collected for this study reveals a history of a trade that was shaped by the decisions of metropolitan manufacturers which left the operation of much of the trade in the hands of the suppliers in Southeast Asia. In that region and in Sarawak, prevailing conditions that included effective trade networks, better local knowledge, and colonial administrations

7 W.R. Tromp de Haas, *The Netherlands India Government Gutta-Percha Plantations in Java*, Tjipetir, Batavia Government Gutta-Percha Plantations, 1908, p. 11 and E.F.A. Obach, 'Cantor lectures on gutta percha', *Journal of the Society of the Arts*, 1897, p. 102.

8 See Appendix 1.

that relied upon local elites, laid the basis for the operation of the trade and power positions within it. The gutta percha trade differed from many other colonial commercial ventures. It was not part of the plantation system of production and did not attract foreign capital or the entry of migrant labor. In Sarawak, the trade was carried out under a unique system of government, where the ruling Englishmen James and later Charles Brooke, known as the White Rajahs of Sarawak, owned the nation as their own personal colony after it was granted to them by the Sultan of Brunei. Aside from the modest impost of export duties, the Brookes did not impose any controls on forest trade so, unlike the trade in many other commodities, the gutta percha trade was largely left beyond direct political intervention. Instead, the Brookes had other interests and concerns which both fostered and benefited from the trade.

While it is difficult to determine exact figures, it seems that even in a typical year of trade, hundreds of thousands of trees were felled by Sarawak gutta percha collectors.[9] The massive effort involved in collecting, carrying, trading and shipping the product from remote forests to trade venues and local ports indicates not only a profitable trade, but one that engaged much of the population. The collector envisaged himself as of the forest, not merely in the forest. For collectors of gutta percha, the forest was an environment that blended the natural and the supernatural – most collectors were animists and some practiced headhunting. The gutta percha trade was used as a means of obtaining both practical and prestige goods of social, cultural and spiritual significance. Chinese traders in Sarawak possessed resources and linkages with local and regional trade networks to connect the hinterland trade with the external market based in Singapore. The trade was built upon the skills and knowledge of each group and on the bonds that were formed among them. This led to material gains for the Sarawak economy – but it was also the catalyst for social change, inter-dependence and hostilities among groups, as well as political opportunism. Local conditions were affected by economic and political regimes in Singapore and Sarawak and those factors need to be considered to understand why actors behaved as they did.

Approaches and Scope

While this is a history of the challenges and opportunities arising from the development of submarine telegraph technology and the associated trade of

9 See Appendix 1 and *Scale of the Trade in Sarawak*, Appendix 2.

gutta percha, it is also a history that maps a process of globalization in the nineteenth century. This is done by drawing out the ways in which technology change and its associated trade linked groups from vastly different communities and cultures. It considers relationships among peoples from these very different environments: those from the world of submarine cable manufacture and the cutting-edge science of its era; the world of the forest collectors who were animists, some even headhunters, and the network of Chinese traders who linked those two very different worlds. The book shows how relationships and trading patterns for those involved in a global trade network were shaped by factors were far more complex than simple commercial gain.

The book has two core sections, followed by a conclusion.

In *Part 1, Submarine Telegraphy: The Forces of Science and Commerce*, the book describes the development of the new technology of submarine telegraphy, and the challenges and opportunities arising from the application of that technology. Chapter 1 begins with an outline of gutta percha itself and its earliest uses. Next follows an overview of the significance of submarine telegraphy – explaining why gutta percha came to be a product of such importance in the nineteenth century world. Chapter 2 then sets out the development of submarine cable connections and technology; the rapid growth and pressures experienced in a conservative industry trying to balance opportunities with the challenges of technology development and commodity supply. What is revealed is that once a reliable method of cable production was developed, few wanted to experiment with what were seen as risky alternatives, resulting in a continued dependence upon gutta percha insulation. Chapters 3 and 4 connect these pressures with the commercial operations of cable producers, and with the economies of the semi-periphery and periphery that were linked in this global trade: from Singapore, where most of the world supply of gutta percha was traded, and then the collecting and trading area of Sarawak. Behaviors and network relationships reveal how the trade was dependent upon those at the periphery. Matching documentary, commercial and trade data reveals firstly the interconnection between colonial and local political and community structures. Secondly, it reveals the speed with which trade and social connections responded to impetus from the metropolitan centres.

In *Part 2, Power Profit and the Periphery*, Chapters 5–8 focus directly on the collectors and traders at the periphery, using Sarawak as a case study. This considers the peculiarities of the commodity, the actors involved in the trade, and the environment in which it was found. This section could be considered a trade ecology: a study of the interaction between a community of actors and the habitat that shaped the commodity trade. The interaction among actors in the trade network is set out, and the interdependence between these actors and the tools of economic change, such as the growth of credit, barter, monetization and the role of bazaars is a key factor in this development. The role of the indigenous communities, seen through the significance of overland trade paths, local cultural and spiritual drivers such as the use of omen, head-hunting and the rise of resource disputes occasioned by the gutta percha trade, are set out.

Parts 1 and 2 are drawn together in a *Conclusions* Chapter. Technical and statistical information is attached in the *Appendices*. Information about gutta percha characteristics and collection is at Appendix 1. Trade statistics collected as the basis of the graphs showing trends in imports or exports are at Appendix 2.

The approach adopted in this book is one that envisages the trade in the context of a set of global to local connections, of mutual dependence among actors be they metropolitan, intermediary or peripheral. The book sets out the strategies adopted by metropolitan manufacturers, and local collectors and traders; these were shaped by commercial, geographic, political, social, cultural and even spiritual factors – and of course by the unique characteristics of gutta percha itself.

Using previously uncollated evidence from commercial sources and trade figures provides a thread of evidence that weaves a picture of the gutta percha commodity chain. That chain is proposed as a set of transnational processes extending from the periphery of 'local' collection in Sarawak, through a regional distribution centre in Singapore (the semi-periphery) and on to production in Britain, Europe and the U.S. (the 'metropolitan core'). Many studies of the period or of the region have focused on trade in minerals or plantation based commodities (such as tin or para rubber) where collection and trade was largely *controlled* by outsiders rather than, as in this case, studying indigenous forest produce where outsiders had almost no role in the collection phase. This study targets the response of local actors to the demands of the global economy where the motives and opportunities were embedded in their traditional methods of resource extraction rather than in introduced systems such as foreign owned companies or mines. Importantly, in Sarawak indigenous collectors had the right to exploit this forest resource – there was no contest with outsiders over who owned those resources. What emerges from the history is that, to comprehend the nature of global trade and its effects, it is essential to understand local relationships and responses as well as metropolitan motivations. That is, global history is not best understood as simply a two-dimensional study of the developed and less developed world, or of basic transactions of demand and supply.

Debates about the definitions, nature and impact of globalization are now long running, cross-disciplinary and long-rehearsed in many publications. Within the field of economic history, definitions such as those put forward by Williamson (with others) focus on what de Vries called 'hard' definitions of globalization as being the process by which commodity and factor prices converge across continents;[10] with O'Rourke and Williamson defining it as 'the integration of markets across space'. This they saw as a phenomenon that became 'economically meaningful' only in the early nineteenth century.[11] For others, changes in economic relations prompt changes additional to price convergence and conversely changes in economic relations are based on a multitude of factors, many of which are not, for example, economic.[12] Flynn and Giraldez propose that globalization is the point at which continents begin to 'exchange products continuously ... directly and indirectly ... on a scale that generated deep and lasting impacts on all trading partners'.[13] The need to embrace other causes of and forms of globalization have demanded other definitions. Held et al. concluded that globalization is best defined as 'a process (or set of processes) which embodies a transformation in the spatial organization of social relations and transactions – assessed in terms of their extensity, intensity, velocity and impact – generating transcontinental or interregional flows and networks of activity, interaction, and the exercise of power'.[14] The definition caters for various forms, causes and outcomes of globalization but is arguably not the smoothest definition with which to grapple.

In response to criticisms about the narrowness of their definition, and their application of it, O'Rourke and Williamson respond that 'The impact of globalization on the world economy will never be properly assessed until we agree on exactly how that impact is to be measured'.[15] This aptly sums up the dilemma – different scholars and different approaches will want to measure

10 J. de Vries, 'The limits of globalization in the early modern world', *The Economic History Review*, New Series, Vol. 63, No. 3, August 2010, p. 714.

11 K.H. O'Rourke, and J.G. Williamson, 'Once more: When did globalisation begin?' *European Review of Economic History*, Vol. 8, No. 1, April 2004, p. 109.

12 See for example: C. Antunes, and K. Fatah-Black, (eds), *Explorations in History and Globalization*, Abingdon, Routledge, 2016; F. Cooper, 'What is the Concept of Globalization Good for?' An African Historian's Perspective', *African Affairs*, Vol. 100, No. 399, April 2001, pp. 189–213; A.G. Frank, *ReOrienting the 19th Century. Global Economy in the Continuing Asian Age*, Boulder, Paradigm Publishers, 2014 and D. Held, A. McGrew, D. Golblatt, and J. Perratton, *Global Transformations: Politics, Economics and Culture*. Stanford, Stanford University Press, 1999.

13 D.O. Flynn, and A. Giráldez, 'Path dependence, time lags and the birth of globalisation: A critique of O'Rourke and Williamson', *European Review of Economic History*, Vol. 8, No. 1 (April), 2004, p. 83.

14 Held et al., p. 16.

15 O'Rourke and Williamson, *European Review of Economic History*, p. 114.

different types of impact because they are interested in different types of impact. Historians, economists and social scientists may never agree on a definition, but it could be argued that the process of debate and counter-debate is in itself adding to the understanding of this vexed yet important question. As Held et al. concluded, any account of a process of globalization must 'acknowledge that, far from being a singular condition, it is best conceived as a differentiated and multi-faceted process'.[16]

The book does not seek to resolve the debate about definitions of globalization. Rather the intention is to add a dimension to the understanding of it by writing about an experience of globalization in a new technology and an associated trade that brought together an unusual combination of challenges and widely disparate economic, social and cultural features. Ultimately, to set a framework for this book, there was clearly a need to embrace a definition of globalization and the one used here is in fact not one from an economic historian.

Globalization in the nineteenth century in the context of *this* study is considered to be a process of change caused in part by the dependence of metropolitan economies on less developed economies for supply of raw materials – in effect, economic globalization – but with the nature and impact of this process linked to the social and cultural experiences in these less developed economies. This can be seen as globalization in its widest sense, as a socio-economic and cultural phenomenon. As will be shown, the speed with which the demand for cable communication was adopted, the fear of failure and the speed with which demand for gutta percha could be met from the semi-periphery and periphery – and the vastly different cultural perspectives of those linked in this trade – make this an interesting case study in both economic, and *socio-economic* forms of globalization. In the broader sense of considering the impact, responses and transformations which arose from interrelationships, Giddens' definition of globalization has much to offer because of its all-encompassing sense of change, and for its succinctness. He defines globalization as an 'intensification of world-wide social relations which link distant localities in such a way that local happenings are shaped by events occurring many miles away and vice versa'.[17] Tucker, in expanding on Giddens, thought that it was the 'concomitant transformation of the local and

16 Held et al., p. 27.

17 A. Giddens, *The Consequences of Modernity*, Stanford, CT, Stanford University Press, 1990, p. 64.

personal contexts of social experience' after influence from distant institutions and events that are a hallmark of globalization.[18]

The point taken up in this work is that it is the shaping of local events by distant influences that is pertinent – it is the concept of network and social (not just economic) linkages over distance, rather than an emphasis on the dominant role of the metropolitan, that holds sway. This study reveals how linkages between both *metropolitan* and *local* institutions and events should not be overlooked when looking to achieve a sense of the experience of globalization. Also, by rethinking the sometimes forgotten aspects of the periphery, by matching historic discourse and trade data, it is possible to gain a useful entry into that understanding. By this means – giving a voice to the periphery – we can lay the groundwork for a better sense of the 'world-wide', of globalization in its broadest sense.

One approach for examining global trade networks – or in the wider sense, global history – has been through world-system analysis, in particular as put forward by Wallerstein.[19] That approach conceives of a world-system linking western developed economies (metropolitan centres) with under-developed 'local' economies considered to be at the periphery of the world-system. The world-system model emphasises the role of the core in dominating the flow of capital and investment, with the periphery supplying resources and labor for the core. The approach is a useful way to highlight the importance of interrelationships, particularly during the era of colonial economic activity; but it does not necessarily assist, on its own, in understanding the impact of trade on the periphery, or the role and responses of local actors. Commodity chain analysis envisages trade as a network of labor and production processes that transform a product from source to a finished product.[20] Commodity chain analysis was initially conceived as part of the world-system approach but

18 K.H. Tucker, *Anthony Giddens and Modern Social Theory*, London, Sage Publications, 1998, p. 144.

19 I. Wallerstein, *The Modern World-System I. Capitalist Agriculture and the Origins of the European World-Economy in the Sixteenth Century*, New York, Academic Press, 1974; *The Modern World-System III. The Second Era of Great Expansion of the Capitalist World Economy, 1730–1840s*. San Diego, Academic Press, 1989; T.K. Hopkins and I. Wallerstein, 'Commodity chains in the capitalist world-economy prior to 1800. Commodity chains: Construct and research', in *Commodity Chains and Global Capitalism*, eds G. Gereffi, and M. Korzeniewicz, Westport, Praeger, 1994, pp. 17–19; and I. Wallerstein, *The Modern World-System I: Capitalist Agriculture and the Origins of the European World-Economy in the sixteenth Century, with a New Prologue*, Berkeley, University of California, 2011.

20 See, for example, Hopkins and Wallerstein, *Commodity Chains and Global Capitalism*, pp. 17–18.

arguably has been utilized and hybridized as an approach in itself.[21] Dougherty proposed that commodity chain analysis is a process that conceptualizes a commodity as a series of sequential nodes that include the design, fabrication and various stages of assembly through to final consumption.[22] Examining the chain helps us to understand, for example, questions about power and the impact of people and organizations in the chain; prevailing economic or political conditions at points along the chain (in this case in Sarawak and Singapore) and, finally, to examine aspects of the modes and strategies adopted by local collectors and traders. The approach used here is to build on that suggested by Frank and Musacchio which proposes that we remain 'agnostic' about the question of distribution of profits between core and periphery, as such categorizations do not necessarily tell us much about the shape and operation of global supply chains.[23] A way of using commodity chain analysis recognizes it as a complement to the more traditional country or regional studies.[24] It is this complementary approach that is adopted herein: that is, an analysis that increasingly focuses on the local actors. The approach envisages the relationships among different actors in the commodity chain, and indeed identifies those actors and their inter-dependencies. Therefore, the commodity chain approach provides a framework for envisaging the connectivity between actors and processes along with the causal connections between demand and supply. The commodity chain approach also needs to be used with caution since it can become linear: merely following processes and networks from one end of the production chain to another will promote understanding only of inter-related network communications, but not a great deal about responses and motivations at the human level. The difficulty, then, with both the world-system and the commodity chain approaches is that they are useful concepts for better understanding systems issues, or power issues within a system, but not necessarily in understanding the role, responses and motivations of actors.

---

21 See J. Bair, 'Global Capitalism and Commodity Chains: Looking Back, Going Forward', *Competition and Change*, Vol. 9, No. 2, June 2005, pp. 153–180.

22 M.L. Dougherty, 'Theorizing Theory: Origins and Orientations of Commodity Chain Analysis', *The Global Studies Journal*, Volume 1, Number 3, 2008, p. 29.

23 Z. Frank and A. Musacchio, 'Brazil in the international rubber trade, 1870–1930', in *From Silver to Cocaine. Latin American Commodity Chains and the Building of the World Economy, 1500–2000*, eds S. Topik, C. Marichal and Z. Frank, Durham, Duke University Press, 2006, pp. 271–299, pp. 290, 293.

24 C. Marichal, S. Topik and Z. Frank, 'Conclusion: Commodity chains and globalization in historical perspective' in *From Silver to Cocaine. Latin American Commodity Chains and the Building of the World Economy, 1500–2000*, eds S. Topik, C. Marichal and Z. Frank, Durham, Duke University Press, 2006, p. 14.

This history is not simply about 'globalization' and systems, but about the people involved in the process of change; it is important to understand not only horizontal connections and relationships, but also to look more deeply into the actors and networks at chosen stages of the chain – Sverrisson suggested this when he called for the need to look 'local' as well as 'global'.[25] The methodology of this book proposes that both global (metropolitan) and local (including intermediary) responses need to be examined; neither has a context without the other. Some suggest there is a risk that the world-system concept could 'entail a fatal disrespect for culture … or agency, or the local'.[26] Edwards concludes world-system thinking is not an overarching template, and domination of the core should not be taken to mean that there is a 'lack of agency' at the periphery.[27] We do have tools to help understand the periphery. Here the linking of trade statistics and historic narrative are key tools; but, at a theoretical level, heuristic models of social capital are also used. Theories put forward by Putnam and others recognize the value (that is, the 'capital') that accrues to individuals and communities from the bonds of trust and sharing of resources that exist among groups.[28] Social capital is useful as a way of visualizing other features of the multi-dimensional aspects of peripheral economies and societies, since it emphasises not only the links that bind groups together but also those that conversely might exclude outsiders. Putnam suggests that social capital refers to the value that arises from social networks and the norms of reciprocity and trustworthiness that arise from them.[29] Conversely, such concepts allow for an understanding of how bonds of social capital might also exclude outsiders. The impact of the gutta percha trade was clearly a material one, but the wider significance in the Sarawak community was not one that might be measured in financial terms alone, but in also terms of changed social arrangements and engagement with the external economy that are considered later. Social capital within groups can have the effect of excluding outsiders: it

25 A. Sverrisson, 'Local and global commodity chains' in *Linking Local and Global Economies: The Ties That Bind*, eds C. Pietrobelli and A. Sverrisson, London, Routledge, 2003, pp. 17–35; and see for example M.R. Dove, *The Banana Tree at the Gate. A History of Marginal Peoples and Global Markets in Borneo*, New Haven, Yale University Press, 2011.

26 D. Palumbo-Liu, B.W. Robbins and N. Tanoukhi, eds, *Immanuel Wallerstein and the Problem of the World: System, Scale, Culture*, Durham, Duke University Press, NC, 2011, p. 5.

27 R.W. Edwards, 'The North American Fur Trade World System', *Field Notes: A Journal of Collegiate Anthropology*, 1 (1), 2009, p. 61.

28 R.D. Putnam, *Bowling Alone*, New York, Simon & Schuster, 2000.

29 Putnam, p. 19, see also M. Woolcock and D. Narayan, 'Social capital: Implications for development theory, research, and policy', *The World Bank Research Observer*, (August), Vol. 15, no. 2, 2000, pp. 225–249.

is a useful concept in examining aspects of hostilities (gutta percha 'wars') and differences that arose in the gutta percha trade, as well as the bonds between various groups of collectors and traders (indigenous and Chinese). Baron, Field and Schuller, in setting out the benefits and limitations of various approaches to the use of 'social capital', draw out the idea that the world is made up of interaction and complexity and that using the concept of social capital 'offers a purchase', in effect a way of taking a snapshot of relationships without needlessly deconstructing them, so they stress the heuristic value as its strongest merit.[30] Such an approach is used here.

The focus of the book is the period that might be considered the 'gutta percha era', that is, from the beginnings of the commercial exports of gutta percha in the early 1840s, particularly from 1851 when the first successful international submarine cable was laid with gutta percha insulation, until the early twentieth century (1903) after which the demand for gutta percha began to fall because of a reduction in the number of gutta percha insulated cables being laid. Although gutta percha was collected and traded in wider areas of Southeast Asia, this book considers the trade that emanated from Sarawak and Singapore – the aim of the approach is to examine a global commodity chain and to highlight the interconnection between indigenous economies and intermediary and global trades.

30 S. Baron, J. Field and T. Schuller (eds), *Social Capital: Critical Perspectives*, Oxford, Oxford University Press, 2000, p. 14.

PART 1

Submarine Telegraphy: The Forces of Science and Commerce

There is no doubt the day will come, maybe when you and I are forgotten, when copper wires, gutta-percha covering and iron sheathings will be relegated to the museum of antiquities.

W.E. AYRTON

Imperial Institute Address

Sixty Years of Submarine Telegraphy, 1897

CHAPTER 1

The Significance of Gutta Percha and the Rise of Submarine Telegraphy

> The bourgeoisie, during its rule of scarce one hundred years, has created more massive and more colossal productive forces than have all preceding generations together. Subjection of Nature's forces to man, machinery, application of chemistry to industry and agriculture, steam-navigation, railways, electric telegraphs, clearing of whole continents for cultivation ...[1]
>
> MARX and ENGELS

∴

What was gutta percha and why did it matter? Put simply, gutta percha mattered because submarine telegraphy mattered. To put these two in context, this chapter begins with a summary of gutta percha, its properties and its early uses. This is followed by an overview of the development and impact of submarine telegraphy: new ways of doing business, even entirely new industries, were made possible and governments came to rely upon this faster and more reliable means of communication. There would be no returning to the old, slow ways of communicating, and no lessening in the demand for gutta percha.

Gutta Percha and Its Significance

> Now that the advantages it confers are appreciated, we should greatly miss its presence if, from some unexpected cause, the supply were cut off, and this universally serviceable material denied to our use.[2]
>
> Society for Promoting Christian Knowledge

1 K. Marx and F. Engels, *Manifesto of the Communist Party*, New York, New York, Labor New Co, 1908, p. 14.

2 Society for Promoting Christian Knowledge, *Caoutchouc and Gutta Percha*, London, 1852, p. 202.

 | DOI 10.1163/9789004357280_003

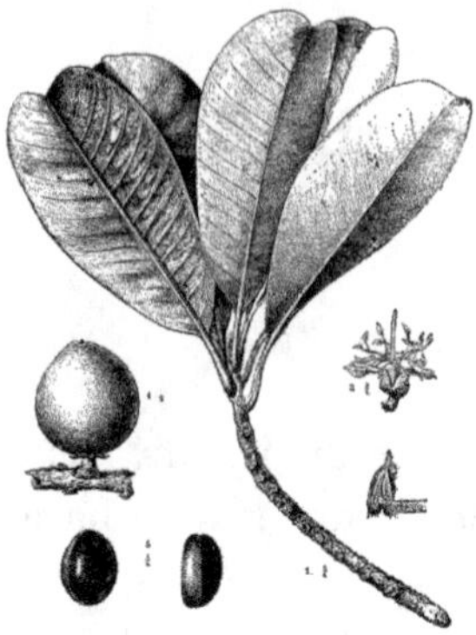

PLATE 2
Palaquium borneense.
Source: H. Jumelle, *Plantes à Latex et à Résines*, Paris, Librairie J.B. Baillière et Fils, 1915. p. 109.

The term *gutta percha* is from the Malay *gĕtah pĕrcha* – *gĕtah* meaning coagulated sap or latex and *pĕrcha* probably from the word for the type of tree from which the sap is obtained.[3] Gutta percha was obtained from particular varieties of trees that grow in parts of what is now Malaysia, Indonesia and, to a lesser extent, the southern Philippines. Gutta percha has been defined as 'a coagulum of the peculiar latex which characterizes the Sapotaceae'.[4]

In the nineteenth and early twentieth centuries, gutta percha used for telegraphic purposes was collected from various species of the *Palaquium* genus of the Sapotaceae family, especially *Palaquium gutta, P. oblongifolia* and *P. borneense* (see Plate 2 above).

As a product that had been traditionally collected in the forest, its malleable qualities had made it suitable for making items such as waterproof containers, handles for blowpipes, knives or axes – but it was of limited interest for subsistence trade. Samples of gutta percha were sent from Singapore to London in 1843 and it became the subject of further study by representatives of the Joint Committee of Chemistry, Colonies and Trade, a Committee that had been formed by the Royal Society for the Encouragement of Arts, Manufacture and Commerce, commonly abbreviated to the Royal Society of Arts. Two members of the Joint Committee decided to pursue the commercial potential of gutta percha. One was Carl Wilhelm (later anglicized to Charles William, known as William) Siemens. Siemens was a brother of Werner Siemens and their work eventually led to the Siemens Company manufacturing and laying submarine telegraph cables. The other member was Christopher Nickels, who contacted

3 See C. Scott, 'The Malayan words in English', *Journal of the American Oriental Society*, Vol. 17, 1896, pp. 93–144 for a detailed exploration of the origin and use of the term.

4 I.H. Burkill, (ed.) with contributions by W. Birtwistle, F.W. Foxworthy and J.B. Scrivenor, *A Dictionary of the Economic Products of the Malay Peninsula*, 2 Vols, London, Crown Agents, 1935, II, p. 1623.

the colonial agency later known as Jewesbury and Co and arranged the first commercial importation of gutta percha from Singapore. Nickels' actions led to the formation in London of the Gutta Percha Co. Together with Jewesbury and Co and a Singapore-based agency, Ker Rawson and Co, these became the first and most significant players to engage in the trade of gutta percha.[5] They dominated the trade for decades and the Gutta Percha Co linked with another company to form the Telegraph Construction and Maintenance Company (Telcon), to dominate the world manufacture of submarine telegraph cables. Singapore remained the major entrepot for obtaining gutta percha from the region and exporting it to the world, but those developments took time.

Initially, gutta percha was used to produce a wide assortment of products. After it is heated, gutta percha softens and can be molded into almost any shape; it then hardens after cooling and retains the molded shape. It was waterproof and impervious to most chemicals so it had an immense range of uses in industrial, decorative, commercial and even medical applications – roughly analogous to various plastics used in the twenty-first century. The Gutta Percha Co produced 'much that is useful, much that is curious, and much that is beautiful' from this substance according to one journal at the time, as the same Journal continued to list:[6]

> Gutta percha has been brought into use in an immense variety of ways. In gutta percha are formed all manner of domestic appliances and ornaments – trays of all sorts and sizes; vases, wash-stands, and plates; bouquet-holders, statuettes, brackets, jugs, mugs, inkstands, and clothes lines; flower-pots and stands, paper-weights, medallions, cornices, doors, mouldings, picture and glass frames, drinking cups, fishing nets, and portmanteaus; skates, policemen's batons, and boats; oil-cans, washing basins, and whips; stethoscopes, splints for dislocations, and curtain-rings; stuffing for horses' feet, mill-bands and stop-cocks; cutting boards, cabmen's hats, and traces; life preservers, bottling boots, and seals; powder-flasks, air-guns, and book-covers; sponge-bags, galvanic batteries, and bandages for broken limbs. For all these, and thousands of other purposes, it has been found of eminent utility, and we think enough has been said to commend it to the reader's attention ... We could go on, but space forbids.

5 E.F.A. Obach, 'Cantor lectures on gutta percha', *Journal of the Society of the Arts*, Vol. 46, 1897, pp. 98–99.

6 *Western Journal and Civilian*, 'Manufacture of Gutta Percha', Vol. XII, No. 2, St. Louis M. Tarver and H. Cobb, May 1854, p. 93.

Its durability in water, and resistance to damage from acids, grease or oil, made it ideal for use in factories such as breweries, bleach and dye-works, mills and brick-yards.[7] Another application was in the manufacture of water pipes; the existing lead pipes 'being discarded as having the risk of causing serious and alarming disorders, such as mania, epilepsy etc'.[8] Dr Thomas Oxley wrote about gutta percha in 1847 with a survey of its applications in surgery and other medical applications, especially in the management of fractures replacing the 'farrago of bandages' and for which purpose 'the Gutta would be no trifling boon to mankind were it to be used solely for this and no other purpose'.[9] At about this time an English dentist began experimenting with the process to purify gutta percha to use in root canal therapy, a procedure which continues to this day and which was in fact to have an impact on both the trade in gutta percha and the practice of submarine cable manufacture. That story will be seen a little later. In the meantime, another product made of gutta percha was police truncheons (or batons as mentioned in the Journal quote above), and one of the first peaks in the price of gutta percha in the key market in Singapore occurred because of demand from Britain '… in consequence of the home government having given an order for 10,000 truncheons for the special constables sworn in for the suppression of Chartism'.[10]

Gutta percha was also greatly in demand to produce soles for shoes. R.J. Dick and Co of Glasgow, for example, made footwear with leather uppers and gutta percha soles and at its peak the company was selling 34,000 pairs of shoes with gutta percha soles every week, with average weeks never lower than 20,000 pairs.[11] The company even traded gutta percha with telegraph cable manufacturers. Gutta percha had naval and military applications as well. By the time of the U.S. Civil War, gutta percha products were an integral part of the military supply catalogue for clothing and many other items such as ground sheets, water bottles and even jacket buttons.[12] Gutta percha became so ubiquitous that the term was adopted into colloquial language: in 1914, for example, the Argentine President described a U.S. government interpretation of its Monroe Doctrine as 'The Gutta Percha Message' because it could be adapted to any

7 G. Dodd, *The Curiosities of Industry and the Applied Sciences*, London, George Routledge & Co, 1852, p. 22.

8 *Western Journal and Civilian*, p. 91.

9 T. Oxley, 'Gutta percha', *Journal of the Indian Archipelago and Eastern Asia*, Vol. 1, 1847, p. 27.

10 *Straits Times and Singapore Journal of Commerce*, Singapore, 1848, 18 August, p. 301.

11 T. Chalmers, *100 Years of Gutta Percha*, Glasgow, R & J. Dick Ltd, 1946, p. 23.

12 See M. Woshner, *India-rubber and Gutta Percha in the Civil War Era*, Alexandria, O'Donnell Publications, 1999.

interpretation that suited U.S. interests.[13] Its wide-ranging applications were so widely recognized that, in George MacDonald's book *Gutta-Percha Willie,* the eponymous Willie was given the name because:

> It was not from his manual abilities alone that his father had given him the name of Gutta Percha Willie, but from the fact that his mind, once warmed to interest, could accommodate itself to the peculiarities of any science, just as the gutta percha which is used for taking a mould fits itself to the outs and ins of any figure.[14]

Products made from gutta percha were shown in London at the Great Exhibition of 1851 that displayed the achievements of contemporary industry, science and design. By this time gutta percha and its many uses were so widely known that they could be targeted by the satirist George Sala in his book describing the exhibition: *The Great Exhibition "Wot is to be"*. The book included a section on gutta percha manufactures; describing an ear trumpet for 'hearing a sermon at twenty miles distance' and gutta percha spectacles that would enable the wearer to 'see thru' a stone wall'.[15] One U.S. writer in 1870, extolling the benefits of gutta percha and india rubber, said that there were possible:

> a greater number of transformations than imagination ever conceived of as being produced by the magic wand of the most potent fairy. It runs, indeed, the whole gamut of the useful and ornamental, from the breastpin on the bosom of the belle to the life-boat in the surges of the ocean[16]

One application for gutta percha that overtook the decorative and industrial uses in leading to the continued demand for this strange product was in the production of golf balls. Because of developments in the late 1840s, 'Gutties' as they were called, surpassed the older leather balls 'in the matter of cheapness and rotundity'.[17] The major telegraph cable producers made golf balls. Silvertown, a cable manufacturer in London, reported in 1895 that it had received a

13 *New York Times*, (hereafter NYT), 24 May 1914. Magazine Section 5, p. 1.

14 G. MacDonald, *Gutta Percha Willie. The Working Genius*, London, Henry S. King & Co., 1873, p. 150.

15 G. Sala, *The Great Exhibition "Wot is to Be"*, London, The Committee of the Society for Keeping Things in their Places, 1850, n.p.

16 L. Abbott, 'Nature's Common Carrier', *Harper's New Monthly Magazine*, No. CCXXXVIII, Vol. XL, March 1870, p. 515.

17 Reported in *India Rubber and Gutta Percha and Electrical Trades Journal* (hereafter IRGPETJ), London, Vol IX, No.9, 8 April 1893, p. 257.

single order for 300,000 balls from G. Spalding and Brothers (New York) in what it estimated to be about half the expected demand for the 'coming season in the States and Canada'.[18] By 1899 it was estimated that the U.S. alone required 2,000,000 balls for the following year's market – the demand for gutta being large because a third of the weight of gutta percha was lost during manufacture.[19] Towards the end of the nineteenth century, the extent of manufacturing with gutta percha and india rubber was sufficiently profitable for a supporting industry of trade literature to be established. In Britain, *The India Rubber and Gutta Percha and Electrical Trades Journal* (IRGPETJ) was established in 1884 after the proprietors saw a gap in British trade literature.[20] The French *Le Caoutchouc & la Gutta-Percha Organe Mensuel*; the *Gummi Zeitung* in Germany and in the U.S., the *India Rubber World* targeted a similar market.

It is difficult now to appreciate the breadth of uses for gutta percha. Demand was both significant and widely based; however, it was the importance of gutta percha in the production of submarine telegraph cables that made gutta percha not just a resource in the production of consumer and industrial products, it brought it into the sphere of global communications, making it a forest product of strategic and economic significance.

The Development and Significance of Submarine Telegraphy

Telegraphy was a system of communication in which a signal was sent via electrical impulses along a conducting (copper) wire between two points. The development of telegraphy was not an overnight event, nor was it the result of one person's research; rather, it was a combination of years of experimentation by many people. Eventually, in 1837 Samuel Morse obtained a U.S. patent for an 'electric' telegraph. He built a trial system in 1843 and an operational system for public use began between Washington and Baltimore in 1844. The commercial beginnings in England also came after 1837, when Cooke and Wheatstone obtained a patent for a telegraphic device and a permanent telegraph line of thirteen miles was laid in 1839. In 1845, the Electric Telegraph Company (ETC) was established with offices in the major commercial centres of England and Scotland. Similar developments followed in North America, Europe and beyond, although in varying formations according to differing economic,

18 IRGPETJ 8 April 1895, p. 266.

19 NYT, 21 November 1899, p. 3.

20 IRGPETJ 3 September 1884, p. 42.

governmental and local conditions.[21] In general, these networks were overland connections until 1851, when the first successful submarine (undersea) cable was laid – insulated with gutta percha. With this cable link in place, the first step was taken on the path towards the near immediate transfer of information along international communication networks. As a result, there was little likelihood of anything other than a continuing demand for gutta percha – at least until an alternative product or technology was developed, but that took more than half a century to occur. Together with transport technologies of railways and steam shipping, telegraphy provided the essential communications for trade and investment that changed the face of the global economy. The changes were popularly and somewhat euphemistically described at the time as the 'annihilation' of time and space.[22] This communication 'revolution' occurred in the context of industrial, social and economic changes – the forces of technology and commerce interacting to change lives across the globe.

Innovations such as steel manufacture, industrial chemistry and electrical industries (the latter enhanced by knowledge arising from developments required by submarine telegraphy) were a significant feature of the new global economy and in the global demand for commodities. Indeed, Kuznets describes this era as the 'modern economic epoch' beginning with 'epochal' innovations achieved by the application of science to problems of economic production.[23] The era was one of organizational and technological revolution that created new opportunities and new industries.[24] Other refinements during the century were the establishment of international co-operation unions such as the International Postal Union and the International Telegraph Union. These bodies set down rules to govern international communications and the formation of globalized networks that commercial and governmental sectors required. Such machinery reflected an increasingly globalized economy and recognized

---

21 See for example K. Beauchamp, *History of Telegraphy*, Stevenage, Institution of Electrical Engineers, 2001; D.R. Headrick, *The Invisible Weapon: Telecommunications and International Politics, 1851–1945*, New York, Oxford University Press, 1991 and W.J. Phalen, *How the Telegraph Changed the World*, Jefferson, North Carolina, McFarland & Company Ltd., 2015 for an overview of this period of telegraph history.

22 For example, G.B. Prescott, *History, Theory, and Practice of the Electric Telegraph*, Boston, Ticknor and Fields, 1866, p. 5, and see I.R. Morus, *Frankenstein's Children: Electricity and Experiment in early Nineteenth-century London*, Princeton, Princeton University Press, New Jersey, 1998, Ch. 7.

23 S. Kuznets, *Modern Economic Growth: Rate, Structure and Spread*, New Haven, Yale University Press, 1966, p. 9.

24 A.D. Chandler with T. Hikino, *Scale and Scope. The Dynamics of Industrial Capitalism*, Belknap Press, Cambridge, Mass., Harvard University Press, 2004, p. 26.

the inevitability of the interdependence among national economies – and the role of the services sectors of economies.[25] As production and wealth – and therefore demand – grew, so too was there a widening need for raw materials and commodities which the developing countries could provide. A feature of the increasing growth of trade was increased flow of capital and labor between countries. This occurred both between the wealthier developed countries which exchanged manufactured goods, capital and technology; and between developed and under-developed but resource rich countries. These latter countries, sometimes described as 'peripheral', provided the commodities that were required by the 'metropolitan' core of the global economy. The distance of countries at the 'periphery' would in the past have limited the extent to which supply chains could easily gain access to these commodities; however, the increasing growth of steam shipping and of railways – facilitated by the telegraph – promoted the capacity and speed with which commodities could profitably be exploited.

The major economies had by this time established, to a greater or lesser extent, outposts of colonial interests that were important as a source of raw materials and an outlet for the investment of capital. To a certain extent also they were a potential market for manufactured goods. Whatever the causes of colonial expansion, its results are constantly re-assessed according to the perspectives held by the analyst – views vary between the extremes of 'triumphalism of the Europeans or the rallying of their opponents'.[26] Much of the capital to fund the world's railways and submarine telegraph cables came from Britain which, despite being the great manufacturing country it was, earned more from 'invisibles' like investment, shipping, finance and insurance.[27] In part this reflected its move towards free trade – and because others, especially Germany, France and particularly the U.S., had overtaken Britain in expanding their own economies and manufacturing capacities. But London remained a central warehousing point for world commodity trade for anything from sugar to ostrich feathers. While not all commodities were shipped through Britain, the arrangements for their trade were handled by the network of agents that were based in London or Liverpool who had connections in the trading ports of the world – such as Singapore. The movement of global investment and commodities was in many ways dependent upon Britain. In that sense, it was not so much the 'British Empire', but the 'invisible' British empire of commerce

25 See P.J. Cain and A.G. Hopkins, *British Imperialism 1688–2015*, Abingdon, Routledge, 2016.

26 N. Tarling, *The Cambridge History of Southeast Asia, Vol. 2 Part 1, From c. 1800 to the 1930s*, Cambridge, Cambridge University Press, 1999, p. xiv.

27 See Cain and Hopkins, p. 172.

that was significant in the nineteenth century. 'Invisible' trade by its very nature depends upon the exchange of information and of course that was the very business of the telegraph. It might be argued that the telegraph was in part a 'tool of empire'[28] but its significance lay more in its application as a tool of 'the invisible empire' – the invisible trade and information exchanges and networks that underpinned the global economy. For many nations, communication with Britain was economically significant and was physically dependent upon the invisible submarine cables making up much of the world's telegraph communication network.[29] This in large measure had also emanated from Britain (Map 1 and Map 2 at the end of this chapter). Further details on the expansion of submarine telegraphic networks are included in this and the next chapter. Statistical details appear in Appendix 2, Table 1.

The first attempt to lay an international undersea cable was made across the English Channel on 28 August 1850. Although the cable broke, just a few days later an Irish newspaper, the *Freeman's Journal*, concluded that news and 'late and speedy intelligence' would probably soon be transmitted via the yet to be laid cables that would stretch from Ireland to England.[30] This very early report reveals how readily and keenly the idea of instant, electrically transmitted information was anticipated; the thirst for speedy information transfer and all that might mean. A successful cable was eventually laid across the English Channel in 1851 and commentary at the time alluded to the electrical basis of the new communication technology by describing this achievement as effectively making it possible to 'send a message by lightning to Paris or Vienna'.[31] It did not take long for the notion to develop that the faster information was received, the better it was. The technology set new standards for what constituted value in information: speed, not just accuracy or relevance, was now a factor in judging its value. Many cable connections soon followed, adding to the first cross-channel linkage. The economies of Britain, Western Europe, and the U.S. were the most dominant and the level of communication between Europe and the U.S. was reflected in the growth of their submarine cable connections.

---

28 See D.R. Headrick, *The Tools of Empire: Technology and European Imperialism in the Nineteenth Century*, New York, Oxford University Press, 1981.

29 See Headrick *Tools of Empire*, and R. Wenzlhuemer, *Connecting the Nineteenth-Century World: The Telegraph and Globalization*, Cambridge, Cambridge University Press, 2012; see also Supplements to Bureau International des Administrations Télégraphiques, *Nomenclature des câbles formant le réseau sous-marin du globe dressée d'après des documents officiels par le Bureau International des Administrations Télégraphiques, Journal Télégraphique*, various years.

30 *Freeman's Journal*, Dublin, 2 September 1850.

31 D. Lardner, *The Electric Telegraph Popularised*, London, Walton and Maberly, 1855, p. 115.

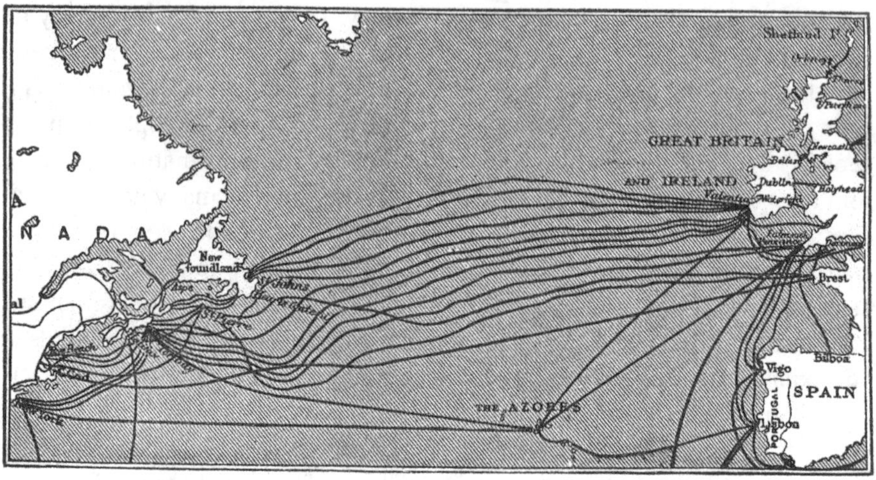

MAP 1 *Map of Atlantic submarine telegraph cable connections by 1903*
Note: Multiple trans-Atlantic cables laid by 1903 reflected the significance of European-North American commercial relations and the role of telegraphic communications. Many connections to North America were linked either to Ireland or England; contact between Europe and North America was most usually made through those linkages, a vulnerability felt in later years, especially during World War I when Britain either cut or monitored many cable communications between even neutral countries.
Source: C. Bright, *The Story of the Atlantic Cable*, London, George Newnes, 1903, p. 217.

Whereas much of the network of nineteenth century global communications relied upon single or sometimes duplicate cable connections, the demand from (and profitability of) trans-Atlantic contact was such that multiple cable links were required. This is evidenced, for example, by the map given by Bright in 1903 (see Map 1), which emphasizes the relationship between trade and submarine telegraph communications.

After the submarine telegraph cable linked Britain and Europe, one of Reuters news agency's first commercial agreements was with the London Stock Exchange for connections with the continental bourses, and later other connections were made between various bourses in Eastern Europe and the Black Sea. In later years, Reuters' market prices and quotations became one of the necessities for merchants in India, China and the Far East.[32] This made it easier to assess prices that traders might obtain and shortened lead times for orders. Inventories and storage requirements could be reduced. The telegraph

32 G. Storey, *Reuters' Century 1851–1951*, London, Max Parrish, 1951, pp. 15, 62.

revolutionized business networks by limiting the role of the middleman in trade, by enabling buyers and sellers to be in immediate contact. Eventually, distant markets, like the Singapore rice market, could be contacted by dealers in Britain or North America, making the telegraph a 'crucial instrument' in the international commodity trade.[33] Traders obtained more accurate information about prices and the state of supplies, with new systems of transport like rail and steamship and the opening of the Suez Canal able to provide more timely shipment.

Britain was the financial capital of the world: it was the engine room of global capital investment, and that engine room could not have been established or operated without the domestic and international network of telegraph communications. By 1875 there were 3,500 miles of telegraph wires underneath the streets of London (most of it insulated with gutta percha because of the damp conditions). By 1896 that amount had increased almost tenfold to 30,000 miles.[34] After the Atlantic connection was established, the London Stock Exchange was at its busiest during those of its opening hours which coincided with the opening hours of the New York Stock Exchange. Because this was usually just the final hour of London trading times, brokers began trading after hours by moving to a position outside the offices of telegraph companies, continuing even in bad weather.[35] This generated an enormous amount of business for the telegraph operators and in similar situations where international time differences could be exploited.[36] In later years, operations in London were improved by the expansion of the Stock Exchange to be near communication links such as the Post Office, or the offices of the Anglo-American Telegraph and Commercial Cable Offices.[37] By 1909 there was a telegram being sent or received every *second* during opening hours between the London Stock Exchange and those of Europe. By 1914 after war broke out, there were 50,000 commercial telegrams being received every day by the cable network in Britain for forwarding on to neutral countries alone.[38] Michie's survey of the London Stock Exchange concludes that the transformation of communications

33 A.J.H. Latham and N. Neal, 'The international market in rice and wheat, 1868–1914', *The Economic History Review*, New Series, Vol. 36, No.2 (May), 1983, p. 273.

34 Obach, 1898, p. 187.

35 R.C. Michie, *The London Stock Exchange: A History*, Oxford, Oxford University Press, 1999, p. 78.

36 C. Bright, *Imperial Telegraphic Communication*, London, P.S. King & Son, 1911, p. 123.

37 Michie, p. 79.

38 *American Journal of International Law*, 'Censorship of telegrams transmitted by Cable and Wireless', Vol. 9, No. 3, Supplement: Diplomatic Correspondence Between the United States and Belligerent Governments Relating to Neutral Rights and Commerce, 1915, p. 288.

destroyed the sovereignty of localized markets and led to competition between markets, such as between London and New York.[39]

Major cities around the world likewise depended upon cable connections: Liverpool and Manchester were connected and their Chambers of Commerce discussed problems of international cable communication as early as the 1860s.[40] One official explained the relationship between the expansion of trade, the empire and the telegraph: 'we had to push on beyond India to the Straits Settlements ... to China and Australia ... looking up all information obtainable as to exports and imports in those Colonies'.[41] Cable communication became an essential feature of shipping and railroad networks around the globe, although a history of such interconnections remains in large measure yet to be studied in depth.[42] Yet information exchanged by sea or overland connections could take weeks, even months, especially on long voyages such as between Australia and Europe. With the submarine telegraph, speed was not the only important difference: the dispersal of information moved from uncertainty to regularity. Letters or business information carried on long sea voyages were not guaranteed to arrive and, by the time they did, the information might not be useful. Ships carrying cargo could obtain updated information cabled to ports, thereby unloading or taking on more profitable cargoes based on updated information, and achieving a more efficient use of shipping that no longer needed to 'hunt' for cargoes, leave cargoes waiting for shipment until vessels arrived, or avoid the unprofitable delays associated with the need

39 *The London Stock Exchange*, p. 71, see also S.M. Muller and H.J.S. Tworek 'The telegraph and the bank: on the interdependence of global communications and capitalism, 1866–1914', *Journal of Global History*, 10, 2015, pp. 259–283.

40 C.T. Bright, *The Telegraph to India and its Extension to Australia and China*, London, William Clowes and Sons, 1866, p. 65.

41 C. Todd, 'Telegraphic enterprise in Australia', *Proceedings of the Royal Colonial Institute*, Seventh Vol., 1886, p. 173.

42 Other than general discussion in texts on the nineteenth century economy, most study to date on the impact of submarine telegraphy relates to diplomacy, social and media changes or the technological or commercial aspects of cable development; certainly nothing on the equivalent scale of Chandler's work on the role of the overland telegraph in changing the U.S. economy in Chandler, 2004; see, however, elements in P.J. Hugill, *Global Communications since 1844. Geopolitics and Technology*, Baltimore, The Johns Hopkins University Press, 1999; D.R. Winseck and R.M. Pike, *Communication and Empire: Media, Markets and Globalization 1860–1930*, Durham, Duke University Press; R.J. Gordon, *The Rise and Fall of American Growth: The U.S. Standard of Living since the Civil War* (*The Princeton Economic History of the Western World*), Princeton, Princeton University Press, 2016.

for ships to await a full cargo before departure.[43] As Mackenzie noted, the extension of telegraphic communication ensured that 'seaborne trade was conducted on the basis of the closed transaction instead of open speculation'.[44] This not only improved commodity trade, but also made for more profitable shipping networks and encouraged investment in shipping. It was said that no steamship service could thrive unless the ports at which it touched were in telegraphic communication.[45] Internationally, shipping networks also used cabled messages to report lost or damaged vessels, problems with crews and local conditions that might have an impact on shipments and trade. Without the telegraph, such issues could go unmanaged for days or even months. Lew and Cater found a direct correlation between the diffusion of the telegraph, the co-ordination of shipping and the growth of world trade, even allowing for the impact of other previously 'well-studied' effects existing at the time (for example, declining tariffs, falling transport costs and so on).[46]

Telegraphic code books were written to simplify and shorten messages, and thereby reduce costs, or to ensure commercial secrecy. They had been developed for overland telegraph users from as early as 1845, by banks, railroads, shipping, trading and manufacturing companies and their increasing use reflects the expansion of telegraph commerce.[47] With the rise of submarine telegraphy, and the cost and global reach of those cables, code books came to be an important feature of nineteenth century commerce, and were used well into the twentieth century. For example, the *ABC Universal Code*, a much-used code book published in several editions from the late nineteenth century was in use until the mid-twentieth century.[48] Such books contained thousands of code words, including words used for international shipping and freight,

43 See for example, V.T. Coates and B. Finn, *A Retrospective Technology Assessment: Submarine Telegraphy. The Transatlantic Cable of 1866*, San Francisco, San Francisco Press Inc., 1979, pp. 68–69; B. Lew and B. Cater, 'The telegraph, co-ordination of tramp shipping, and growth in world trade, 1870–1910', *European Review of Economic History*, Vol. 10, 2006, pp. 147–173.

44 C. Mackenzie, *Realms of Silver. One Hundred Years of Banking in the East*, London, Routledge & Kegan Paul, 1954, p. 40.

45 *Blackwoods Edinburgh Magazine*, 'The All-British Trans-Pacific Cable', (February), 1897, p. 272.

46 Lew and Cater, p. 147.

47 See D. Kahn, *The Codebreakers: The Comprehensive History of Secret Communication from Ancient Times to the Internet*, New York, Scribner, 1996, p. 189 and Ch. 6 for an overview of telegraph codes and cryptography.

48 W. Clauson-Thue, *ABC Universal Commercial Electrical Telegraphic Code. Specially adapted for the use of Financiers, Merchants, Shipowners, Brokers, Agents Etc.*, London, 1873. Other

machinery, local conditions, commodities, goods and weights, currency and so on. These and similar code books reveal just how extensively global shipping and trade used cable communication, with their details providing a snapshot of the commerce and supply chains they represented. So entrenched was cable communications in global commerce that many commodity trades also used specialized cable code books.[49] Code useage was so prevalent that even as early as 1868 the International Telegraph Union began formulating regulations for code word design, application and costing, and this continued at future Union conferences well into the twentieth century.[50]

It was not simply that the spread of worldwide telegraph connections achieved greater efficiency but that it led to developments that would not otherwise have eventuated. It was not only profitability and timeliness that changed, but the entire architecture of commerce. Chapman concludes, for example, that the most far-reaching change in Asian finance came with the extension of the telegraph cables to the region[51] and Cain and Hopkins note the impact of global telegraphic links on British trade networks.[52] Mackenzie gives numerous examples of how the availability or absence of cable connections between London and Asian bankers influenced banking and the growth of all types of trade of many eastern enterprises.[53] Gordon has considered whether today's 'New Economy' based on e-commerce and the technology revolution measures up to the industrial revolution of the latter half of the nineteenth

---

editions through to *The Seventh Edition Five-Letter ABC Universal Commercial Telegraphic* Code, William Droege (ed.), London, Sir Isaac Pitman & Sons, 1936.

49 For example, cotton: A.B. Shepperson, *Telegraphic Cipher Code Especially Adapted to the Cotton Trade*, Shepperson Publishing Company, New York, 1878; and well into the twentieth century – T. Buenting, *Buenting's International Cotton Code*, Second Edition, Bremen, Theodor Buenting & Co., 1929: mining industry, such as C.A. Moering and T. Neal, *The New General and Mining Telegraph Code*, London, William Clowes and Sons, 1891.

50 See J.C. Hemmeon, *The History of the British Post Office*, Cambridge, Mass., Harvard University, 1912, p. 212 and W.F. Friedman, *Report on the Use of Codes and Code Language, The International Telegraph Regulations Pertaining thereto, and the Bearing of this History on the Cortina Report, International Radiotelegraph Conference of Washington 1927*, Washington, U.S. Government Printing Office, 1928 for a useful history of telegraph code useage and its regulation.

51 S. Chapman, *The Rise of Merchant Banking*, London, Routledge, 2010, p. 114.

52 P.J. Cain and A.G. Hopkins, for example p. 186.

53 Mackenzie, *Realms of Silver*: in providing more effective Head Office control, enhancing bank stature in China and altering the business of currency exchange see p. 58; funds settlements and the rise of telegraph transfers pp. 69–70; improved banking management p. 74; providing international telegraphic funds transfer during local currency crisis in the Straits Settlements p. 191.

century and notes that no development in communication has achieved a change in communication speed comparable to the telegraph.[54]

Commodity trades provide other examples of the impact of submarine telegraphy. One of the early advantages of the Atlantic cable connection was that Manchester cotton-good manufacturers and Lancashire spinners could obtain accurate information about U.S. cotton prices, rather than being subject to manipulation by Liverpool merchants.[55] It was said that no Bombay cotton dealer would sell his cotton before he had read Reuters' overnight quotation.[56] In the U.S., the cotton trade was altered by domestic and international telegraphic connections. For example, agents in the port cities lost considerable business as the telegraph linked inland growers and merchants with the external market, bringing 'remote producer mills and exporters' closer together. The telegraph was used to such an extent that U.S. mill buyers, brokers and even English buyers had access to hourly updated information about price and availability in distant markets.[57] When a cable connection was put in place to link South America with New York (and thus with London), this enabled a globally integrated coffee market to be formed – it established the exchange of information on prices, leading to 'further institutionalized access to information and standardized commercial practices'.[58] A coffee exchange then followed in New York in 1882, followed by a futures exchange; such an exchange would have been impossible prior to the era of international telegraphic communications. Mercantile exchanges that were established to deal in commodity markets were near centres of finance and telegraph offices, even though these were at quite a distance from the 'site of production' of the commodities.[59]

54 R.J. Gordon, *Journal of Economic Perspectives* p. 68.

55 C. Smith and M.N. Wise, *Energy and Empire. A Bibliographical Study of Lord Kelvin*, Cambridge, Cambridge University Press, 1989, p. 673; see also A. Engel, 'Buying time: futures trading and telegraphy in nineteenth-century global commodity markets', *Journal of Global History*, Vol. 10, Iss. 2, July 2015, pp. 284–306.

56 Storey, p. 64.

57 H.D. Woodman, *King Cotton and His Retainers: Financing and Marketing the Cotton Crop of the South, 1800–1925*, Washington, Beard Books, 2000, p. 275.

58 S. Topik and M. Samper, 'The Latin American commodity chain' in *From Silver to Cocaine. Latin American Commodity Chains and the Building of the World Economy, 1500–2000*, eds S. Topik, C. Marichal and Z. Frank, Duke University Press, Durham, 2006, pp. 118–146, Topik and Samper, 2006, p. 136, also Engel, *Journal of Global History*.

59 R.R. John, 'Recasting the information infrastructure for the industrial age' in *A Nation Transformed by Information: How Information Has Shaped the United States from Colonial Times to the Present*, eds A.D. Chandler and J.W. Cortada, New York, Oxford University Press, 2000, p. 83.

Instead, exchanges were established where traders themselves could access the telegraph and form personal networks and relationships. At the local level, people were indirectly affected by the telegraph as international trade connections came to rely upon it. Costa Rican coffee growers used telegraphed information in local newspapers to mobilize and force better prices for their produce.[60] The access to telegraphed information not only achieved better prices for growers, but empowered them by revealing both to them and to the processing firms that the growers were no longer ignorant of wider commercial forces. In Australia, the regional wheat trade was stimulated, for example, by news such as a 'firmer market' telegraphed from the key market in Melbourne.[61] Submarine telegraphy became a means of empowerment for local actors who now better recognized their role in international trade and how to make use of market information. This was a result of their access to something they had not had – commercial information previously only in the hands of factors, middlemen or shipping agents. That is, an unintended consequence of submarine telegraphy was a change in power relations between actors in commodity chains because now more actors in the chain had access to useful information, rather than sometimes being excluded from it.

The impact of cable telegraphy went beyond its economic significance. The telegraph also became 'a concrete definition of Imperial unity ... the controlling instrument of national aggrandisement and individual enterprise' while Britton and Ahvenainen point to the 'deployment of advanced technology in lesser-developed nations' as a means of replacing 'the traditional imperial device of territorial acquisition' and as a means of gaining advantages of trade and diplomacy.[62] Global telegraphic connections were a significant tool in colonial administration and a means of managing both diplomatic and military operations. The telegraph could be used to deal with problems in a timely fashion, make better use of resources, and remove 'the possibility of misunderstanding'.[63] Submarine telegraph connections were more secure than overland connections which could be intercepted or destroyed by civil or military forces. Governments could use the telegraph for diplomatic purposes

60 Topik and Samper, p. 132, see also Engel, *Journal of Global History.*

61 *The Argus*, Melbourne, Australia, 12 August, 1859, p. 4.

62 F.E. Kappey, 'Electric Telegraph Service' in *The British Empire Series, Vol. v*, London, Keagan Paul, Trench Trubner & Co Ltd, 1902, p. 332, and J.A. Britton and J. Ahvenainen, 'Showdown in South America: James Scrymser, John Pender, and United States-British cable competition', *Business History Review*, Vol. 78, 1, (Spring), 2004, p. 24.

63 A.J. Herbertson and O.J.R. Howarth (eds), *The Oxford Survey of the British Empire*, London, Oxford University Press, 1914, p. 23.

and for linking with overseas posts. The British Prime Minister concluded that '… we positively exist by virtue of the electric telegraph … it is a phenomenon to which nothing in the history of our planet up to this time presents anything equal'.[64] Adas puts this in a broader context by concluding that this superior technology not only made the administration of 'vast areas possible; but further, it gave an assurance of the 'right' even the 'duty' to police, arbitrate or demand tribute' from colonial regions.[65]

Britain and its imperial interests and cable connections were more complex and diverse than those of any other country, but the submarine telegraph and the newspapers that used it also caused aggravations and difference of opinion among its administrators. For instance, by 1902 it was felt that the great increase in publicity given to issues was a cause of frequent exchange of opinion between London and its diplomatic staff in India.[66] One official of the Indian frontier so hated telegraphic interference that he famously remarked that 'I might have been a great man, but for the telegraph'.[67] Still, as Headrick concludes, although the telegraph might have had an impact upon diplomacy, it did not necessarily improve international relations.[68] While the telegraph might not prevent hostilities, governments recognized that telegraph connections were important and so provided financial support for many submarine cable ventures, either to protect colonial interests or to ensure they were not dependent upon linkages passing through territory belonging to other powers. Subsidies could be extremely costly. For example, between 1879 and 1893, the British, German, French and Portuguese governments subsidized cable networks around Africa for up to a period of 23 years to a total value of £4,174,000.[69] The strategic significance of submarine cable communications was most notably evidenced during war time. One of the earliest such uses was during the Crimean War of 1853–56 when a British cable was laid temporarily to transmit official and military messages. By 1895 a total of 100,000 miles

64 Marquis of Salisbury [Prime Minister], 'Address to Dinner of the Institution of Electrical Engineers', *The Electrician*, Vol. 24, (November 8), 1889, p. 13.

65 M. Adas, *Machines as the Measure of Men: Science, Technology, and Ideologies of Western Dominance*, Ithaca, NY, Cornell University Press, 1990 p. 204.

66 In R.L. Greaves, *Persia and the Defence of India, 1884–1892 A Study in the Foreign Policy of the Third Marquis of Salisbury*, London, Athlone Press, 1959, p. 43.

67 A.L.P. Tucker, *Sir Robert G. Sandeman K.C.S.I. Peaceful Conqueror of Baluchistan*, London, Society for Promoting Christian Knowledge, 1921, p. 6.

68 D.R. Headrick, 'Review: "Under the Wire: How the Telegraph Changed Diplomacy"', *Victorian Studies*, (Spring), Vol. 47, no. 3, 2005, p. 493.

69 C. Bright, p. 68.

of military cables had been put to use.[70] Bright noted that the British Afghan campaigns of 1878–1880 were well monitored – for military purposes and in the newspapers – because of submarine and overland telegraph links. In comparison, he said, the absence of such links with South Africa was 'a disaster' in 1879 and failures there went unknown for weeks in Britain.[71] Bright goes on to conclude that it was those failures that promoted government support for a submarine cable to connect with South Africa, the achievement of which greatly promoted negotiations during the Boer War.[72]

Although all telegraph cables carried a proportion of government messages, it was not just the transmission that made international telegraph connections important; the cable lines were themselves strategic assets. It was the potential to use the cables (or withhold useage rights or censor messages) during times of hostility that made submarine telegraphic resources so important. In wartime, the submarine telegraph remained a tool of diplomacy, but it was an instrument of war as much as any armament. Indications of its significance are reflected in the steps that governments made to strengthen or protect submarine telegraph resources. The International Telegraph Convention of 1884 recognized the protection of submarine telegraph cables, but retained a protective clause to ensure that this did 'not in any way restrict the freedom of belligerents' to cut them.[73] The British Cabinet had decided as early as 1898 that if necessary it would cut an enemy's cables.[74] Peacetime talk of such measures, however, described this as an act of 'barbarism'.[75] The U.S. cut several cables near Cuba and the Philippines during the Spanish-American war in 1898. The accidental destruction of the incorrect cable revealed to the U.S. its vulnerability in Pacific cable communication. After the annexation of Hawai'i because of that same war, U.S. government interest in a Pacific cable connection increased. At the outbreak of World War I, the very first hostile act undertaken by Britain was to cut all the German undersea cables that passed through the English Channel. The act took place within just five hours of war being

70 Obach, 1898, p. 186.

71 C. Bright, *Submarine Telegraphs. Their History, Construction and Working*, London, Crosby Lockwood and Son, 1898, p. 174.

72 Ibid.

73 See C. Bright, p. 151.

74 P.M. Kennedy, 'Imperial cable communication and strategy', *The English Historical Review*, Vol. 86, no. 341 (October), 1971, p. 740, see also J.R. Winkler, *Nexus. Strategic Communications and American Security in World War I*, Cambridge, Mass, Harvard University Press, 2008, p. 5 ff.

75 J. Anderson, *The Colonial Office List Comprising Historical and Statistical Information Respecting the Colonial Dependencies of Great Britain*, London, Harrison and Sons, 1892, p. 312.

declared.[76] The significance of submarine telegraphy was recognized in the peace negotiations following World War I. One U.S. official urged that a role for the newly established League of Nations should be that it take control of the world's submarine cables or that 'arrangements must be made by which gutta percha ... could be obtained by all nations'.[77] Germany's overseas submarine cable connections, even its cable-laying ships, were all recognized as valued resources and were forfeited by that country as part of its post-war reparations.[78]

Yet it was through its role in the print media that submarine telegraphy arguably most influenced warfare and diplomacy. Regular and timely reports from 'the front' or from reporters in foreign outposts informed the public of government and military failings; the very speed of the telegraph put pressure on both governments and officials to act more promptly in order to be seen to be in control of the situation.[79] The telegraph and the news media enjoyed a particularly close relationship; each expanded with the commercial benefits gained from the other. The potential for newspapers to use this new technology was observed as early as 1844 when the New York Herald concluded satirically that so significant were the changes arising from telegraphic transfer of current affairs that 'There is nothing now left for invention to achieve but to discover news before it takes place'.[80] Postal services were already being used for the brokering of information to newspapers, but the timeliness of and distances over which world telegraphic cable links could be formed, provided a more productive arrangement for the sale and dissemination of information, and could do so effectively on a wider scale. In the early years, the newspapers' most profitable use of the telegraph was in publishing timely information of a commercial nature, rather than from information about current affairs.[81] Commerce and government both interacted with the print media so that all were in some fashion mutually dependent. Gradually, the demand for information grew. Reuters news agency, the Associated Press and other similar

76 C. Bright, *Telegraphy, Aeronautics and War*, London, Constable & Company, 1918, p. 30.

77 B.E. Schmitt, 'The Peace Conference of 1919', *The Journal of Modern History*, Vol.16, no. 1 (March), 1944, p. 54.

78 F.J. Brown, *The Cable and Wireless Communications of the World*, 2nd edn, London, Sir Isaac Pitman & Sons, 1930, p. 38.

79 See D.P. Nickles, 'Telegraph diplomats. The United States' relations with France in 1848 and 1870', *Technology and Culture*, Vol.40, no.1, (January) 1999, pp. 1–25 and D.P. Nickles, *Under the Wire: How the Telegraph Changed Diplomacy*, Cambridge, Mass, Harvard University Press, 2003.

80 Reported in NYT, 1 May 1894, p. 4.

81 See also R.B. Du Boff, 'Business demand and the development of the telegraph in the United States, 1844–1860', *Business History Review*, Vol. LIV, no.4 (Winter), 1980, pp. 467–468.

agencies used the telegraph to receive and transmit information that was then sold to newspapers: by this means information was commoditized and made available to smaller regional papers. The scale and timeliness of these network arrangements had not been possible under early methods of sending information by horse or rail. That is, the telegraph made possible new systems in the brokering of the news – this was an entirely new way of conceiving current events, information on which became commodities in their own right. In 1851, a British Parliamentary enquiry was told that telegraphed information was 'the great point' that enabled one newspaper to achieve market advantage because it could bring the quickest news.[82] It was said that Reuters moved from Paris to London after the first cable connection was established because the company's policy was to 'follow the cable'.[83] Reuters even commissioned submarine cable connections, including one crossing the Atlantic.[84] Newspapers had special arrangements to send messages more cheaply because of the great volume of work they generated.[85] In 1876 it was said that foreign telegraphic news was a prominent feature of all English newspapers[86] and some Australian newspapers were each willing to pay Reuters' London bureau the very considerable sum of £2000 per year for daily telegrams of just forty words as it provided them with 'more prestige at home'.[87] As a result, readers obtained new information: the telegraph contributed to changing ideas about national identity, colonial linkages and about time itself.

Individuals were empowered by submarine telegraphy by using the telegraph to transmit funds across the globe, for example, people who moved to new areas as part of the global movement of labor. Remittances have been described as the 'human face' of globalization.[88] The Commercial Bank of Sydney in 1896 'frequently' had people telegraphing even small sums to relatives 'on the other side', a service for which they were willing to pay up to £2 per

82 *Select Committee on Newspaper Stamps: Report, together with Proceedings of the Committee, Minutes of Evidence, Appendix, and Index,* 1851; Reports of Committees (558) Vol. XVII.1, p. 448.

83 See T. Standage, *The Victorian Internet*, London, Weidenfeld Nicholson, 1998, p. 142.

84 C. Bright, *Submarine Telegraphs*, 1898, p. 107 and Headrick and Griset, *Business History Review*, p. 551.

85 Lardner, 1855, p. 88.

86 Storey, p. 58.

87 Storey, p. 69.

88 M. Orozco, 'Worker remittances in an international scope', *Inter-American Dialogue, Research Series*, Remittances Project, (March) 2003, p. 11.

message.[89] In Europe, the telegraphic transmission of foreign money orders was regulated in 1898 with an arrangement between the British Postal Telegraph system and that of Germany, followed by connections with other major European countries.[90] Funds could be transmitted from thousands of postal and telegraph offices around the world, or from banks which also offered that service. There was a large market for remittances from Singapore and the Malay Peninsula as Chinese who moved to the region sent money home to China; with Singapore important not just because of the size of the Chinese population, but because of its banking and telegraphic transfer facilities. The operation relied upon timely information of telegraphed exchange rates and market conditions, in part because its owners ran the operation by engaging in the practice of hedging.[91]

Cable communication was also an important means of promoting debate that challenged imperialist ideals. Chinese language newspapers in Singapore, for example, used telegraphed updates from Chinese newspapers as well as those from London in support of either pro-British or non-Anglophone sentiments. They obtained updates from China about the Sino-Japanese War in 1898, and used information to promote debate about the future of Confucianism.[92] Political concerns also found a voice through the new means of communication: on one occasion, a telegraphed message with the names of 1,000 supporters was sent from Singapore to the Empress in China, and messages of a political nature were sent on other occasions.[93] So, while the submarine telegraph supported European ideals and trade, it was also adopted by indigenous and intermediary groups or what Anderson called 'imagined communities' that recognized and exploited its benefits.[94] It created a sense of belonging to an empire, or conversely a community within but apart from an empire.

Developments resulting from submarine telegraphy touched populations in even the most remote regions and changes occurred almost immediately upon connections being made. The operation and design of business changed

89 G. Johnson, *The All Red Line. The Annals and Aims of the Pacific Cable Project*, Ottawa, James Hope & Sons, 1903, p. 227.

90 Hemmeon, p. 213.

91 K.F. Lian and K.W. Koh, 'Chinese enterprise in colonial Malaya: the case of Eu Tong Sen', *Journal of Southeast Asian Studies*, *35* (3), 2004, pp. 426–427.

92 See M.R. Frost, 'Transcultural diaspora: The Straits Chinese in Singapore 1819–1918', *Asian Research Institute Working Paper Series*, no. 10, National University of Singapore, 2003, for a more detailed study of the effect on Straits Chinese.

93 Frost, *Asian Research Institute Working Paper Series*, p. 31.

94 R. O'G. Anderson, *Imagined Communities: Reflections on the Origins and Spread of Nationalism*, London, Verso, 1991.

because it moved from being based largely on the *physical* articulation of goods and transport to being facilitated by the *abstract* and immediate exchange of information. The administration of governments and colonies, the conduct of wars and railways, timely information from newspapers and the remittance of money, were all changed. The impact of submarine telegraphy was both direct and indirect, and long lasting.

The continued functioning of all these achievements was dependent upon the supply of a latex with a strange sounding name – gutta percha.

CHAPTER 2

Gutta Percha and the Challenge of Submarine Telegraphy

> … if every person who, on the introduction of gutta percha into this country tried its insulating properties, even in a basin of water, be admitted as one of those claimants, their name would be legion[1]
>
> WILLOUGHBY SMITH, *Manager, Gutta Percha Co*

Achieving the most from the opportunities made possible by this new global communications technology was not without challenges, some of which related directly to gutta percha (and therefore to its trade). The development of submarine telegraphy was by no means straightforward; there were many scientific and production problems to solve including issues relating to the study and use of gutta percha. After many expensive cable failures, a reliable method of submarine cable production was developed with gutta percha insulation; as a result, few wanted to experiment with risky alternatives. A conservative industry relied upon a commodity that had significance beyond its immediate commercial worth. One of the challenges of the new system of global communications was to study and test gutta percha, and then to maintain a consistent standard of quality control during the manufacture and laying of the cables. How this challenge was met and global communication networks grew is considered in this chapter.

Gutta Percha and Submarine Telegraphy – The Early Years

Most of the developments described in the previous chapter did not occur until after the first successful cable was laid in 1851. Before that time, telegraph networks were confined largely to overland connections because an effective means of insulation for wires laid in damp or underwater conditions had not

1 W. Smith, *The Rise and Extension of Submarine Telegraphy*, London, J.S. Virtue & Co, 1891, p. 11.

 DOI 10.1163/9789004357280_004

been found. Yet overland connections also had problems. Wires could be damaged by bad weather, accidents or deliberate human acts. When telegraph lines crossed national boundaries, this could require messages to be re-transmitted or translated and then re-transmitted. In 1865, for example, the overland telegraph connection from Britain to India via Persia, usually required messages to be translated from English into Russian, then into Persian and back into English.[2] Another problem with messages being sent on overland wires was that they were not secure. It was relatively simple to cut wires or to connect to wires to obtain messages being relayed for others, or indeed to transmit false messages. This was a sensitive issue for government communication, but commercial information also needed a secure medium of communication. The best way to achieve this was to send messages on undersea cables that were more difficult to intercept and did not necessarily require changes in transmission from one country to another. Aside from these problems, the simple fact was that in many cases the most direct route for telegraph lines happened to be ones that crossed water. So, for truly flexible, speedy and reliable international communications to develop, telegraphic communication across the sea was needed. Submarine telegraphy if it could be achieved would incorporate 'the three great ideals of moral improvement, a dramatic triumph over nature, and return on capital'.[3] The Directors of the Gutta Percha Co recorded in their first Minutes that the manufacture of submarine telegraph cables was 'absolutely necessary for the purposes of our social system, and a highly remunerative investment'.[4] Yet these developments all took time and many experiments took place before the submarine telegraph could be established.

Gutta percha was first used on underground cables because of the damp conditions that existed there. Werner Siemens tested a gutta percha insulated cable along the track of the Anhalt Railway in Prussia in 1846 and the following year he laid an underground wire for the Prussian Telegraph Commission. In 1847, Siemens' company developed a machine with which to apply continuous and symmetrical coatings of gutta percha to copper telegraph wires.[5] Other companies later developed machines similar to that invented by Siemens. The use of gutta percha for underground telegraph wire insulation continued, and

2 C.T. Bright, *The Telegraph to India and its Extension to Australia and China*, London, William Clowes and Sons, 1866, p. 25.

3 V.T. Coates and B. Finn, *A Retrospective Technology Assessment: Submarine Telegraphy. The Transatlantic Cable of 1866*, San Francisco, San Francisco Press Inc., 1979, p. 126.

4 BICC I/12/103, 1864, p. 5 April.

5 C. Bright, *Submarine Telegraphs. Their History, Construction and Working*, London, Crosby Lockwood and Son, 1898, p. 250.

its use had become relatively common in Britain in the 1850s.[6] Samuel Morse had experimented with using insulated wires in underwater conditions in 1842 in New York harbour[7] but U.S. investment in the telegraph in subsequent years centered upon its growing domestic economy. So, the focus there was on overland telegraph networks, or across rivers or lakes rather than on building a capability to manufacture oceanic submarine cables. Instead, submarine telegraphy grew from that island with the greatest sources of capital, access to industry and science, and with a global colonial and trading network that could benefit most from an overseas cable network. That nation was of course Britain, which also had the largest trade in services: that is, its economy was based on the exchange of information, the very business of the telegraph. As a result, a few British companies came to dominate submarine cable manufacture, but with other European firms also occupying important positions as the industry developed through the latter part of the century.

The unsuccessful 1850 cable attempt was made by Englishmen John and Jacob Brett who had a 25 nautical mile copper wire cable with a thin gutta percha covering manufactured by a company that had already established itself as the foremost producer of other gutta percha products – the Gutta Percha Co.[8] The cable was laid between Dover and Calais – this was to be the first of tens of thousands of miles of telegraph cable to be manufactured by the Company, and it was the beginning of its long domination of the submarine telegraph cable trade. The cable was manufactured without any specification of electrical or physical standards – it was not known at the time that such standards were required, what they ought to be, nor indeed how they would be measured. The cable failed after a few messages. The second attempt made in 1851[9] became the first successful international submarine cable and it was the basis for future submarine telegraph cables for many decades to come, although improvements were made after long and expensive experimentation.

Submarine telegraph cables consisted of two parts. The first part was the inner 'core' which was made up of a conductor of copper wires (which carried the electrical impulse, the actual 'message' itself) and the insulation of gutta percha. Second was the outer, protective layers that are referred to as the 'armoring'.

6 See for example, *Report of the Joint Committee Appointed by the Lords of the Committee of the Privy Council for Trade and the Atlantic Telegraph Company to Inquire into Construction of Submarine Telegraph Cables, Together with the Minutes of Evidence and Appendix*, Command Papers, [2744] LXII. 591, HMSO, London, 1861, (hereafter Jt. Comm.), p. 105.

7 C. Bright, *Submarine Telegraphs*, p. 4.

8 C. Bright, *Submarine Telegraphs*, p. 252.

9 C. Bright, *Submarine Telegraphs*, pp. 10–12.

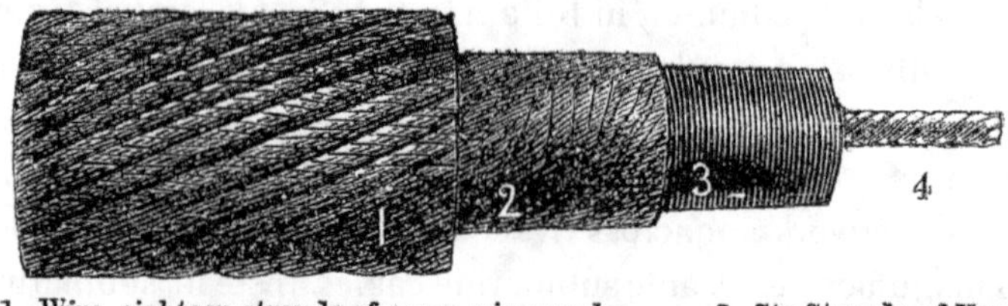

1. Wire, eighteen strands of seven wires each. 2. Six Strands of Yarn.
3. Gutta Percha, three coats. 4. Telegraph Wires, seven in number.

SMALL CABLE, 11-16THS OF AN INCH IN DIAMETER, SHOWING THE VARIOUS COVERINGS.

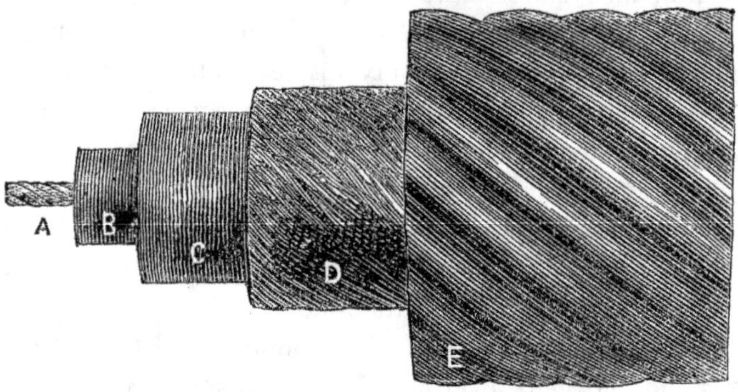

A. Telegraph Wires. B. Gutta Percha, three coats. C. Gutta Percha and Sawdust, two coats. D. Spun Yarn. E. Twelve Solid Wires.

LARGE CABLE, 1½ INCH IN DIAMETER.

PLATE 3 *Design of submarine telegraph cables.*
Source: *Illustrated Times*, London, 1 August 1857, p. 93.

The armoring could be layers of material such as hemp, and outer layers of wire wrapped in a spiral formation. All future submarine cables essentially consisted of these two parts, i.e. core and armoring, although the manufacturing and electrical features improved over future years. The cables shown at Plate 3 are two early examples – the core was always of copper, the outer layers varied according to where the cable was to be laid and manufacturer's and contract specifications.

Werner Siemens wrote that, 'after the successful Channel cable was installed, a veritable cable fever broke out especially in England'.[10] Within two years of the successful Channel cable, the Gutta Percha Co had produced 1300 miles of insulated cable.[11] The Siemens family and associates also began manufacturing

10 G. Siemens, *History of the House of Siemens Vol. 1: The Era of Free Enterprise*, (Trans. Rodger, A.F.), Freiburg/Munich, Karl Alber, 1957, p. 37.

11 G.L. Lawford and L.R. Nicholson, *The Telcon Story*, London, The Telegraph Construction & Maintenance Co. Ltd, 1950, p. 41.

cables (as *Siemens and Halske*), in both London and Germany. Another Company, known as Silvertown, became a successful London based manufacturer. German companies such as Norddeutsche Seekablewerke (NSW) and a Danish, French and an Italian company made cables. Another British Company called Henley's produced cables, including for the niche market, manufacturing strong cables for the 'shore end' cables used where cables came to shore in rocky or difficult situations. The Gutta Percha Co was by far the largest manufacturer, with Siemens and Silvertown the next most significant manufacturers in the nineteenth century. Some of the original companies exist today in the submarine cable business (after some strategic realignments) including Siemens and NSW. Despite the continuing expansion of submarine telegraphy, the technology was not the result of fully developed scientific knowledge or skilled manufacture at the time. Enough of the early ventures developed along short routes were, however, successful and as a result thoughts turned to the possibility of laying a cable across the Atlantic or on longer routes. There were two international submarine cable endeavors that were pivotal in the history of global communications – and in turn they altered the course of cable manufacturing practice and the trade in gutta percha. These two projects were the first (two) Atlantic cable projects and a Red Sea Cable which was to have been part of a new submarine telegraph link between Britain and India. These endeavors became causes célèbres – not because they were a success, but because they failed.

Telegraphic communications networks had developed between Britain and Ireland and between those countries and Europe, and within North America. Instant communication between North America and Britain/Europe had not been achieved yet it was an objective which had commercial, political and social attractions. By comparison, information otherwise took around fourteen days by steam shipping to reach the other side of the Atlantic (or much longer under sail).[12] A project to lay a trans-Atlantic cable was commenced and, although the project was conceived by an American (Cyrus Field), it was almost entirely British capital that funded the venture. The Atlantic Telegraph Co was formed and an order was given to the Gutta Percha Co for preparation of the cable core. The core consisted of 107 lb of copper wire and 261 lb of gutta percha insulation per nautical mile – or 300 tons of gutta percha – costing £40 per mile.[13] The cable was produced in relative haste and without adequately developed specifications. The result was a somewhat poor quality cable although it

12 P.J. Hugill, *World Trade Since 1431: Geography, Technology and Capitalism*, Baltimore, Johns Hopkins University, 1995, p. 127.

13 C. Bright, *The Story of the Atlantic Cable*, London, George Newnes, 1903, p. 49.

was not clear at that time what a *good* quality cable was. The cable venture was the first to be attempted across such a length and at such depth. There were many technicalities that might have been addressed but these were not recognized in the state of knowledge at that time.

The first Atlantic cable attempt was one born in part of haste and naiveté – it was at risk of failing – and it did.

Field and other investors decided to make a second attempt in 1858, despite views that the attempt was the product of stubborn ignorance.[14] Again, the core was made by the Gutta Percha Co. A new, more robustly constructed cable with increased gutta percha insulation was successfully laid and, as a result, direct communication across the Atlantic was established. Jubilation followed: songs were written, public firework displays held, crowds and newspapers proclaimed the beginning of a new world of peace and understanding. Then, after just six weeks, the Atlantic telegraph cable failed. The *New York Leader* described those involved with the cable as 'a pack of asses' and, as Willoughby Smith, Manager of the Gutta Percha Co put it: 'Of course it was thought that the gutta percha had proved wanting'.[15] The failure of the cable was a shock and a humiliation after the preceding euphoria; it shattered the confidence of the public in not just telegraphy but also the sense of man's ability to control nature through achievements in science and commerce. An explanation was required.

Then, in 1859, another important cable connection failed. This was a cable via the Red Sea to enhance communication between Britain and India. Telegraphic communication with India was vital because:

> It meant no less than the intelligence, the wealth, and the resources of thirty millions of British subjects at one end of the wire, and the wants, the interests, and the industries of upwards of one hundred millions of the Queen's subjects at the other end of the wire… [16]

Again, problems with the quality of manufacturing, and a lack of understanding about what was needed to make a good cable, lay at the heart of the problem. The British government was sensitive to memories of the then recent Indian Mutiny (proving the importance of telegraphy to colonial and commercial life

14 C. Bright, *The Story of the Atlantic Cable*, p. 113.

15 26 September 1858, in G.P. Oslin, *The Story of Telecommunications*, Macon, Mercer University Press, 1992, p. 172; W. Smith, *The Rise and Extension of Submarine Telegraphy*, London, J.S. Virtue & Co, 1891, p. 49.

16 C.T. Bright, 1866, p. 42.

in Britain) and the failure of the Atlantic cable connection (a warning about the as yet unproven nature of submarine telegraphy); but it had still agreed to underwrite part of the costs of a submarine telegraph line between Suez and Karachi and had lost more than £1,000,000 on these projects.[17] Clearly, the Government wanted the technology to work and it could not continue providing financial guarantees for cable projects until the technology and manufacturing practice had been objectively assessed and a way found for it to work. Moreover, the commercial future of international submarine cable manufacture – a largely British commercial endeavor – required such certainty. Britain's international standing and commercial reputation were at stake.

A Joint Committee was established by the British Board of Trade to investigate the entire question of submarine telegraphy. At the time of the Committee's survey, it estimated that a total of 11,364 miles of submarine cables had been laid, of which just over 3,000 miles was working (the total distance included 2,200 miles for the failed Atlantic cables and 3,499 for the failed Red Sea and India cables).[18] If the Committee could resolve these problems it would not only reassure investors, but also 'secure the technical basis for future success, by discovering where past inadequacies could be remedied'.[19] The Committee's Report was one of the earliest studies to integrate scientific, manufacturing and commercial investigation. It addressed questions related to scientific measurement, particularly the electrical and physical properties of cables (including requirements for the use of gutta percha). Moreover, it put such measurement and scientific study into a commercial context. It laid down ground rules for quality manufacture and practice. The Committee concluded that 'the greatest precaution is necessary in employing any new material'.[20] The Committee was referring to an important development that occurred while it was still taking evidence: an improvement in the processing of the gutta percha used for cable manufacture was adopted by Willoughby Smith of the Gutta Percha Co. Oddly, the earlier history of experimentation with gutta percha in many fields of endeavor appears to have had unexpected returns: in large part the new processes adopted by the Gutta Percha Co built upon a patent bought by the company from Edwin Truman, a dentist who had purchased gutta percha from the company for his 'professional purposes'.[21]

---

17 Jt. Comm. 1861, p. x.

18 Jt. Comm. 1861, pp. v–vi.

19 C. Smith and M.N. Wise, *Energy and Empire. A Bibliographical Study of Lord Kelvin*, Cambridge, Cambridge University Press, 1989, p. 675.

20 Jt. Comm. 1861, p. xxiv.

21 Smith, p. 97.

Truman experimented with ways to clean and treat the product to make it suitable for dental applications. Because of his attempts to find a way to purify gutta percha so that it could be injected into the human dental root canal, Truman unwittingly helped to solve the technological challenge of producing reliable insulation for undersea cables. Indeed, he became an active, if somewhat forgotten figure in the history of gutta percha and its telegraphic application (see Appendix 1.2). The Committee noted that 'the Company ... sorted the gum with the most elaborate care, and soon produced a tolerably pure material'.[22] This gave reason for optimism about what material might be used for insulation: in turn, it provided for an optimistic outlook for the gutta percha trade.

By endorsing the 'newly introduced pure gutta percha' as the best insulator for submarine cables,[23] the Committee's Report ensured that the demand for considerable quantities of gutta percha would continue for many decades, and that the quality of gutta percha would be addressed by cable manufacturers who would establish rigorous testing regimes for its use.[24] The Committee's work reflected the great enhancement that scientific study brought to commerce and economic development in the nineteenth century. In the case of gutta percha, this led to it becoming what Hunt has termed the most intensively studied and tested substance in the world.[25] In later years, specifications for cable manufacture and laying became far more detailed; in using them, contractors and clients had agreed standards which had not previously been available. A result of setting out such standards, however, was that the role of gutta percha in the communications industry was reinforced, making it difficult for other products to be put forward as alternatives. The Committee boosted the science and commerce of submarine telegraphy but at the same time, in identifying what it considered to be reliable guidelines for the future, the Committee's report also resulted in a conservative approach to future experimentation, including the important question of cable insulation. The implication was that alternatives to the use of gutta percha could be risky. Cable manufacturers would have required a good reason to veer from these guidelines when

22 Jt. Comm. 1861, p. xviii; see also G. Cookson and C.A. Hempstead, *A Victorian Scientist and Engineer. Fleeming Jenkin and the Birth of Electrical Engineering*, Aldershot, Ashgate, 2000; B.J. Hunt, 'Insulation for an empire. Gutta percha and the development of electrical measurement in Victorian Britain', in *Semaphores to Short Waves*, ed. F.A.J.L. James, Royal Society of the Arts, London, 1998, pp. 85–104 and W. von Siemens, *Scientific and Technical Papers*, 2 Vols, London, John Murray, 1892, Vol. 1, p. 138.

23 Jt. Comm. 1861, p. xxv.

24 Jt. Comm. 1861, pp. xxv and xxxiv, and Smith, *The Rise and Extension of Submarine Telegraphy*.

25 B.J. Hunt, *Semaphores to Short Waves*, p. 86.

promoting their own cables to investors or clients. Clients themselves would be hesitant to seek to invest in a project that might involve an expensive failure, so would also be unlikely to take the risk of looking for novel, if cheaper, alternatives. In its conclusions, the Joint Committee endorsed the concept of telegraphy, particularly submarine telegraphy, and set out recommendations which it considered would ensure that:

> ... if regard be had to the principles we have enunciated in devising, manufacturing, laying, and maintaining submarine cables, this class of enterprise will prove as successful as it has, hitherto, been disastrous.[26]

Even as late as 1931 it was noted that 'no manufacturing process requires a higher degree of insurance against occasional defects than does the submarine insultion art, a fact that has engendered a strong conservatism in the industry'.[27]

In the context of such uncertainty, manufacturers and their clients sought the safety of a conservatism that was seen to bring reliability. After this period the production of submarine cables saw relatively few changes to the physical design of cables; over three decades later, similar guidelines, and requirements for gutta percha insulation, were still being followed by cable customers, including Governments.[28] The specifications included extensive requirements for scientific measurement, particularly in relation to electrical matters as applied to quality controls both for cable manufacture generally, and in the demands placed on quality gutta percha. Such standards were not known in the early years of telegraphy; nor would it have been possible to measure them in any case – progress in the understanding and measurement of electricity and its application in a commercial setting were required to achieve this. Whilst the demand for telegraphic communication grew and cable manufacture continued to be refined, relatively little occurred in the technological sense after the Joint Committee issued its Report. No one wanted to take the risk of designing a cable that might fail. In any case, most cable specifications called for cables that matched the kind suggested by the Joint Committee – not just in Britain but elsewhere – causing one U.S. rubber industry writer in 1893 to complain:

26 Jt. Comm. 1861, p. xxxvi.

27 A. R. Kemp, 'Paragutta, A New Insulating Material for Submarine Cables', *Bell System Technical Journal*, Vol. 10, Iss. 1, 1931, p. 133.

28 For example, the 1894 Atlantic Cable Specification described in C. Bright, *Submarine Telegraphs*, p. 514.

> The Government has never sanctioned any attempt to insulate its submarine cables with India-rubber. It does not allow the rubber manufacturers to make specifications. It insists on its own specifications, and will have nothing but gutta-percha. … It is prejudice – nothing else.[29]

Beyond these political and commercial machinations, at a more prosaic level, the simple manufacturing demands of the telegraph cable industry meant that the demand for quality and quantity of gutta percha continued. Submarine telegraph cables remained very expensive to produce – between five and seven times as much as a land line, or around £150 per nautical mile in 1898.[30] Quality cables required expensive materials. In 1898, for example, when copper cost around 7d per pound, gutta percha could vary between 3s to 6s (shillings) per pound (that is, between five and ten times the cost of copper). If it is recalled that heavy cables required between 320 lb to 400 lb of gutta percha per nautical mile,[31] it is not surprising that anyone with an interest in telegraphy was concerned to maximize the likelihood that a reliable cable was produced – no wonder Bright observed that, when it came to new ideas in cable technology, 'everybody preferred to wait for somebody else to make the experiment first'.[32] Cables were also heavy, dangerous, cumbersome and expensive to lay – the cost being about half as much again as the cost of producing the cable.[33] It was not only the cable repairs that were very expensive. When an Aden-Bombay connection was broken, it was out of service for 251 days – 148 days of that delay were caused by a monsoon which prevented the repair being undertaken.[34] These figures underline why cable manufacturers and customers were so cautious about investing in new processes or products to be used in cable manufacture. It was not likely that governments or companies commissioning new cables would want to speculate on new design or materials – it was, then, only gutta percha, good quality gutta percha, that was sought. Cable manufacturers determined what they thought was the best type of gutta percha for their needs. One of the judgements made was that only particular types of gutta percha were suited for the needs of cable manufacture, and these were priced accordingly. Interestingly, Edwards told me that such views continued even

29 *India Rubber and Gutta Percha and Electrical Trades Journal* (hereafter IRGPETJ), London, 8 December 1893, p. 131.

30 C. Bright, *Submarine Telegraphs*, p. 163.

31 See E.F.A. Obach, 'Cantor lectures on gutta percha', *Journal of the Society of the Arts*, 2, 46, 1898, p. 102.

32 C. Bright, *Submarine Telegraphs*, p. 111.

33 C. Bright, *Submarine Telegraphs*, p. 163.

34 C. Bright, *Imperial Telegraphic Communication*, London, P.S. King & Son, 1911, p. 71.

into the 1950s when customers of Telcon continued to insist on gutta percha of only the 'best Pahang White type'. This was despite the fact that gutta percha was at that time little used for cable purposes, but more for industrial and golf ball production; and even though customers had often (unknowingly) been supplied with what Telcon produced. The company mixed the varieties of gutta percha *it* considered best met standards required for their customers, and frequently this was in fact a blend of types, not necessarily containing 'Pahang white' at all.[35] In the nineteenth and early twentieth century the gutta percha collected in Sarawak, for example, was amongst the better regarded – although not the most expensive – of the gutta percha types on the market.

Cable manufacturers were interested in more than just the insulating qualities of gutta percha, as can be seen in the detailed measurements below in Plate 4.

From Obach's table it can be seen that the gutta percha consisted of both gutta percha and resin (and dirt and water). Too high a level of resin produced

CHEMICAL COMPOSITION, PHYSICAL, MECHANICAL, AND ELECTRICAL PROPERTIES OF VARIOUS BRANDS OF GUTTA PERCHA.

Description of Material.			Percentage Composition.					Quality.	Physical Properties.			Mechanical Properties.		Electrical Properties (per cube knot.)	
Group.	Name of brand, and Grade.	No. of Lots tested.	Gutta G.	Resin. R.	Dirt. D.	Water W.	Ratio G/R	q. (r/10)	Temperature when commencing to soften ° C.	Temperature when becoming pliable. ° C.	Time hardening. Mins.	Tensile strength lbs. per sq. in.	Elongation. percent.	Insulation. 75° F. 2nd min. Meg-ohms.	Inductive capacity. Micro-farads.
I.—GENUINE.	Pahang ... I.	19	80·0	17·7	1·4	0·9	4·52	1c	48·8	66·1	2¾	5,067	444	1,077	·0511
	Pahang ... II.	9	77·8	19·0	1·8	1·4	4·10	2	48·3	64·4	4	4,138	414	2,662	·0559
	Pahang ... III.	12	74·8	22·0	1·5	1·7	3·41	2a	45·5	65·5	3½	3,940	410	4,326	·0558
	Average	40	78·1	19·2	1·5	1·2	4·08	2	47·7	65·5	3⅜	4,577	427	2,408	·0535
	Banjer red I.	27	70·5	26·9	1·4	1·2	2·62	2b	45·0	67·2	5	4,123	417	3,723	·0542
	Banjer red II.	22	67·5	29·6	1·5	1·4	2·28	3	44·4	65·5	5	3,546	377	6,113	·0559
	Banjer red III.	21	62·0	35·0	1·6	1·4	1·77	3c	42·2	65·0	11	2,932	384	10,150	·0580
	Average	70	67·0	30·2	1·5	1·3	2·22	3	43·8	66·1	6¾	3,584	394	6,400	·0559
	Bulongan red I.	10	73·4	24·2	1·4	1·0	3·03	2c	46·1	64·4	4	4,200	440	4,511	·0537
	Bulongan red II.	7	67·7	29·8	1·4	1·1	2·27	3	45·5	62·7	7	3,400	415	5,292	·0554
	Bulongan red III.	5	60·3	37·4	1·3	1·0	1·61	3b	43·8	62·2	15¾	2,494	383	17,200	·0579
	Average	22	68·6	29·0	1·4	1·0	2·37	3	45·5	63·3	8	3,557	419	7,643	·0552

PLATE 4 *Extract of gutta percha analyses from Obach.*
Note: The data reveals the extent to which gutta percha was studied from a scientific or manufacturing viewpoint and the demanding requirements that were set down for gutta percha by cable manufacturers.
Source: E.F.A. Obach, 'Cantor Lectures on Gutta Percha', London, *Journal of the Royal Society of Arts*, 1898, p. 179.

35 L. Edwards, Manager of the Telcon Balata [and Gutta Percha] Processing Plant, Greenwich; *Telcon and Gutta Percha, Discussion with Author*, 19 July 2007.

cables with greater strength but less flexibility and reduced dielectric (insulating) capacity. Too high a level of gutta percha at the expense of low levels of resin led to improved dielectrics but resulted in physically weaker cables. Obach noted that the commercial value of gutta percha depended principally upon the relative amounts of these two components.[36] The ideal was to firstly understand *what* the balance should be, secondly to *measure* it in any given supply of gutta percha. This is a very simple example of some highly detailed analyses of gutta percha undertaken over several decades by many scientists engaged in the scientific and commercial aspects of submarine telegraphy. The tensile strength (seen in Obach's chart) of gutta percha was very significant from the perspective of loading the cable on board ship, and more importantly of 'paying out' the heavy cable from ship to ocean floor. Poorly produced cables could snap during laying; moreover, they could be inclined to twist which could damage the cables or cause large tangles, once described as resembling a large cargo of live eels.[37] Problems with twisted, knotted or damaged cables could cause parts of the cable to be destroyed altogether, or require expensive attempts to recover and replace the cable. Cable strength was also important in that any future attempts to raise cables from the sea floor put great strains on cables as they were hoisted to shipboard level for repair. Obviously, the cable armoring (the protective wire covering) was an important aspect of this capability. All of these processes could be put at risk by the use of poor quality gutta percha, poor manufacturing methods, and of course failures of the technology behind that manufacturing.

The detailed measurements set out in Obach's table (which represents just a minor example of much wider analytical research from the period) and the Contractor Specifications noted above underline the considerable degree of quality control in manufacturing of submarine cables and the level of scientific study and measurement that lay behind it. It was, then, not just the availability of gutta percha that enabled global submarine communication links, but the application of scientific research to understand the latex and to determine how its properties were to be measured and managed during cable manufacture. The study and testing of gutta percha had begun from the period when Werner Siemens first took an interest in it, with extensive work carried on over decades to come by many of the other scientific minds associated with telegraphy including Whitehouse, Thomson, Jenkin, Maxwell and Faraday to name just a few.[38] Faraday never claimed to have been the first to suggest the potential of gutta percha for submarine telegraph applications, but his diary entry in

36 Obach, 1898, p. 23.

37 A.C. Clarke, *Voice Across the Sea*, New York, Harper & Brother, 1958, p. 45.

38 See B.J. Hunt, *Semaphores to Short Waves*, pp. 85–104 for a useful overview of this period in relation to gutta percha.

1848 showed him to have been one of its earliest analysts. It was here that he wrote that 'The Gutta Percha is probably a complicated substance' quoted at the beginning of this book.[39] In 1848 he raised in the *Philosophical Magazine* its potential both as an electrical insulator and as a material of interest for study by scientists (as an aside, it is clear that he sourced some of his experimental gutta percha supplies from shoemakers).[40] In later years, these systematic testing and manufacturing practices became industry standard during cable production. The submarine telegraph cable in Plate 5 below is being manufactured at Henley's North Greenwich, London in 1892.

Hunt has written of the scientific and technical challenges facing submarine telegraphy and concludes that after the 1851 cable many cables were laid

PLATE 5 *Cable manufacture.*
Source: *The India Rubber and Gutta Percha and Electrical Trades Journal*, 'Homes of Industry', 8 September 1892, p. 39.

39 M. Faraday, *Faraday's Diary, Volume 5, September 6, 1847 – October 17, 1851*, London, G Bell and Sons Ltd, (Reprint 1934), 1848, p. 45.

40 M. Faraday, 'On the use of gutta percha in electrical insulation', *The Philosophical Magazine*, xxxii., 1848, p. 167.

with 'more enthusiasm than expertise'.[41] Cookson and Hempstead, in considering the early years of cable telegraphy concluded that 'some of the bitter feuds within the industry reflect the fact that there was no clear agreement on what constituted competence'.[42] Over several decades, considerable scientific effort brought the rewards of competence to the industry. Understanding and measuring – as seen in the examples above in the case of gutta percha – was just one facet of this developing competence. Such measurement was possible only after agreement had been reached upon what to measure, the units of measurement to be adopted, and a technique for measurement. Gooday sets out the problems that physicists, telegraphers and 'electricians' overcame – including the 'material, moral and managerial' problems and theoretical matters, such as 'measurement', that underwent a 'form of industrialization'.[43] Thomson's address on the 1865 Atlantic cable is a useful context of the scientific, practical and commercial complexities of telegraphic cable ventures.[44] As Hunt notes, later scientific efforts owed a great debt to the work and resources that the companies and individuals who worked in the cable industry allocated to resolving such problems. William Thomson (later Lord Kelvin), for example, worked closely with the cable companies and took out patents for much of his work in the telegraphic industry.[45] His contribution to the success of submarine telegraphy was in large degree due to his work on theoretical matters at the heart of the 'electrical' industry. For him, the problems of retardation and propagation of electricity through submarine telegraph wires were:

> the most beautiful subject possible for mathematical analysis…the differences between the insulating power of gutta percha and the coatings of tar and pitch around it…[46]

---

41 B.J. Hunt, *The Maxwellians*, Ithaca, Cornell University Press, 2005, p. 53, see also B.J. Hunt, 'Doing science in a global empire: cable telegraphy and electrical physics in Victorian Britain', pp. 312–333, in *Victorian Science in Context* edited by B. Lightman, Chicago, University of Chicago Press, 1997.

42 G. Cookson and C.A. Hempstead, p. 36.

43 G.J.N. Gooday, *The Morals of Measurement: Accuracy, Irony, and Trust in Late Victorian Electrical Practice*, Cambridge, Cambridge University Press, 2004, p. xx.

44 W. Thomson, *Atlantic Telegraph Cable. Address Delivered before the Royal Society of Edinburgh December 18, 1865, with other Documents*, London, William Brown & Co, 1866.

45 See C. Smith and M.N. Wise, especially Ch. 19.

46 S.P. Thompson, *The Life of Lord Kelvin*, 2 Vols, Chelsea Publishing Company, New York, 1910, as 2nd edn Reprinted New York, Chelsea Publishing Company, 1976, Vol 1, p. 336.

The development of submarine telegraphy, and achieving an understanding of the chemical, physical and electrical properties of all aspects of submarine cables and gutta percha insulation represented a major scientific and commercial challenge in the nineteenth century. This continued throughout the latter part of the century.

Growth Years of Submarine Telegraphy and Its Dependence upon Gutta Percha

> Oceanic telegraphy is absolutely dependent on this substance. Its supply is a matter of interest to the whole civilised world.[47]
>
> SIR WILLIAM THISELTON-DYER Director, Royal Botanic Gardens Kew

Those who had been involved with the previous, unsuccessful, Atlantic cable effort tried again in 1865 – when the cable broke – but succeeded in 1866 – during which attempt they also successfully repaired the 1865 cable.[48] After the Atlantic cable was finally proven viable, others followed. The success of the Atlantic cable connections was not just important in itself and to trans-Atlantic telegraphy, but also because it proved the feasibility of long distance, deep ocean telegraph projects generally. Trans-Atlantic communications once established did not end with one cable connection – there was a growing demand for increasing communication and commercial reasons to add more cable connections. By 1903, there were 15 Atlantic cable connections, totaling approximately 40,000 nautical miles of cable (see Map 1 above).[49]

The most ambitious international cable connections were, however, achieved in 1902 and 1903 when trans-Pacific cable connections were established. A British connection known as 'The All Red Line' and a nominally U.S.-owned cable connection to Honolulu and the Philippines were laid. The British cable connection crossed the Pacific from Canada to Australia, thereby

47 W.T. Thiselton-Dyer, Correspondence to C.P. Lucas (Colonial Office), in *Kew: Malaya, Gutta Percha, 1876–1904*, Royal Botanic Gardens Kew, Archives, 1899, f.45.

48 See W.H. Russell, *The Atlantic Telegraph*, London, Day & Son, 1865.

49 C. Bright, *The Story of the Atlantic Cable*, p. 218.

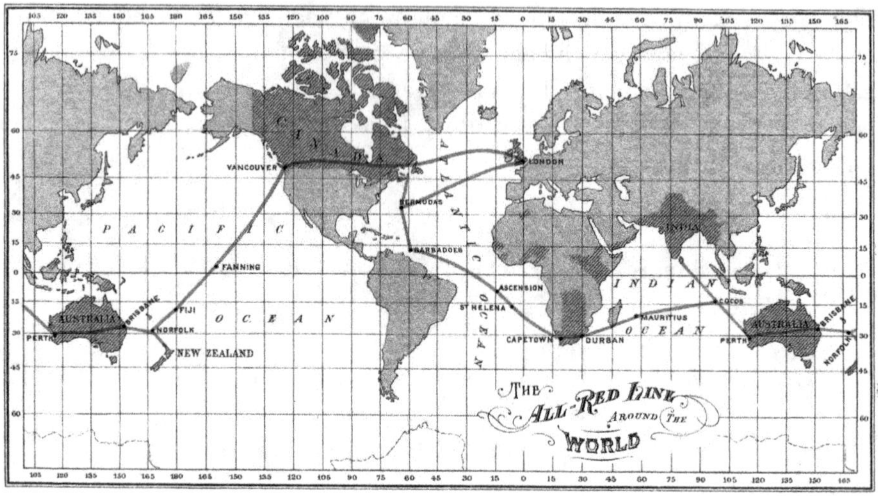

MAP 2 *The All Red Line.*
Note: When the *All Red Line* was laid across the Pacific it coincided with the laying of another Pacific cable and other cable ventures. The ventures caused a massive demand for gutta percha that had a significant effect on price and the collecting and trading activities of the Southeast Asian region, including the entrepot of Singapore and the forests of Sarawak.
Source: G. Johnson, *The All Red Line. The Annals and Aims of the Pacific Cable Project*, Ottawa, James Hope & Sons, 1903, Preface.

completing a ring around the globe to connect all those countries of the British Empire – countries marked in red on the world map, the 'All Red Line' as seen in Map 2.

The *All Red Line* completed a global link of the commercial and strategic interests of the scattered outposts of the British Empire. It ensured that during a time of war there would be not just one, but multiple connections around the globe – for example, 49 separate cables would need to be cut to isolate Britain because of the network linkages the *All Red Line* would provide.[50] All of those connections would be on British or imperial territories: that is, the network would be entirely independent of 'foreign' countries.[51] The need for such a cable was felt not only in Britain: pressure and funding for it also came from the governments and businesses of Canada, Australia, and New Zealand.[52] The

50 See P.M. Kennedy, 'Imperial cable communication and strategy', *The English Historical Review*, Vol. 86, No. 341 (October), 1971, p. 741.

51 Kennedy, p. 733.

52 See Kennedy 1971; D.R. Winseck and R.M. Pike, *Communication and Empire: Media, Markets and Globalization 1860–1930*, Durham, Duke University Press, 2007, Ch. 5.

All Red Line became, in a symbolic sense, the hallmark of both British imperial dominance and of the worldwide telegraphic cable production and networks that underpinned it. It also reflected the importance of British global trade networks that relied on those resources. It is beyond the scope of this work to expand on the implications of the British cable communications network, but it must be said that the British cable capacity as 'crowned' by the *All Red Line* would have been noted by other nations, especially since so many nations were dependent upon British cable industry and the cable linkages.

As if to respond to that emphasis, the U.S. also finalized its own Pacific cable network at this time. While it lacked the British capacity to produce submarine cables, the U.S. had commercial and strategic concerns both globally and in the Pacific, that emphasized the need for it to maintain international cable connections not fully dependent upon, or vulnerable to, British or other interests. Some of those interests had been made prominent after the U.S. war against Spain brought territorial gains such as the Philippines and Guam, in addition to its interests in Hawai'i: thus making 'speedy cable communication between the United States and all these Pacific islands... imperative' according to the President in a special address to Congress in 1899.[53] Winseck and Pike note that the *All Red Line* and the U.S. Pacific Cable 'campaigns' seemed almost mutually reinforcing.[54] Both cable networks were based upon commercial and strategic interests and the need for communications self-reliance and the protection from vulnerability to attack from belligerents during hostilities. U.S. colonial and trade interests in the Pacific made cable communications vital: from there, cable connections could be made with important trade points in Japan, China and Southeast Asia.[55]

The *All Red Line* was the longest gutta percha insulated cable project ever undertaken. It had been discussed and argued over for more than a decade previously and, because of these discussions in the few years prior to its achievement, cable manufacturers had commenced building up large stocks of gutta percha to cope with the possibility of being the winning contractors to lay the cable. The U.S. Pacific Cable project was also a massive venture and, although

53 G.O. Squier, 'An American Pacific Cable', Paper Presented at Thirteenth Meeting of the *American Institute of Electrical Engineers*, New York and Chicago, December 27, 1899, Vol. XVI Transactions, p. 603.

54 Winseck and Pike, p. 168.

55 See also Squier; and Winseck and Pike, p. 170 for an overview of the cable's history; J.R. Winkler, *Nexus. Strategic Communications and American Security in World War I*, Cambridge, Mass., Harvard University Press, 2008, Ch. 1 for an overview of the U.S. strategic communications vulnerabilities.

it was laid in 1903, the cable itself was manufactured in Britain in 1901.[56] These two Pacific cables came soon after two Atlantic cable connections (1897–8 and 1900). The demand for gutta percha to produce these cables ensured that cable manufacturers were active in the marketplace; so, in the period of the end of the nineteenth and the early twentieth century, there was a significant demand for gutta percha that was felt throughout the collecting region including Sarawak.[57]

With the completion of the two Pacific cables, the global telegraphic network laid with gutta percha insulation totaled just over 230,000 nautical miles or around 426,000 kilometers (to put that figure in another context: this was more than enough cable to stretch from the earth to the moon and part of the way back again).[58] If the unknown total of underground cables insulated with gutta percha is added to that figure, then the scale of the coverage was even greater. Up to this time no other insulating material was found suitable other than the limited use of india rubber (the likely use of it being of the order of 15,000 nautical miles).[59]

---

56 *India Rubber World*, New York, 1 February 1902, p. 136; J. Ahvenainen, *The Far Eastern Telegraphs, The History of Telegraphic Communications between the Far East, Europe and America before the First World War*, Helsinki, Suomalainen Tiedeakatemia, 1981, p. 173.

57 See H.L. Terry, *India-Rubber and Its Manufacture. With Chapters on Gutta-Percha and Balata*, London, Archibald Constable & Co. Ltd., 1907, p. 280.

58 *The Telegraph Construction and Maintenance Co. Ltd. On the Occasion of the Visit of the Delegates to the International Telegraph Conference*, London, Telegraph Construction and Maintenance Co. Ltd, 1903, p. 22, and see Appendix 2 here for calculations made for this book.

59 Produced by two companies, one an English company called *Hooper's* (the company changed ownership at various times). While exact figures are not certain, the total length of cables Hooper's produced was probably less than 10,000 n.m – known lengths can be calculated to just under 8,000 n.m. (see Glover, 2017). Between 1901 and 1904, an American company, the *Safety Insulated Wire & Cable Co* produced rubber insulated cable networks totaling around 5,000 nautical miles. These included a network in the Philippine archipelago and another to connect the contiguous U.S. States with Alaska (making up the undersea part of what became known as WAMCATS – the Washington Alaska Military Cable and Telegraph System, see for example D. E. Jessup, Connecting Alaska: The Washington-Alaska Military Cable and Telegraph System, *The Journal of the Gilded Age and Progressive Era*, Vol. 6, No. 4 (October) 2007, pp. 384–408. The Company also produced a network in the Gulf of Mexico for the Mexican Government in 1902–3. There is no known evidence that it produced cables after 1904. Details at W. Glover, *The Safety Insulated Wire & Cable Company*, New York, http://atlantic-cable.com/CableCos/SafetyCable/index.htm, 2013; and United States Department of Commerce and Labor, Bureau of Census, *Manufactures 1905, Part IV, Special Reports on Selected Industries*, Washington, Government Printing Office, 1908, p. 206.

While commercial, strategic and imperial in their intent, the result of these two Pacific cable endeavors in terms of the gutta percha trade was the same: an extraordinary level of demand. The global telegraphic network connected by the Pacific cables was the crowning achievement and the finale of the great era of the gutta percha trade. The manufacture of all these cables, and the consequent demand placed on supply of gutta percha, was unprecedented – and never repeated. Figure 1 is a summary of submarine telegraph cables produced from the (first attempted) 1850 English Channel cable through to the 1903 Pacific cables.

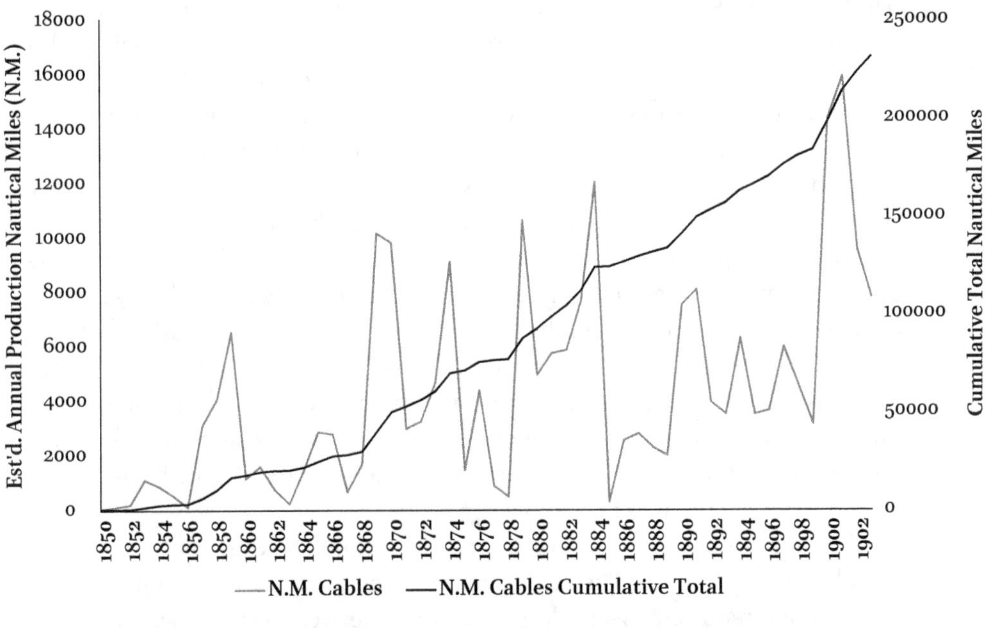

FIGURE 1 *Estimated world submarine cable production 1850–1903.*
Sources: Estimates calculated from data in *Pall Mall Gazette*, 'The West Indian Telegraph Expedition', 1870, July: n.p.; United States Hydrographic Office, *Submarine Cables*, Washington, Dept of Navy, Bureau of Navigation, 1892, pp. 41–59; India *Rubber World*, New York, 1 February. 1902, p. 136; Telcon, *The Telegraph Construction and Maintenance Co. Ltd on the Occasion of the Visit of the Delegates to the International Telegraph Conference*, London, Telegraph Construction and Maintenance Co. Ltd, 1903, Appended; C. Bright, *The Life Story of Sir Charles Tilston Bright Civil Engineer, with Which is Incorporated the Story of the Atlantic Cable and the First Telegraph to India and the Colonies*, London, Constable & Company, 1908, p. 313 and W. Glover, *Cable Timeline 1845–1900* and *Cable Timeline 1901–1950*; 2017/Continuing, http://www.atlantic-cable.com. See Appendix 2, Table 1 for data. The data is a calculation, as far as possible, for the years in which the cables were *manufactured* – rather than the year when *laid* (if different), to more closely parallel likely years that gutta percha was used.

One estimate calculated that the indigenous collecting method probably yielded 250 grams of gutta percha from a 30-year-old tree. Even using a generous estimate of 400 grams yield per tree, and a very conservative allowance of 25 kilograms per kilometer of submarine cable, the estimate calculated it would have required more than 62 mature trees to provide the gutta percha needed for each kilometer of cable. This would put the use of gutta percha by this date, just for submarine telegraphic purposes, at slightly under 27 million trees.[60] The exports of gutta percha from Singapore and Sarawak, especially in the last few years of the nineteenth and early years of the twentieth century, had a significant economic impact and led to a great deal of activity on the part of collectors and traders that is discussed in later chapters. In the previous decades, other peaks in exports were sometimes related to other major cable ventures, especially Atlantic cables or long-distance cables.

With the completion of the Pacific Cables, by a combination of overland and undersea connections, much of the world had at least some form of link to the global network of communications. Many of these cables were laid by one company formed after the 1865 Atlantic cable had been achieved – the Gutta Percha Co was joined with another company to form the Telegraph Construction and Maintenance Company, or Telcon.[61] Telcon became the largest manufacturer of submarine telegraph cables in the nineteenth and early twentieth centuries. Although it did not manufacture all the world's submarine cables, Telcon produced a very large proportion of them. The map below at Map 3 was produced in 1903 to illustrate the extent of Telcon's achievement and market domination. It gives a sense of the broad coverage of the global cable network – insulated with gutta percha.

In the mid-nineteenth century submarine telegraphy became a key feature of global communications. The achievement of the scientific and manufacturing competency required to develop this global network was considerable – a major factor in this achievement was the quest to understand and optimize

---

60 W.R. Tromp de Haas, 1908, *The Netherlands India Government Gutta-Percha Plantations in Java*, Tjipetir, Batavia Government Gutta-Percha Plantations, 1908, p. 3, based on Obach 1897; see also Appendix 1.1.

61 Sometimes referred to as TCM; although employees and those associated with the company called it Telcon. Edwards, who joined the company in the 1930's, told me 'we always called it Telcon' and that even the 'old' employees had always used the term 'Telcon' (L. Edwards, Manager of the Telcon Balata [and Gutta Percha] Processing Plant, Greenwich; *Telcon and Gutta Percha, Discussion with Author*, 19 July 2007). The term Telcon may hark back to nineteenth century share listings: the company appears, for example, in the *Pall Mall Gazette* as 'Tel. Con. and Main. (Limited)'; *Pall Mall Gazette*, London, J.K. Sharpe, 7 November 1866, p. 8, and in later issues.

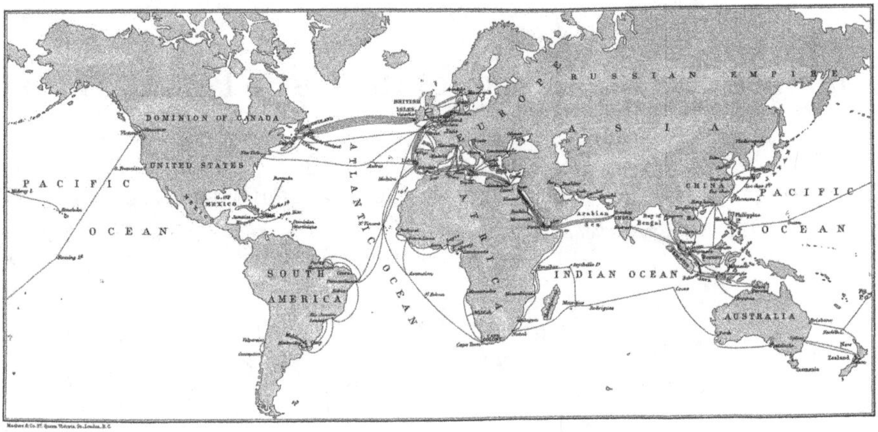

MAP 3 *Global network of submarine telegraph cables constructed by Telcon up to 1903.* *Source*: Telcon, *The Telegraph Construction and Maintenance Co. Ltd on the Occasion of the Visit of the Delegates to the International Telegraph Conference*, London, Telegraph Construction and Maintenance Co. Ltd, 1903, p. 22.

the use of gutta percha. These challenges were complex, but solutions to the problems were developed; and such solutions assumed a continuing supply of quality gutta percha – a sense of which can be gleaned from the cumulative production of cables shown on the graph at Figure 1.

The spread of global trade and economic imperialism could almost be mapped by the spread of the world's submarine telegraph network. The result was a requirement for the continual production of submarine cables from the first international attempt in 1850 onwards. Kuznets saw the geographic spread of modern dynamic growth to less developed areas as being facilitated by changes in transport and communication. The various characteristics of modern economic growth were interrelated and one induced another.[62] Submarine telegraphy was, therefore, integral to global economic growth and the changes that growth brought to most regions of the world. As a result, gutta percha was a globally significant commodity the supply of which was of immense significance beyond its immediate commercial worth.

For the those who supplied it, the possession of gutta percha provided a significant economic asset and integrated them more closely into the world economic system. How this process occurred is examined in the following two chapters.

62 S. Kuznets, 'Modern economic growth: Findings and reflections', *The American Economic Review*, Vol. 63, no. 3, (June.), 1973, p. 250.

CHAPTER 3

The Rise and Challenges of the Gutta Percha Trade

For gutta there is a good deal of business doing in barter[1]
The Weekly Market Reports, Singapore

In previous chapters, it was seen how gutta percha became a product of strategic and economic significance and was the subject of intensive scientific and commercial testing. It was a commodity in constant demand. Essentially, there were two aspects to the European hunt for gutta percha; the first aspect was the search for information and understanding – in a technical and a botanical sense – as attempts were made to understand how to identify and use the best varieties of the gutta percha for cable production. Much of that quest was undertaken in the laboratories and factories of Europe, discussed in the previous chapter. The second approach adopted in the European quest were direct, field expeditions – scientists, traders, adventurers and colonial administrators – went to the potential sources of supply to obtain knowledge or, hopefully, actual sources of supply. These efforts and an overview of the region that supplied gutta percha are examined in this chapter. The survey explains why ultimately it was the regional trade networks in Southeast Asia that were best placed to ensure the supply of gutta percha. The survey begins with a short overview of the region and its history.

The Lands of Plenty: Southeast Asia and Regional Trade

In 1897, a series of lectures on gutta percha was presented at the Royal Society of Arts in London by Dr Eugene Obach, Chief Chemist at Siemens cable factory at Woolwich, London.[2] As part of his lectures he produced a map of the region of the world that supplied gutta percha for submarine cables. Broadly

1 *Straits Times and Singapore Journal of Commerce*, Singapore, 21 February 1849, p. 406.

2 E.F.A Obach, 'Cantor lectures on gutta percha', *Journal of the Society of the Arts*, Vol. 46, 1897, pp. 97–114, 117–133; 1898, pp. 137–164; 1898, pp. 169–197.

 | DOI 10.1163/9789004357280_005

speaking, he was discussing what we now know as Southeast Asia – areas colonized by western powers, all of which were to be investigated by European governments, officials and adventurers in the quest for gutta percha. What had been determined was that the preferred forms of gutta percha came not from all Southeast Asia, but from only a limited area within that region – marked as 'Orbach's proposed area' on Map 4 below. Those areas fell within the area of six degrees north or south of the equator, and 99 to 119 degrees latitude. The area depicted in the map serves to emphasize the very limited part of the world that supplied the commodity underpinning global communication – and the potentially limited number of people engaged in the trade at the supply stage.

By the time gutta percha came to European attention in the nineteenth century, the external trading relationships of Southeast Asia were not new, but the scale of them was. Trans-regional trade networks among Southeast Asia, China, and Arab traders were centuries old, but by the latter part of the nineteenth century, colonial powers – especially Britain, France, the Netherlands, Spain and the U.S. – metropolitan economies – had largely split up the region with

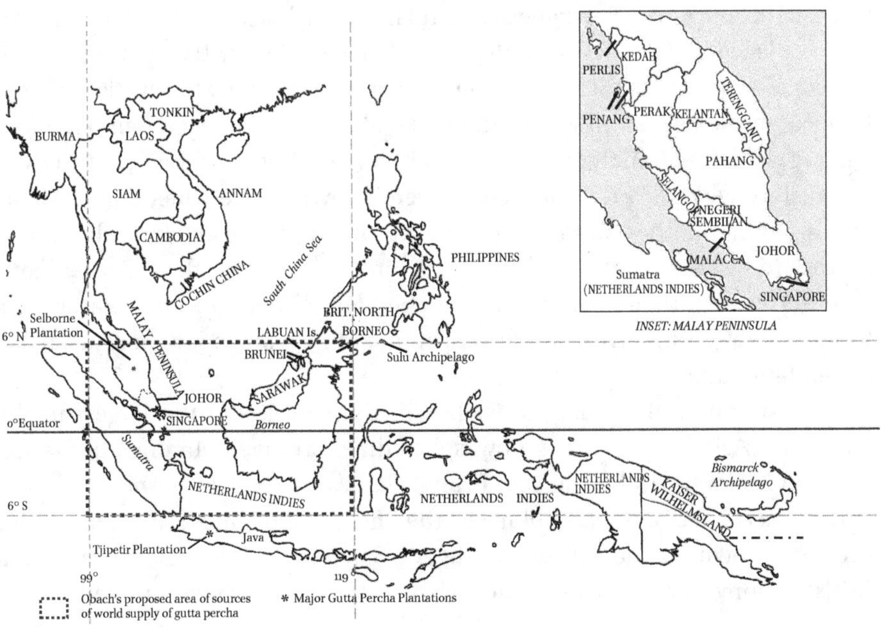

MAP 4 *Historic map of sources of world supply of gutta percha in the nineteenth century.* *Note:* Obach's outline of the region that provided gutta percha for the world's submarine telegraph cables is marked. In fact, British North Borneo and the southern Philippines also supplied gutta percha, although the latter product, justifiably or not, was not well regarded by the telegraphic industry.

varying forms of indirect and more formalized types of administration.[3] Each aimed to maximize the freedom of trade and influence over indigenous populations and to minimize incursions from other colonial powers. Britain, France, and the U.S. (and Germany in the near Pacific region of east New Guinea) were also interested in submarine telegraphy and therefore in the supply of gutta percha, while the Netherlands maximized the trading potential of its colonial possessions in the region, some of which included areas that produced gutta percha. Each of these metropolitan nations was involved in the quest for gutta percha.

The heart of the gutta percha export trade was Singapore, which was part of the British colonial possessions known as the Straits Settlements, on the Malay Peninsula. Penang and Malacca on the Malay Peninsula made up the remainder of the Straits Settlements. The Straits Settlements were to some extent fusion settlements in that they were a result of trade and migration that sprang from British rule and the expansion of global trade networks in the region. The growth of tin mining, and the influx of Chinese laborers and merchants, leads Osborne to describe the Straits Settlements as being 'in but not of the Malay world that surrounded them'.[4] Elsewhere on the Malay Peninsula, Malacca had come under Portuguese control in the early sixteenth century until it was displaced by Dutch control 1641, and then ceded to British administration in 1824. Several Malay Sultanates signed agreements with the British in 1895 to become the Federated Malay States (Perak, Selangor, Negeri Sembilan and Pahang). Other Sultanates under British influence, but not direct control (Kelantan, Terengganu, Perlis and Kedah) were known as the Unfederated Malay States. Johor, another Sultanate on the tip of the Peninsula and adjacent to Singapore was also under indirect British influence. From these Settlements, together with parts of Borneo, the independent Federation of Malaysia was formed in 1963, initially including Singapore. In 1965 Singapore became an independent State.

Borneo is one of the largest islands in the world, and the largest in the Southeast Asian region. It is now split administratively among the nation states of Malaysia, Indonesia and Brunei. Sarawak is part of Borneo and from 1841 it was ruled as a private colony by the Englishmen James and later Charles Brooke, known as the 'White Rajahs' of Sarawak. While not technically a British colony, Sarawak was part of what was loosely termed 'British Borneo'.

3 For earlier Southeast Asian history see for example A. Reid, *A History of Southeast Asia: Critical Crossroads. Blackwell History of the World*, Chichester, Wiley-Blackwell, 2015.

4 M. Osborne, *Southeast Asia. An Introductory History*, Tenth Ed., Crows Nest, Allen & Unwin, 2010, p. 130.

One of the factors that made Sarawak unusual in the gutta percha trade was the fact the British owned trading company, the Borneo Company Limited (BCL), had important sources of trade there, and a direct involvement in the business of submarine telegraphy. This promoted the gutta percha trade in Sarawak in a manner and to an extent not seen elsewhere. Another unusual characteristic of Sarawak was the indigenous Iban traders who were very actively involved in collecting gutta percha – to the point that they went on collecting expeditions throughout, in effect, the whole region mapped by Obach. No other collectors engaged in this regional pattern of trade. Also part of British Borneo was an area called British North Borneo, run under Chartered Company rule; it is now known as Sabah. Both Sarawak and Sabah are part of the Federation of Malaysia. The island of Labuan, off Brunei, was ceded to Britain by the Sultan of Brunei in 1846, and is also now part of Malaysia. The Sultanate of Brunei, in the north of Borneo, is an independent state but during the nineteenth century was subject to British influence and became a Protectorate in 1888. These areas were an important part of the gutta percha trade. The remainder of Borneo that is now part of the Republic of Indonesia (Western, Southern and Eastern Kalimantan) was one of the Dutch colonial possessions.

The Dutch possessions in Southeast Asia were established under the Dutch East India Company (*Vereenigde Oost-Indische Compagnie* or VOC) in the sixteenth century during the great era of spice trading. The Dutch government took over the possessions in 1800 when it became known as the Dutch East Indies, or Netherlands Indies as it was often termed in English language records at the time. Java, with its capital at Batavia (now Jakarta), was the centre of economic and political administration; with agricultural and commodity trade, harnessing local labor, with local elite and Chinese intermediaries. The colony included the island of Sumatra, Celebes (now Sulawesi), the Kalimantan areas of Borneo, and across into Netherlands New Guinea (now Western New Guinea). Between 1830 and 1870 a Dutch policy known as the *Cultivation System* was implemented in Java and Sumatra whereby local farmers and landholders were forced to set aside a proportion of their land and labor to grow cash crops for the export market.[5] Although profitable trades, the cultivation system was regarded as harsh and was abandoned in the 1870s, except for the coffee and sugar trades which, because of their profitability, remained under

5 See for example C. Fasseur, The Cultivation System and its Impact on the Dutch Colonial Economy and the Indigenous Society in Nineteenth-Century Java, in *Two Colonial Empires. Comparative Essays on the History of India and Indonesia in the Nineteenth Century*: Comparative Studies in Overseas History 6, C.A. Bayly and D.H Kolff, (eds), Dordrecht, Martinus Nijhoff Publishers, 1986, pp. 137–154.

government control until the twentieth century.[6] Official interest in plantation growth and botanical research into the commercial exploitation of tropical produce (both indigenous and introduced) continued, with official plantations and research stations established.[7] One such product investigated was gutta percha.

Dutch interest in Borneo evolved, beginning in Pontianak in 1814 in Western Borneo, in part as a response to the development of external interest in Borneo.[8] This external interest was largely the increasing British presence in Borneo, which had grown because of British relationships with the Brunei Sultanate, the Brookes position in Sarawak (albeit independent but still British in the diplomatic sense) and finally with the establishment of what became British North Borneo. Eventually, an Anglo-Dutch treaty was signed in 1891 which notionally set in place borders between the Borneo States, although it was decades before such borders were eventually surveyed.[9] Dutch Borneo was an important source of gutta percha, some of it traded overland through Sarawak, but with large amounts shipped directly to Singapore because of the Chinese trade network there and British linked sources of demand from submarine cable manufacturers.

While British and Dutch policies in Borneo were broadly accommodative as the century progressed, British policies in Borneo and the degree to which Britain wished to be responsible for Sarawak, Brunei and British North Borneo were complicated by British concerns to exclude likely intervention by other powers with an interest in the region. Coal resources on the island of Labuan off the Borneo coast signaled its economic importance, and its politico-strategic location was of concern. However, compared with the Straits Settlements, British Borneo was of lesser significance and this may have contributed to what has been described as an 'element of improvisation'[10] in British policy responses to difficulties that arose in British Borneo over the latter part of the century.

6 J.G. Taylor, *Global Indonesia*, London, Routledge, 2013, p. 68.

7 See for example A. Goss, *The Floracrats. State-sponsored Science and the Failure of the Enlightenment in Indonesia*, Madison, University of Wisconsin Press, 2011; P. Boomgaard, 'The making and unmaking of tropical science: Dutch research on Indonesia, 1600–2000', *Bijdragen tot de Taal-, Land- en Volkenkunde*, Vol. 162, No. 2/3 (2006), pp. 191–217 and L. Pyenson, *Empire of Reason: Exact Sciences in Indonesia, 1840–1940*, Leiden, Brill, 1989.

8 See M.F. Somers Heidhues, *Golddiggers, Farmers, and Traders in the 'Chinese Districts' of West Kalimantan*, Indonesia SEAP Publications, Ithaca, Cornell University, 2003, p. 47.

9 Reid, p. 244.

10 N. Tarling, *Imperialism in Southeast Asia: 'A fleeting, passing phase'*, London, Routledge, 2001, p. 86.

Chapter 8 will explain how these diplomatic tensions were to have an influence upon, and in turn be influenced by, the gutta percha trade of Sarawak.

Elsewhere in the near region, the countries now known as Vietnam, Cambodia and Laos made up French Indochina. Although French traders and missionaries had been a presence in Vietnam, it was in 1861 that an attack on Christian missionaries was used as justification for the French takeover of Saigon on the extreme southern coast in 1861, with control extending over the whole country by the 1880s. For Osborne, the French were interested in Vietnam in large part as a staging point for trade with China, and as a counterfoil to British presence in the region.[11] The country was then formed into three sections: Cochin China, Tonkin and Annam. Nearby, Vietnamese and Thai rulers had been making inroads into Cambodia in the seventeenth century but the country became a French protectorate in 1864. After periods of conflict in the seventeenth century, Laos (as it is now known) was brought under French control between 1885 and 1899 when it became part of French Indochina with Vietnam and Cambodia.[12] France, with its scientific and manufacturing capacity, as well as a network of colonial outposts, took an interest in both submarine cable manufacture and in the gutta percha trade. French gutta percha hunters went there or were sent there on official expeditions; however, gutta percha of the type required for telegraphy was not indigenous to the French outposts in Southeast Asia.

The Philippines had a patchwork history of imperial rule in the nineteenth (and early twentieth) century. Notionally a Spanish colony from the sixteenth century (Manila was taken over in 1568), with large areas remaining beyond official Spanish administration. While trade and international standing were an aspect of Spain's quest to establish a presence in the region, the impetus to spread Catholicism was a significant factor. Reid concludes that the Spanish presence was the least commercially active of all the major European networks in the region.[13] The Sulu archipelago in the southernmost area of the Philippines had long been a site of Chinese trade in products such as mother-of pearl, birds' nests and trepang.[14] The region was also to develop a trade in gutta percha in later years, but the product was of limited interest to submarine cable producers (see Obach's outline in Map 4 above), although Sarawak Iban gutta percha collectors travelled to the southern Philippines to collect gutta percha. Despite attempts to achieve independence in the 1870s,

11 Osborne, p. 115.

12 See Osborne Ch. Five for an overview of this period of regional history.

13 Reid, p. 73.

14 See Reid, p. 235.

Spanish rule continued until 1898 when the Philippines were ceded to the U.S. under the Treaty of Paris after the end of the U.S. war with Spain. Local insurrections against the U.S. continued until 1902 when the U.S. gained control. Apart from the Japanese occupation in World War II, U.S. administration continued until 1946 when the Philippines became an independent nation. U.S. officials investigated both the local and regional sources of gutta percha with the idea of developing U.S. access to gutta percha, but there was no significant American-sourced gutta percha made available for telegraphic use.

The final colonial power of note in the region in the European hunt for gutta percha was Germany, with its colonial outpost of German New Guinea which existed for what was a short time by the standards of imperial history of the era – the thirty years from 1885 to 1915. Somewhat mirroring the European colonial and trading history of Southeast Asia, German interest in New Guinea began initially with the establishment of a colony with a commercial charter. It was developed and administered under an Imperial Charter by the *Neuguinea Kompagnie* (1885–1899), and then from 1899–1914 under administration of the Reich represented by the Imperial Government of German New Guinea. The territory grew during this period, from the north-eastern section of New Guinea proper (called *Kaiser Wilhelmsland*) and what was known as the *Bismarck Archipelago*, sections of the Solomon Islands (now part of Papua New Guinea), and islands including the Mariana, Caroline, Marshall etc in Micronesia (known as *Island Territory*). These areas are now known as the Federated States of Micronesia, the Marshall Islands, Nauru, the Northern Mariana Islands and Palau. Both the Kompagnie, and later the Imperial Government, issued Annual Reports.[15] While it was under the Kompagnie rule, commercial objectives were important, and even under official rule the colony was judged 'primarily on the success of the private sector'.[16] By the end of 1914 with the outbreak of World War I, the colony was taken over by military forces from Australia (Kaiser Wilhelmsland), New Zealand (Samoa), and Japan (the Marshalls and Caroline islands). As can be seen from the map above, the German territories were all, however, outside the zone that Obach had identified as being within the area which would likely possess forest sources of the varieties of gutta percha suitable for cable telegraphy.

---

15 This summary based on P. Sack and D. Clark, (eds), *German New Guinea. The Annual Reports, Edited and Translated.* Canberra, Australian National University Press, 1979, Introduction.

16 P. Sack and D. Clark, p. xii.

The European Hunt for Gutta Percha

Nothing but Gutta Percha[17]

India Rubber and Gutta Percha and Electrical Trades Journal, 1893

Europeans began looking for gutta percha soon after it became an item of trade and particularly so after its adoption for use in submarine telegraphy – 'manufacturers, merchants, travellers, scientific men, and even governments' were all engaged in this quest.[18] While these expeditions could themselves be the topic of a single study, an overview here gives a sense of the European hunt for gutta percha in the context of the history of submarine telegraphy.

One of the earliest expeditions to find gutta percha was commissioned in 1845 by the French Ministry of Commerce in an unsuccessful attempt to locate supplies in French Indochina.[19] British expeditions also began relatively early and specimens sent to the Royal Botanic Gardens at Kew in 1847 were already accompanied by warnings that gutta percha would soon become a very scarce article.[20] Various reports were sent to the Royal Botanical Gardens and were copied to the Colonial Office because of the importance of gutta percha as a strategic resource (hence Thiselton-Dyer's comment about gutta percha being of 'importance to the whole civilized world').[21] Leonard Wray explored the British Malay territories to obtain specimens and information about gutta supplying trees in 1883, raising concerns about the future of supply due to the destructive methods of collection then taking place.[22] Several reports arose from numerous Dutch expeditions searching for gutta percha in the Southeast Asian colonies of the Netherlands Indies – de Vriese's *De Handel in Getah-Pertja*

17 *India Rubber and Gutta Percha and Electrical Trades Journal*, London (hereafter IRGPETJ), 8 December 1893, p. 131.

18 T. Seeligmann, L. Torrilhon and H. Falconnet, *Indiarubber and Gutta Percha. A Complete Practical Treatise*, 2nd English edn., London, Scott, Greenwood & Son, 1910, p. 296.

19 Seeligmann et al., 1910, p. 290.

20 T. Oxley, 'Gutta percha', *Journal of the Indian Archipelago and Eastern Asia*, Vol. 1, 1847, p. 23.

21 For example, J. Collins, *Report on the Gutta Percha of Commerce*, London, W.H. Allen, 1878; W.T. Thiselton-Dyer, Correspondence to C.P. Lucas (Colonial Office), in Royal Botanic Gardens Kew, Archives, *Kew: Miscellaneous Reports, Gutta Percha*, 1899, 26 December, f.45.

22 L. Wray, 'Gutta-producing trees', *Journal of the Straits Branch of the Royal Asiatic Society*, no.12. (December), 1883, pp. 207–221.

was an early example that set out both botanical and commercial information, including on local collecting practices and the potential risk to long term supply due to non-sustainable methods of collection.[23] Another Dutch report raised similar findings, but the writer of this article said, overly optimistically, that planting had started on Java which would 're-assure society' with regard to future supply; while other Dutch reports over several decades from colonial and forest/plantation officials report on location of trees, botanical features and varieties, local collecting practices and repeatedly talk of depletion of supplies, or the greater difficulties in finding trees ('getah-pertja' discussion appears frequently in issues of *Tijdschrift Voor Nijverheid* throughout the latter half of the century and early twentieth).[24] Dr William Burck, an Assistant Director of the Botanical Gardens at Buitenzorg, Java studied gutta percha producing trees and yields on the island of Sumatra and his 1884 report to the Government was later considered a 'standard' on the subject.[25] Several specimens he obtained were planted at the gardens at Tijkeumeuh near Buitenzorg. There were many other reports and private and official expeditions.[26]

While much important information was obtained, there were still problems that remained to be solved and Europeans did not take over the sourcing of gutta percha for commercial export from the region to any great degree. Not all expeditions were well managed and recommendations were not always acted upon; personal disagreements were a feature of the European hunt for gutta percha. In 1885, Wray, by then Superintendent of the Perak Forest Conservancy, had written to Thiselton-Dyer about problems in funding for gutta percha research and management due to a lack of support for 'science'; Kathirithamby-Wells sets out a history of ill-feeling and differences of opinion that existed about the 'gutta percha crisis' among the Colonial Office, the Royal Botanic Gardens at Kew and various officials in the Straits Settlements and the Federated Malay States. For example, the observation that it was not an issue for the Straits Settlements since there were no more gutta percha exports coming from there in any case![27] Similar efforts and mishaps can be seen elsewhere in the historic record.

---

23 W.M. de Vriese, *De Handel in Getah-Pertja* (Gutta-Percha), Leyden, A.W. Sythoff, 1856, pp. 15–16.

24 'Getah Pertjah van den Nederlandsch Indischen Arcipel', *Tijdschrift voor Nijverheid in Nederlandsch-Indië*, Volume IV, 1858, pp. 396–410; other issues of the journal are readily found online.

25 Obach, 1897, p. 114.

26 See Seeligmann et al., p. 296 ff for a summary; also, Obach.

27 L. Wray, in Kew Director's Correspondence, 1896, 21 April, f.165/262 *Kew Director's Correspondence*, Royal Botanic Gardens Kew, Archives 165; 166; 168 and J. Kathirithamby-Wells,

There were renewed French efforts to find sources of gutta percha in their Southeast Asian colonial outposts after the failure of the French government to find local tenderers for two of its cable projects – as none could supply sufficient pure gutta percha. In 1881, the French Minister of Posts and Telegraphs sent an engineer, M. Séligmann-Lui, on a mission to determine suitable varieties of trees that could supply gutta percha and to identify a possible means of propagating them in French colonies in the region.[28] In one sense the move to send an engineer, rather than a botanist, to undertake the task was at least recognition that the problem was the need to manage the future of (French) telegraphic communication rather than simply re-engaging in discussions of theoretical botanical quandaries about the nature and sources of gutta percha supply. Séligmann-Lui submitted a report to the Minister, and arranged collection of several varieties that were sent to Saigon (then the centre of French colonial Cochin China and now Ho Chi Min City) for propagation; with orders for a botanical garden to be constructed for experimentation and acclimatization of the plants. It is difficult to gauge how serious the Ministry was about solving the problem or perhaps they lacked influence in the colonial administration: the specimens were left in the cellars of the Government offices, forgotten about, and 'nothing more was heard of the affair'.[29] Séligmann-Lui did not reach an optimistic conclusion in his report on the future situation of the trade and the electrical industry: 'the present' he wrote, 'is without remedy and even the future is compromised, it is urgent to decide what to do; if measures are not taken quickly, the industry will soon lose a material that, up to now, it has not known how to replace'.[30]

Another series of ill-fated efforts came after the French government sent a different scientist, this time an economic botanist, to try to identify options to develop gutta percha resources. M. Sérullas went to Southeast Asia, studying and hunting for gutta percha sources and spent several years trying to develop options for cultivation.[31] Over several visits to the region, his efforts were plagued by personal health problems, lack of government funding and a series of bureaucratic bunglings and arguments. In the end, the test seedlings he finally gathered were at first rejected, then accepted and then finally neglected and left to die.[32] In the end, all of his endeavors were 'wasted' after the whole

---

Nature and Nation Forests and Development in Peninsular Malaysia, Singapore, Nias Press, 2005, pp. 72–73 see also Ch. 2 of that book.

28 G. Séligmann-Lui, *Rapport sur les Origines de la Gutta-Percha*, Paris, Libraire des Corps Nationaux des Ponts et Chaussées, des Mines et des Télégraphes, Dunod, 1883, p. 5.

29 IRGPETJ 8 November 1890, p. 88.

30 Séligmann-Lui, p. 80.

31 See IRGPETJ 8 January 1891, p. 162; IRGPETJ 8 November 1890, p. 88.

32 Summarised in IRGPETJ 8 April 1891, pp. 248–9.

issue was 'abandoned by the French Government which...entirely ceased to interest itself in the matter'.[33] At the turn of the century the French government commissioned reports from the region with a 'Gutta Percha Commission' in Saigon without any apparent result, and in 1899, the French scientist Lecomte, in a report to the French Colonial Ministry, observed that all the many studies that had preceded his report 'n'ont pas encore pu jeter une lumière complete sur cette importante question'.[34] In 1902 d'Abbans, the French consul in Singapore summarized the gutta percha resources of the Malay region as a whole.[35] The Government tried again with some minor but unsuccessful attempts including in the Caribbean, but these numbered just a couple of hundred plants.[36]

The United States possession in the Philippines presented it with a potential source of gutta percha. In 1901, Penoyer Sherman, Chemist at the Government Laboratories in the Philippines, was sent out as Special Agent for the Philippines Forestry Bureau to investigate gutta percha resources in the Philippines and to study the trade of the region more widely.[37] There was an existing trade in the southern Philippines and a concern that such resources were being lost due to the destructive means of collection, and that traders were not paying proper dues to the Government.[38] Sherman returned with several samples for record purposes and others for laboratory testing. He found a brisk trade in gutta percha in Mindanao and the Sulu Archipelago, but expressed concerns about the need to control the non-sustainable means of collection which would threaten future supplies. Gutta percha collected from the Philippines was usually shipped to other areas, such as Labuan near Borneo, and eventually sold in Singapore as North Borneo or Sarawak gutta percha so that it would

33 Obach, 1897, p. 114.

34 M. Lourme, 'Rapport sur les travaux de la commission de la gutta-percha', *Bulletin Economique de l'Indochine*, Vol. 2, No. 10, 1 April 1899, pp. 320–328; H. Lecomte, *Les Arbres a Gutta-Percha. Leur Culture. Mission Relative A L'Acclimatation de ces Arbres Aux Antilles et a La Guyane*, Paris, Georges Carré et C. Naud, 1899, p. 2.

35 J. d'Abbans, 'La Gutta-Percha et Le Caoutchouc en Malaisie. Conference de M. Le Comte de Jouffroy d'Abbans', L'Agriculture Pratique des Pays Chauds, *Journal du Jardin Colonial et des Jardins d'Essai Des Colonies Françaises*, Juillet 1902-Juin 1903, pp. 586–603.

36 See Lecomte and L. Leon, 'L'Industrie de la gutta-percha en Indo-Chine', *Bulletin de la Société Académique de Brest*, Deuxième Série, Tome XXV, 1899–1900, pp. 51–109.

37 P.L. Sherman, *Report of the Special Agent of the Forestry Bureau Sent to Investigate Gutta-Percha and Rubber in the Straits Settlements, Java and Sumatra*, Manila, Forestry Bureau, 1901, p. 339.

38 P.L. Sherman, *The Gutta Percha and Rubber of the Philippine Islands*, Manila, Bureau of Public Printing, 1903, p. 5.

achieve a better price as its origin could be obscured.[39] Sherman's mission to find gutta percha was not surprising but, aside from gathering information about the operation of the trade in the Philippines and the near region, there was no resulting wide scale plantation development on U.S. territory. The U.S. uncertainty over its potential vulnerability in cable communications, and the role of gutta percha supply in that vulnerability, continued during and even after World War I. By then, so concerned was the U.S. administration about its lack of a cable producing capacity, and the need to access sources of gutta percha, that it sent out secret missions to Southeast Asia to try to identify options for offering highly attractive prices to Chinese traders, thereby taking over the perceived British monopoly of the world gutta percha market. In the event neither the mission to control the gutta percha market, nor the production of cables in the U.S. with this insulation ever eventuated.[40]

As another colonial power in the region and, unlike the U.S., a submarine cable producing country, it is not surprising that German colonies were investigated for possible sources of gutta percha. Samples were sent from forests in German New Guinea to Obach at Siemens' London factory for testing, but none proved anything like suited to telegraphic needs.[41] The best known of the German expeditions were made by Dr Rudolf Schlechter who undertook three missions to try to find sources of rubber and gutta percha (or any other latex that might be of general commercial interest). These missions were promoted by the Kolonial-Wirtschaftliches Komitee [Colonial Economic Committee]. His first expedition was in fact to southern Africa, undertaken in part with support of various rubber manufacturing concerns including Felten & Guilleaume Carlswerk,[42] producers of submarine cables; but it did not lead to any significant results for gutta percha. His two later expeditions were to German New Guinea and in 1902 the Committee awarded him a prize for finding gutta percha on the first of these expeditions. A report he produced suggested that the gutta percha he found was of use for cable producers when blended with other better quality gutta.[43] Obach had already written that this locality was

39 Sherman, *The Gutta Percha and Rubber of the Philippine Islands*, pp. 16, 18.

40 Examined in J.R. Winkler, *Nexus. Strategic Communications and American Security in World War I*, Cambridge, Mass., Harvard University Press, 2008, especially pp. 230ff.

41 Obach, 1897, p. 112.

42 R. Schlechter, *Westafrikanische Kautschuk-Expedition 1899/1900*, Berlin, Kolonial-Wirtschaftlichen Komitees, 1900.

43 R. Schlechter, Über die neue Guttapercha von Neuguinea, *Der Tropenpflanzer Zeitschrift für Tropische Landwirtschaft*, Oktober 1903, Nr 10, p. 467.

an unlikely source of gutta percha suited to telegraphic requirements.[44] Obach proved to be correct: the product sent by Schlechter was of relatively low grade and of limited interest to cable users. A German cable producer, the Norddeutsche Seekablewerke (NSW) was sent 900 kg for testing but it was felt that the likely market price it would achieve would not justify the cost of collection, the product not being suited to cable production.[45]

Schlechter's second expedition was in 1906–1909. This expedition was also funded by the Committee, the German rubber manufacturing industry and the German Colonial Company, which operated in German New Guinea.[46] These two expeditions have been described as, in turn, exploratory and then developmental.[47] Schlechter's two expeditions were thought moderately successful at the time but had little lasting impact, his collection of orchids regarded as probably the true success, based on his 1902 mission.[48] Schlechter's second expedition was intended to determine the amount of rubber and gutta percha in the colony, the best means of preparing it for export or cultivation, and finally to instruct the local population in how best to gather the products.[49] While these endeavors took the quest for gutta percha more seriously, and there were exports, the amounts were small, especially when compared with other exports from the colony. It was discovered that collecting gutta percha in the colony was difficult due to the problems of labor shortage and skilling, and the difficult collecting and transport environment, especially given the product's relatively poor quality and its likely price in the market place. The local population, it was thought, could not compete with the 'growers' in Malaya and elsewhere.[50] Four 'Malays' were brought in to train 150 villagers in how best to collect gutta percha and rubber, and two facilities to store gutta percha before export were also built. The hope was that collecting gutta percha would provide a means for local people to earn cash to pay their head tax, rather than having to undertake compulsory labor which was very unpopular. The training ceased in 1912 when funds ran out and the district office ceased purchasing

44 Obach, *Journal of the Society of the Arts*, 1897, p. 112, and see Obach's map of the gutta producing regions, at Map 4 above.

45 Y. Péhaut, *Les Plantations allemandes des Mers du Sud avant 1914*, CRET, Bordeaux, Presses Univ. de Bordeaux, 1990, p. 139.

46 *India Rubber World*, New York (hereafter IRW), 1 January 1913, p. 192.

47 D.G. Frodin, 'The Biological Exploration of New Guinea' in *The Ecology of Papua Part 1*, Marshal A.J. and Beehler, B.M. eds., Singapore, Periplus Publishing, 2007, p. 41.

48 See Frodin, p. 42.

49 IRW 1 January 1913, p. 192.

50 P. Lawrence, *Road Belong Cargo: a Study of the Cargo Movement in the southern Madang District, New Guinea*. Manchester, Manchester University Press, 1964, pp. 44, 48, 227.

gutta, leaving any trade in the hands of the New Guinea Company – the trade was left to survive based on the marketplace rather than as a means of looking to underpin German telegraphic objectives.[51] German cable producers, like those of other nations, could obtain forest sourced gutta percha in the marketplace, such as from agency or auction houses in country or the Netherlands. Norddeutsche Seekablewerke (NSW), for example, also had agents in Singapore that supplied them with gutta percha[52] and NSW sent an engineer to Sarawak to attempt to find likely other suppliers of gutta percha.[53]

Other Expeditions and Ventures in the European Quest for Gutta Percha

As the centre of the global gutta percha trade, Singapore became a centre for those hunting for sources and information on gutta percha. Many visited Singapore and Henry Ridley (Director of the Botanic Gardens for the Straits Settlements from 1888–1912) was critical of scientists like Schlechter and Sérullas – and other 'adventurers' as he called them – interested in gutta percha.[54] When Ridley talked of gutta percha 'adventurers' he no doubt meant it as a disparaging term for those seeking easy profit or fame by finding new sources of a valued resource.

Perhaps a true gutta percha 'adventurer' did exist. Adolphe Combanaire was a French explorer who travelled to Singapore, Sarawak and Borneo in 1899 hunting for gutta percha to find a solution to the problem about to face global telegraphy, the disappearance of sources of supply of gutta percha. This was his second visit to the region, having previously found gutta percha in Sumatra in 1894. His book *Au Pays des Coupeurs de Têtes: A travers Bornéo*, [In the Country of the Headhunters: Crossing Borneo], is today a useful historical resource, including for the gutta percha trade in Borneo. He discusses the nature of the trade and difficult collecting conditions, even the role of fruit bats in the propagation and distribution of gutta percha trees – noted elsewhere in the region as being important in the dissemination of seeds over

51 Sack and Clark, pp. 330, 346; IRW 1 January 1913, pp. 192,193, Schlechter 1911 in R.E.P. Dwyer, 'Rubber Production in New Guinea and Papua (Continued)', *New Guinea Agricultural Gazette*, Vol. 7, No. 4., November 1941, p. 265.

52 See *Singapore Free Press and Mercantile Advertiser*, Singapore (hereafter SFPMA), 29 September 1924, p. 3 and see TCM/21/53: Jewesbury to the Gutta Percha Co, 28 April 1914.

53 C. Schutze, *Kabel, unser Leben, Narrative Identitätskonstruktionen eines Unternehmens und seiner Belegeschaft*, Isensee Verlag, Oldenburg, PhD Dissertation, 2012.

54 *Kew Director's Correspondence*, Royal Botanic Gardens Kew, Archives 168, 12 July and 2 August 1902, 2 May f.168/71–2 and 1891, 20 October, f.166/593–4.

'considerable distances'.[55] He also gives details of some of the natural resources of Borneo, writes of headhunters and astute Chinese traders. Combanaire went prepared with the practical necessities but, to ease the burdens of his travel, he took with him a supply of wine, beer, cognac, liqueurs, and Dutch cigars in hermetically sealed tin boxes; along with tobacco and gin to appeal to the local population.[56] D'Abbans, the French Consul in Singapore (a gutta percha savant himself), described Combanaire as a 'fanatique de la gutta' who 'il les sentait, les flairait, les palpait, les caressait…' the gutta percha trees he found in the forest.[57] Combanaire was more than just a gutta percha 'scientist', adventurer and recordist. In the late 1890s he obtained a patent for a proposed means of obtaining gutta percha from the leaves and branches of trees and it was his aim to find suitable gutta percha resources such as seeds or seedlings to establish plantations in French Indochina – thus offering both profit and the benefit to France of its own source of supply.[58] He and another French colleague took out a patent for an 'Improved Process for the Purification of Gutta-Percha'.[59] In the event, such plantations and techniques were not established, but Combanaire seems to have kept in mind his colleague d'Abbans' exhortation about the hunt for gutta percha 'Oh! Messieurs, les joies de l'exploration en pays équatorial' and he went on to other explorations and adventure in French colonial territories.[60]

Another European who hunted for gutta percha has possibly become the best known of the gutta percha hunters in popular record because of a presence in the literary world – it may be only through the writing of Joseph Conrad that today some may be familiar with the words *gutta percha*:

> Many tried to follow him and find that land of plenty
> for gutta-percha and rattans, pearl shells and birds' nests[61]

55 A. Combanaire, *Au Pays des Coupeurs de Têtes: A travers Bornéo*, Singapore, Editions Pagode, 1902, Reprint 1993, p. 198 and *Agricultural Bulletin of the Straits and Federated Malay States*, 'The Dissemination of Seeds by Natural Means – Gutta Percha', New Series, 1902, Vol. 1, No.4, p. 154.

56 Combanaire, p. 34.

57 d'Abbans, *Journal du Jardin Colonial*, p. 598.

58 See Combanaire, Introduction and p. 302.

59 A. Combanaire, and J. de la Fresnaye, Patent Application 'Improved process for the purification of Gutta Percha', English Patent Record No. 22,758, November 11, 1902, in IRW 1 February 1902, p. 149.

60 d'Abbans, ibid; see B. Sellato, 'Preface', Adolphe Combanaire, *Au pay des Coupeurs de Têtes. A travers Bornéo*, Singapore, Editions Pagodes, 2003 for a summary of Combanaire's life.

61 J. Conrad, *Almayer's Folly. A Story of an Eastern River*, London, T. Fisher Unwin, 1896, p. 13.

Conrad's so called 'eastern' fiction drew upon his experiences in the Malay Archipelago in the 1880s and is widely recognized as incorporating both factual and thematic features of the period with discourse on the trade and traders in gutta percha and other 'exotic' produce featuring as part of conveying this life.[62] The quotation is from *Almayer's Folly* where Conrad presents a character known as Tom Lingard, an acknowledged 'king' of the schooner captain-traders in the region. As a Rajah-Laut ('King of the Sea'), Lingard was known for his considerable wealth – trading 'Manchester goods [fabric, shirting], brass gongs, rifles and gunpowder' in hinterland Dutch Borneo which he traded for local produce, achieving 'enormous profits' from his adventures.[63] The *Almayer* of the title is based upon William Olmeijer an actual east Borneo trader who dealt in produce including gutta percha in the East Netherlands Indies part of Borneo.[64] Conrad's Lingard was in turn based upon Captain William Lingard, an actual gutta percha trader, considered at the time probably the greatest expert on the commerce of gutta percha. William Collins, the Government Economic Botanist and Secretary of the Raffles Museum at Singapore, refers several times to Captain Lingard in his *Gutta Percha: Its History, Commerce and Supply* in 1878 as the source of much of the information in his pamphlet. Interestingly, unlike most writers of the time, Lingard was of the view that the best gutta percha came from the area not 6° north and south of the equator (see Obach map above), but just 4 degrees North and 1 degree south, an area considerably smaller than that generally reported elsewhere.[65] Lingard features in another study by John Dill Ross, a shipper-trader-adventurer of the period in his roman-à-clef biography (of his father's and his life as traders), *Sixty Years: Life and Adventure in the Far East*. They were well known Borneo traders and appeared in Colonial/Foreign Office records of the period.[66] Ross senior and son also traded in local produce including gutta percha, and Ross the younger was a key trader of Labuan produce. In his work, Ross refers to Captain Lingard (whom he calls 'Captain Lugard') as 'an awfully nice fellow ... everybody thought of the piles of money that Lugard made of his gutta-percha

62 See for example A. Francis, *Culture and Commerce in Conrad's Asian Fiction*, New York, Cambridge University Press, 2015.

63 Conrad, p. 13.

64 J. Baines, *Joseph Conrad. A Critical Biography*, Harmondsworth, Penguin, 1971, p. 116.

65 J. Collins, *Report on the Gutta Percha of Commerce*, London, W.H. Allen, 1878, see p. xiii and p. 8.

66 See for example N. Tarling, *Britain, the Brookes and Brunei*, Kuala Lumpur, Oxford University Press, 1971.

trade'.[67] Elsewhere Ross writes of merchants in Singapore who met and traded gambier, pepper and gutta percha – but that such speculations did not always end well.[68] Ross dominated the trade between Singapore, Labuan and onward connections to North Borneo and the Sulu region of the Philippines for several years, trading also in opium and arms.[69] His links could have facilitated the trade of the less marketable Philippine gutta percha to Singapore with its origin being disguised as having come through Borneo. Burbidge, another recordist of the time, describes how in the local market square in Singapore, it was common to hear discussion about the price of gutta percha, and listen to Captain Lingard after another of his successful trips to the southeast coast of Borneo, with few men knowing more than this 'Rajah Laut' and that he and Ross senior were considered to be celebrities.[70] In their day, they were part of the profitable European quest for gutta percha, but, as Francis notes in his study of commerce in the era, traders like Lingard and Ross were overtaken by global changes – for example, steam shipping, telegraphic communication, and an expanding Chinese trade network. That is, the type of traders considered later in this book.[71] Warren also notes the important role of Singapore Arab traders who exploited the gutta percha trade in Borneo and took over regions of East Borneo from traders like Lingard (see Warren, who also sets out some gutta percha trade data from East Borneo for these shippers in the late 1860's – early 1870s[72]). The history of submarine cable manufacture considered in the previous chapter shows that traders such as Lingard and Ross could certainly have made good profits from gutta percha during this time. They were trading during the period of some of the greatest demand ever recorded.

European Ideas about the Future of Gutta Percha

While the above sections represent a brief overview of the European traders and expeditions engaged in the quest for gutta percha, much of the information

67 J.D. Ross, *Sixty Years: Life and Adventure in the Far East*, 2 Vols, London, Hutchinson & Co, 1911, Vol. I, p. 81.

68 Ross, Vol. I, p. 74.

69 See J.F. Warren, *The Sulu Zone 1768–1898: The Dynamics of External Trade, Slavery and Ethnicity in the Transformation of a Southeast Asian Maritime State*, Singapore, NUS, 2007, p. 110.

70 F.W. Burbidge, *The Gardens of the Sun or A Naturalists Journal on the Mountains and in the Forests and Swamps of Borneo and the Sulu Archipelago*, London, John Murray, 1880, pp. 18, 19.

71 See Francis, Introduction.

72 J.F. Warren, 'Joseph Conrad's fiction as Southeast Asian History: Trade and politics in East Borneo in the late nineteenth century', *Brunei Museum Journal*, 1977, pp. 21–34.

gathered still had not led to the finding of new or sustainable sources of supply for metropolitan users. Instead, by 1891:

> Confusion now commenced in nomenclature. The native terms were recklessly employed, with the result that the same gum came to European markets under various designations, while absolutely different perchas were labelled with the same name...[73]

The confusion was described decades later in 1910, by which time little had changed:

> Scientific men discovered ... trees the juice of which may yield a good, a mediocre, or a bad gutta percha. They determined, with greater or less care and exactitude, the physical and chemical properties of these different products, but their investigations ended there, whilst manufacturers continued to receive most diverse substances, through numerous savage or civilized merchants, who could not or *would not* (for a very good reason if we consider the point well) give exact indications upon the botanical origin, nor even upon the exact geographical origin of these products.[74]
>
> [Original emphasis]

When adventurers and the agents of various governments began the quest for gutta percha, they were in most cases not seeking to develop new, sustainable sources of supply, but rather to find existing but as yet undiscovered forest resources. Yet cable producers and those with an interest in the electrical sciences expressed concerns about the need to find a way to protect the future supply of this precious resource:

> Submarine and underground telegraphy...are threatened with the entire disappearance at an early date of a material for which an effective substitute has hitherto been vainly sought – gutta percha[75]

In 1848, in one of the very earliest reports on the resources and trade in gutta percha, Logan noted that the trees in Johor were already becoming scarce – well before the demand from cable telegraphy had commenced; and, while his reports were perhaps based on limited details compared with what came from

---

73 IRGPETJ 9 March 1891, p. 221.

74 Seeligmann et al., p. 296.

75 IRGPETJ 8 November 1890, p. 88.

later reports, it highlighted the level of early demand and extent of resource extraction taking place on the Malay peninsula.[76] In 1852, the British Association for the Advancement of Science heard a report from one of its Committees that urged extreme vigilance in protecting the gutta, noting that even 'admitted financial value has not been sufficient to protect them from thoughtless waste'.[77] Decades on, in 1878, William Siemens, as President of the Society of Telegraph Engineers, raised the issue again, stressing the need to find an alternative compound that could substitute for gutta percha as demand would soon 'outrun the supply'.[78] A few years later, the *Telegraphic Journal & Electrical Review* referred to the limited supply of gutta percha.[79] The issue was the subject of debate at an international conference that considered the need for European action to protect and develop the supply of gutta percha to ensure the future of submarine telegraphy – see, for example, the British representative to the Congrés International des Électriciens, who noted that without some form of control, the supply of gutta percha could fail within thirty years.[80] Continuing the debate in 1898, Bright wrote that the 'gradual scarcity, and threatened extinction of the guttifers, has for some years been the cause of much anxiety'.[81]

The only 'salvation' was thought to be Government control: forcing collectors to take out licences and collect in a manner designed to prevent destruction and waste.[82] Attempts were made in the Straits Settlements to limit or totally ban the forest collection of gutta percha in an effort to ensure supply.[83] This too was ineffective – in part because of the near impossibility of controlling the

76 J.R. Logan, 'Range of the gutta taban collectors, and present amount of imports into Singapore', *Journal of the Indian Archipelago and Eastern Asia*, Vol. II, 1848, p. 532.

77 H. Cleghorn, J.F. Royle, R.B. Smith and R. Strachey, 'Report of the Committee appointed by the British Association to consider the probable effects in an economical and physical point of view of the destruction of tropical forests', *Report of the British Association for the Advancement of Science,* London, John Murray, 1852, p. 95.

78 W. Siemens, 'President's Address', *Journal of the Society of Telegraph Engineers*, Vol. VII, No. 21, January 23, 1878, p. 10.

79 *Telegraphic Journal and Electrical Review*, Vol. XI, No. 241, July 8, London, Henry Gillman, 1882, p. 1.

80 Webber in Congrés International des Électriciens 1881, *Comptes Rendus des Travaux*, Paris, Ministère des Postes et des Télégraphes, 1882, p. 328.

81 C. Bright, *Submarine Telegraphs Their History, Construction and Working*, London, Crosby Lockwood and Son, 1898, p. 258.

82 H.L. Terry, *India-Rubber and Its Manufacture. With Chapters on Gutta-Percha and Balata,* London, Archibald Constable & Co. Ltd., 1907, p. 275.

83 See J. Kathirithamby-Wells, 2005, p. 70 ff.

level or location of collection without a large and expensive force to police it, a problem recognized at the time.[84] Potter summarizes various attempts to either promote tapping or to prohibit the trade altogether in the Malay peninsula – this simply led to collectors crossing borders and smuggling the material out, or else completely ignoring the restrictions because it was most unlikely the 'offenders' would be caught in forest areas where poor roads and little reliable transport prevented easy access by those wanting to police the areas.[85] In this regard, the 'Dyaks of Sarawak' collecting gutta percha were known for being able to avoid controls even in the late 1920s.[86] The issuing of collection permits was tried in the Straits Settlements, and in British North Borneo, but collection of fees for permits and enforcement of appropriate techniques for collection were difficult to monitor (see Burn-Murdoch on the problems and failures of 'Measures of Protection'[87]). It is notable that the British administrators in the Straits Settlements who were willing to ban collection of gutta percha in their own colonies could do so only with the knowledge that their home country's need for continued supply would come from elsewhere – the Dutch controlled colonies – or from Sarawak, and the Philippines. That is, British interests were happy to protect their own gutta percha resources, but were only able to do so because the market was supplied by others.

The British publication, the *India Rubber and Gutta Percha and Electrical Trades Journal* (*IRGPETJ*), to the extent it was the mouthpiece of the British cable manufacturing industry, reflected repeatedly over several decades a sense of bewilderment that more progress was not achieved in ensuring a sustainable supply of gutta percha. The Journal had asked:

> If the Government deems itself not called upon to take further action in this matter, why is it not taken up as a commercial speculation?[88]

In the January 1901 issue of the IRGPETJ, a writer records the 'swollen' exports of gutta percha that had been recorded for the previous decade, and that 'those

84 Terry, p. 274.

85 L.M. Potter, 'A forest product out of control. Gutta percha in Indonesia and the wider Malay World 1845–1915', in *Paper Landscapes: Explorations in the Environmental History of Indonesia* eds P. Boomgaard, F. Combijn and D. Henley, Leiden, KITLV Press, 1997, pp. 290–291.

86 S.W. Kirby, 'Johore in 1926', *The Geographical Journal*, Vol. 71, no. 3 (March), 1928, p. 256.

87 A.M. Burn-Murdoch, 'Some facts about gutta percha', *Agricultural Bulletin of the Straits and Federated Malay States*, Vol. IV. no.2 (February), 1905, p. 35.

88 IRGPETJ 8 November 1890, p. 88.

who look with anxiety to the continuance of the supply will be seen to have cause for their disquietude'. He then wonders:

> How long the producing districts can stand the strain is a speculative matter of great importance, and it is regrettable that the system of gathering by natives, as practiced in Borneo, cannot be effectively controlled by a responsible person. As things are at present, the natives do pretty much as they like, and there seems to be no way of enforcing the regulations as to cutting down trees…

In other words, the journal was noting the collecting practices as part of the problem. What the writer did go on to conclude, however, was a suggestion that western interests should do something to ensure a sustainable future supply:

> We are promised a supply of gutta-percha in the future from French and German colonies…but it will probably be some time before supplies from the new sources affect the Singapore market – if, indeed, they ever do so. Still … our own Government might do well to take some decided action in the matter.[89]

One factor holding back action on the problem was that the trade was, in fact, continuing. Even if it was known that in the longer term there could be a problem, it would be tomorrow's problem. It may also have been that it was thought that the local collectors were the real cause of the problem and it was their behavior that ought to change in order to protect the future of the trade. Comments resting the blame on collectors were not uncommon. The British Consul for Trade's Report for Sarawak for 1869 described 'the reckless way in which trees are felled and the gutta extracted'.[90] Dutch reports reached similar conclusions, such as that 'the tree is killed because of the ease of doing so, or out of recklessness, not out of necessity'.[91] Sérullas' compatriot, Séligmann-Lui, had concluded in his 1883 report that the indigenous collectors were so lazy, and had so little idea of the impact of supply on price, that they would rarely rouse themselves to make the extra effort involved in searching longer distances for supplies of gutta percha as resources were being cut down – even if they could have received a better price by doing so as shortage of supply might bring them

89 IRGPETJ 21 January 1901, p. 55.

90 Consular Report on the Trade of Sarawak for 1869, BSP 1871 [C.343-LXV], p. 7.

91 C. de Haan, Report in *Tijdschrift voor Nijverheid in Nederlandsch-Indië*, Volume XII, 1866, p. 67.

a better price.[92] Sherman warned that 'stringent instructions' to protect the trees were having little effect on collectors throughout the region, whom he concluded would not 'give up the present method of gathering…unless a better method can be shown them; but apparently no better method is known which will appeal to people who can not see the folly in killing the goose that lays the golden eggs'.[93] Others report of the indigenous collectors being more 'greedy of immediate lucre than anxious to ensure the future of their production', and stuck to their 'ruinous method of collection'.[94] At the 1881 International Congress of Electricians, one of the Congress' committees, prompted by a British delegate, noted a report that countries producing gutta percha should be '*told*' to 'regularize' the operation of the management of the trees.[95]

Yet these various claims came after reports as early as the 1850s that tapping a tree produced a considerably smaller amount of gutta percha than was obtained from felling.[96] This was reported at the time and again in a summary presented to the *Indisch Genootschap* in 1900.[97] It is true that indigenous collectors did not practice sustainable collection methods. There were good reasons for felling the tree – not just because tapping was slow and produced limited amounts of product, but because of the danger and difficulty of living in the forest during lengthy collecting expeditions, lack of transport, and the likelihood that other collectors might find and collect the resource in the meantime. In addition, most collectors had other seasonal agricultural responsibilities (for example, swidden rice production) and often were fitting in collecting expeditions around other obligations. Local spiritual and omen restrictions were also an important factor, though these did not usually feature in European considerations at the time. There were also certain forest areas in which collectors were prevented from both entry and collection of produce because the areas were preserved as sacred or inhabited by spirits.[98] The European quest for gutta percha in the scientific sense made great progress in the

92 Séligmann-Lui, p. 42.

93 Sherman, *Report of the Special Agent of the Forestry Bureau Sent to Investigate Gutta-Percha and Rubber in the Straits Settlements, Java and Sumatra*, p. 351.

94 Seeligmann et al., 1910, pp. 292, 322.

95 *1881 Congrés International des Électriciens, Report*, 1882, p. 328.

96 de Vriese, 1856, p. 15 and in Von Gaffron 1858, p. 226.

97 *Indisch Genootschap*, 'De Gouvernements getah pertja-kultuur', *Indisch Genootschap Algemeene Vergadering* van, 20 November 1900, pp. 212–213.

98 See for instance P. Boomgaard, 'Oriental nature, its friends and its enemies: conservation of nature in late- colonial Indonesia 1889–1949', *Environment and History*, Vol.5, No.3, October 1999, pp. 257–292.

study and analysis of gutta percha in its application for submarine telegraphy; but the search for understanding the sources of supply and trade in the regions of supply was embedded to some extent in the cultural values of the day. The understanding of the indigenous collection practices was based in part on a tendency to overlook the practical and cultural issues associated with forest life and collecting. In that context, and given the dependence of global communication on the supply of gutta percha, it remains noticeable that even in 1903, just after the manufacture of the lengthy Pacific cables, Clouth concluded that it was a duty of civilized governments to investigate the issue, in particular because indigenous collectors were more interested in immediate gain than in the future, and that they tried neither to improve collecting methods nor increase their annual production by improved cultivation.[99]

Extraction of Gutta Percha from Leaf Product

While discussions and botanical expeditions continued, there had in fact been some attempts to bring metropolitan scientific and commercial initiatives forward to solve or profit from the need to find new or sustainable ways of sourcing gutta percha. This was through various attempts to extract commercial grade gutta percha from leaves and branches of trees sourced from the forest, rather than from latex collected by the indigenous collectors felling the trees. If the process could be made to work, collectors would simply need to remove leaves or small branches, rather than cut down whole trees to collect gutta percha and this product would then be processed in European owned factories. Potentially, this would also have the benefit (to European business) of cutting out the intermediary and Chinese traders. The concept, if it could be made to work, could in theory be both profitable and sustainable, but in the end proved to be neither.

One system that aimed to function in this manner was that patented by Sérullas and another French national, a M. Hourant, for a commercial endeavor in the Straits Settlements, but in the event in 1895 a factory managed by Hourant's nephew was established in Sarawak, based upon the Sérullas and Hourant patent. The company was a commercial failure due to lack of leaf supply, the cost of production and of transporting the final product, and the poor quality of the final product.[100] Two of Britain's leading scientific minds also joined in the hunt for profits arising from the Serullas and Hourant patent. One was William Thomson, as noted above, one of the leading scientific

99 F. Clouth, *Rubber, Gutta-Percha and Balata*, London, Maclaren & Sons, 1903, p. 166.

100 Samples were sent to Paris for testing in 1894; *Sarawak Gazette*, Kuching, 1 April 1895, p. 57; *Agricultural Bulletin*, FMS, July 1903, p. 227, and see later in Ch. 3 here.

minds working to solve the technical issues associated with the development of submarine telegraphy. While Thomson appreciated the 'most beautiful subject' of mathematical analysis associated with the study of gutta percha,[101] he was fully aware of its commercial potential as well. The other was Sir William Ramsay, Professor of Chemistry at University College. Both lent their scientific reputation and financial backing to a consortium and the somewhat confusingly named Gutta Percha Corporation.[102] The Corporation was formed in 1897 with the purpose of extracting gutta percha in a factory to exploit the Sérullas and Hourant patents. The prospectus for the company set out extensive details about the patents for processing gutta percha – endorsed by Thomson and Ramsay and other leading experts in the industry which made the corporation 'splendidly equipped technically' but its commercial structure was thought to be inadequate and underfunded.[103] In fact, under-investment and problems obtaining leaf gutta percha supplies led to the company's failure and a high-profile court case.[104] Had these investors seen the trade figures from the Hourant Sarawak factory, they may have thought better of investing in the venture at all. From 1894, through to 1897 when the Corporation was formed, just 2,720 pikuls, or $1,418.18 worth of leaf gutta percha had been exported from Sarawak.[105] In the event, collectors soon discovered that it was far simpler to cut down small trees to collect the leaves and branches to obtain the product, so the method proved to be not a sustainable means of collection and in Sarawak the practice was banned.[106]

Sérullas' endeavor extended to the public arena after he became involved in an exchange of critical correspondence in the Singapore *Straits Times*. The acrimonious exchanges were with another gutta percha (leaf) extraction patentee, a Singapore-based French national Dieudonné Rigole; each man criticizing the patented processes and even the honesty and personal qualifications of the other.[107] Rigole had a factory in Singapore which ultimately was also a commercial failure due to the problems in quality, high production costs and

---

101 S.P. Thompson, *The Life of Lord Kelvin*, 2 Vols, New York, Chelsea Publishing Company, 1910, as 2nd edn Reprinted New York, Chelsea Publishing Company, 1976, Vol. I, p. 336 and see previous chapter, above.

102 Not to be confused with the Gutta Percha Co, the cable producing arm of Telcon.

103 IRGPETJ 12 July 1897, p. 303.

104 See for example *The Times*, London, 12 March 1901, p. 4.

105 Sarawak Gazette, Kuching, 1895–1898 various issues, see Appendix 2.

106 Sarawak Gazette, Order No. VII, 1895, in SG 1 October 1895, p. 178.

107 For example, especially *The Straits Times*, Singapore (hereafter ST) 14 March 1893, p. 2; but also in ST 11 October 1892, p. 9 – in this account the writer thought the gutta percha obtained by the Sérullas process was useless for electrical (ie., telegraphic) purposes; rather

difficulties in obtaining sufficient supply of gutta percha leaves. Rigole had, however, sold patent rights to another company, the British owned Borneo Gutta Percha Extraction Syndicate. Little known in the historic record, what is interesting about the Syndicate is that it claimed to have purchased a 99-year lease for 93 square miles of territory in Brunei for the extraction of local forest products including (but not only) gutta percha.[108] It may have been that both Rigole and Sérullas achieved more personal financial profits from selling their patents than from the commercial operations that used techniques that they had patented.

Other efforts in the region to obtain and process leaf product had some modest but short-lived success. These were largely Dutch initiatives in the late nineteenth and early years of the twentieth century, with small steamers used to transport gutta percha leaves sourced in the region. These were run by the *Nederlandsche Gutta-Percha Maatschappij* (Netherlands Gutta Percha Company). The company reported losses and problems at its Singapore gutta percha processing factory for several years, often occasioned by difficulties in obtaining sufficient leaf product, with the factory closing for up to sixty days in some years.[109] This company had used yet another patented process, known as the Ledeboer process after its patentee, but again it was viewed with suspicion about its suitability for telegraphic use, the factory at least in its early days not even employing a chemist.[110]

The commercial and practical difficulties of a forest-sourced leaf trade were well summarized in an issue of the *Straits Times* in 1892. Again reflecting the emotion that the matter conjured up in the region, the writer demanded 'who is this idiot who would believe in the possibility of getting a picul of the leaves laid down' from Borneo to Singapore at a commercially attractive price?[111]

---

it was of the quality used for the manufacture of tennis shoes; and 8 November 1892, p. 674.

108 Durnford and Co Prospectus, (TCM) *Telcon Archives*, National Maritime Museum, Greenwich, London TCM/21/18.

109 IRW 1 July 1904, p. 345.

110 See A.G.N. Swart, *Rubber Companies in the Netherland East Indies*, Amsterdam, The Netherland Commission for the International Rubber Exhibition, London, 1911, p. 145; IRW 1 July 1903, p. 337; IRW 1 July 1904, p. 345; Sherman 1901, p. 345; ST 28 August 1900, p. 2. The other plantation established by Dutch interests was in the Riau Archipelago, off the Malay peninsula – the *Indragiri Sumatra Rubber and Gutta Percha Company Limited*. It also planted thousands of trees, but appeared to have more success in the rising trade in rubber, see Swart, p. 93.

111 ST 11 October 1892, p. 3.

As in Sarawak the problem associated with the leaf gutta process was that rather than being a more sustainable process, it in fact encouraged depletion as it was simpler for collectors to completely cut down small trees to collect the leaf and twigs used in the trade.[112] All factories in the region and in Europe, based upon similar extraction processes, had problems obtaining leaf supply – for local collectors the difficulties in carrying this bulkier form of gutta percha and the price it attracted (although it was easier to collect), meant that the trade did not repay the labor involved.[113]

A regular supply (and guaranteed quality) of product for this form of extraction would require investment in plantations if it were to be a practically viable, and a longterm solution for the future of cable telegraphy. In the meantime, unless or until such plantations came into production, even the involvement of highly regarded scientific minds and thousands of pounds of investments proved insufficient to proceed to financial success: the hunt for gutta percha continued to rely upon a trade network with local knowledge and regional networks.

Gutta Percha Plantations

It might be thought that nations engaged in cable production might have been keen to establish gutta percha plantations either for reasons of commercial gain (even despite the long-term investment needed) or as a means of ensuring future supply for the cable industry. Governments from nations such as Britain, Germany and France – all cable producing countries and each with colonial possessions in or near the areas where gutta percha was naturally occurring, or might be readily established on plantations – might have looked to develop plantations.

The stalled and vexed French attempts at collecting seedlings for possible plantations have been noted above and French success in the quest for gutta percha appears to have been more focused on individuals working to develop patents for extracting gutta percha from forest (leaf) supply rather than from official plantations.

As the nation with the greatest capacity to produce submarine cables, Britain could be expected to be at the centre of the development of plantation resources of gutta percha. This proved not to be, at least not in terms of supplying commercial levels of gutta percha to the market during the period of

112 Ibid.

113 *Agricultural Bulletin of the Straits and Federated Malay States*, 'Gutta Percha From Leaves', New Series, 1902, Vol. 1, No.4, p. 153.

this study. Even by the end of the century there were considerable arguments among various British officials – in the Straits Settlements, the Malay Peninsula, and in the U.K. at the Foreign and Colonial Office and the Royal Botanic Gardens, Kew – about the need both to protect gutta percha resources and to initiate a platform for future development of the resources. The situation and arguments became so heated as to be described as a 'Gutta Percha Crisis'.[114] Thiselton-Dyer at the Royal Botanic Gardens, Kew in a lengthy letter to the Colonial Office set out the frustrations, and differing opinions about efforts to protect or develop knowledge and resources in the region, with comments such as 'Could any system of management be made more futile?' and referred to his earlier statement that not five sizeable gutta trees could be found in the Straits Settlements in 1894'.[115] A few years later, Thiselton-Dyer noted, however, that 41,000 saplings that could supply gutta percha were planted in the Straits Settlements in 1901.[116] If the calculation that replacing the world's cables might have required 27 million trees was even roughly correct,[117] then the plantings mentioned by Thiselton-Dyer emphasize the extent to which the hunt for gutta percha would require rather larger plantations than those being established at that time. Following a recommendation in a report in 1902 from the Secretary of State for the Colonies, the British began relatively limited forest (as opposed to plantation) planting. In 1905, three attempts were made – at Penang, Malacca and Singapore – but it was said none succeeded.[118] Burn-Murdoch (Conservator of Forests for the Straits Settlements) summarised progress in 1905 when he wrote that the question of planting was also discussed, 'but not in a very practical manner'.[119] In the meantime, Ridley thought that in fact Burn-Murdoch knew nothing about gutta percha, and that he took only an idle interest in the work being undertaken, 'merely stopping by to drink tea'. Ridley also thought that some decisions opposing gutta percha planting were being

114 Kathirithamby-Wells, 2005, p. 68ff.

115 W.T. Thiselton-Dyer, Correspondence to C.P. Lucas (Colonial Office), in Royal Botanic Gardens Kew, Archives, *Kew: Miscellaneous Reports, Gutta Percha,* 1899, 26 December, ff.39–46.

116 CO 273/263:553.

117 W.R. Tromp de Haas, *The Netherlands India Government Gutta-Percha Plantations in Java,* Tjipetir, Batavia Government Gutta-Percha Plantations, 1908, p. 3, after Obach 1897.

118 I.H. Burkill, (ed.), with contributions by W. Birtwistle, F.W. Foxworthy and J.B. Scrivenor, *A Dictionary of the Economic Products of the Malay Peninsula,* 2 Vols, London, Crown Agents, 1935, II, p. 1631.

119 A.M. Burn-Murdoch, 'Some facts about gutta percha', *Agricultural Bulletin of the Straits and Federated Malay States,* Vol. IV. no.2 (February), 1905, p. 35.

made by people who knew more about 'grey shirting and golf'.[120] In 1909, the *Straits Times* reported that the Director of the Botanic Gardens in Singapore – [Ridley] – offering a reflection on the general European attitude to the study of gutta percha advised that while there were a 'couple of gutta plants at Kew ... these looked so miserable that they were usually kept out of sight...[and] two were said to be at Paris, but they could not be found when wanted'.[121] After Sherman concluded his study of the gutta percha resources of the region, he reported that the gutta percha nurseries in Singapore, Malacca and Penang were on a small scale owing to lack of funding or adequate assistance to the various Directors of the botanic gardens.[122] Arguments about long-term objectives and preferred management practices continued for decades – the Conservator of Forests for the Federated Malay States in his report as late as 1923 noted that disagreements about aims and practices for gutta percha resources existed between the Conservator and other senior staff.[123]

With regard to Germany, reports from the 1890s of gutta percha plantations in German New Guinea suggest that while some had been established, they were very modest in scale. The focus was placed instead on profitable plantations such as copra and coffee. In the 1901–02 reporting year, for example, 3950 hectares of new plantations of various products were established in the colony, most of which was coconut (copra), with just one hectare of gutta percha.[124] In 1909 the *Frankfurter Zeitung* reported that the cultivation of gutta percha in New Guinea was not undertaken on any noteworthy scale and that naturally occurring resources of the product were too difficult to collect and there were no skilled laborers to do so.[125] It was expected that there would be no great increase in production or exports of gutta percha. According to Ohff, some small plantations were established but these activities were conducted initially under license as part of efforts to develop the local economy.[126] Schlechter had urged the government to establish plantations until private plantations could

120 Cotton cloth, one of the most important items of the regional barter trade sought by hinterland traders; in *Kew Director's Correspondence*, Royal Botanic Gardens Kew, Archives 168: 12 July and 2 August 1902, 168/f76 and f77.

121 ST 14 October 1909, p. 7.

122 Sherman, *Report of the Special Agent of the Forestry Bureau*, p. 345.

123 FMS Govt. Gazette Supplement, June 13, 1924, p. 103.

124 Sack and Clark, p. 226.

125 In *The Sydney Morning Herald*, Sydney, Charles Kemp and John Fairfax, 27 September 1909, p. 9.

126 H.J. Ohff, *Empires of Enterprise: German and English commercial interests in East New Guinea 1884 to 1914*, Ph.D Dissertation, Adelaide, University of Adelaide, School of History and Politics, 2008.

be proven profitable,[127] but by 1915 it was 'clear that the government had decided to ... concentrate on copra production'.[128] By 1913–14 it was noted that the market for this gutta percha was poor, with the New Guinea Company buying all the produce, but this was just 299 kg. and even the Company had ceased purchasing.[129] By 1915 the gutta percha trade in German New Guinea was described as 'very slight' in comparison with other trades from the colony (copra and bird-of-paradise skins apparently were more profitable).[130] From that time judgments about any gutta percha plantations in the German colonies were influenced by several other developments. These included the restrictions on German trade that arose during and after World War I, and the rise of wireless telegraphy (and the uncertainty that generated for both the future of cable telegraphy and the demand for gutta percha). Germany had also been placing an increasing emphasis upon the strategic significance of wireless as opposed to cable telegraphy.[131]

It was the forces of commerce, rather than the outcomes emanating from within cable producing countries, that had the greatest impact in prompting an interest in gutta percha plantations. These endeavors came not from a cable producing nation, but from a regional economic force, the Netherlands and its interest in developing the resources of its colonies in Southeast Asia. Such efforts did not begin auspiciously: Obach reports a Dutch plantation established in Java in 1847, but half the plants died within two years, though the remainder did provide two trees which produced seeds for future plants.[132] While there had been plantings in various parts of the Dutch colonial territories, many were either experimental, such as at Buitenzorg in Java, or small local efforts such as at Pontianak in the western region of Netherlands Indies part of western Borneo[133] and were not sources of commercial levels of production.[134] The *Nederlandsche Gutta-Percha Maatschappij* that had the leaf processing plant at Singapore also began gutta cultivation on Java in 1902, but ceased

127 Schlechter, 1911 in Dwyer, p. 266.

128 Dwyer, p. 268.

129 Sack and Clark, *German New Guinea. The Annual Reports, Edited and Translated,* p. 308; P. Sack and D. Clark, (eds), *German New Guinea. The Draft Annual Report 1913–14, Edited and Translated.* Canberra, Australian National University Press, 1980, p. 90.

130 SFPMA 30 March 1916, p. 12.

131 See for example H.J. Evans, 'The path to freedom? Transocean and German wireless telegraphy, 1914–1922', *Historical Social Research*, 35 (1), 2010, pp. 209–233; and Winkler, pp. 15 ff.

132 Obach, 1897, p. 117.

133 de Vriese, 1856, p. 16.

134 See summary in Obach, 1897, pp. 117–118, and IRW 1 October 1901, p. 13.

new plantings in 1906, when it began planting rubber (*hevea*) instead. Another Dutch company had a plantation on the Riau Archipelago, off the Malay peninsula. It also planted thousands of seedlings but likewise converted to rubber after a few years.[135] Both companies' gutta plantations were established in a time of increasing rubber trade, so commercial considerations and the likely faster returns from rubber could have had an impact on these ventures; how suited their product was for cable uses may also have been uncertain. It was an interesting reflection of the trade and the apparent emotions it raised in the region that the matter of plantation growing was raised again in the *Straits Times*, when arguments for and against the plantation growing of gutta percha, and its commercial possibilities and practicalities, again raged in the letters to the Editors.[136]

Finally, in 1900, the Dutch Ministry of Colonies decided to expand a small plantation at Tjipetir in Java, with the use of seedlings from Dutch and British colonial sources. Java was not an indigenous source of gutta percha but growing conditions there proved to be very suitable, subject to proper management. Even then – after all those decades of research – botanists and telegraph experts disagreed about which varieties should be grown. Some incorrect varieties were chosen and 160 acres of seedlings that had been planted had to be removed after arguments arose that they were entirely unsuitable for cable manufacture.[137] Somewhat mirroring the British disagreements about gutta percha resource development, there were very public and personal attacks made about the management of Tjipetir. Van Romburgh, an organic chemist who had been a Director of the Government Chemical Laboratory at the Dutch Botanic Gardens in Buitenzorg, Java, publicly accused the Tjipetir management of not paying attention to the characteristics of the plants (that is, not growing the correct variety for the telegraphic market), of poor forest and cultivation management, and a lack of good record keeping (including data on yield of each variety, costs of growing, market price and so on). The Inspector of Forestry, S.P. Ham, published a response that van Romburgh had not understood the intention behind the growing of the particular varieties (defending the growing of those later removed) or recognized the good management of the plantation generally. Even further, Ham said that not only had Tjipetir been mismanaged previously by the Forestry Department but 'nothing

135 See A.G.N. Swart, p. 145 for The *Nederlandsche Gutta-Percha Maatschappij* plantation. The plantation established by Dutch interests in the Riau Archipelago was the *Indragiri Sumatra Rubber and Gutta Percha Company Limited*, see Swart, p. 93.

136 Such as ST 22 March 1902, p. 2; ST 14 October 1909, p. 7 and ST 16 October 1909, p. 8.

137 Tromp de Haas, 1908, p. 8.

worth mentioning' had been achieved during that period. After further criticisms of van Romburgh he added that the Forestry Department prior to Ham's own time, had shown little or no interest in Tjipetir, none of their staff having ever even visited. Further, while it was all very well for van Romburgh to call for the garden to have supervision by staff with broader education and more expertise, van Romburgh himself had no education in the culture and subjects required and did not know what was involved in forestry.[138] Van Romburgh returned to similar criticisms, while another participant in turn criticized van Romburgh for his opinion that 'a forester only produces wood'.[139] Others thought a forester, not a 'detached overseer' was needed, while van Romburgh still argued for a botanist and so on.[140]

In 1907, Terry wrote that little had been achieved by way of plantations in any location to that date and that, compared with rubber, they required a far longer period of investment before they could return a profit.[141] By 1909, questions were being asked about the advisability of continued funding for Tjipetir since it was known that the plantation and the smaller gutta percha resource at Buitenzorg were 'a financial disaster'.[142] Concerns about the role of government expenditure, the commercial viability of the plantations, and the lack of likely future demand from the telegraphic industry meant that the Tjipetir plantation was restricted to a 'primitive experimental factory'.[143] The Tjipetir plantation was assessed against commercial objectives – not against the objective of ensuring a future supply of the commodity that underpinned global communication. Then again, gutta percha was, after all, largely a commercial issue for the Dutch. The Netherlands was not itself a telegraph cable manufacturer and there was no particular reason why it should shoulder the burden of ensuring the needs of global communications and its associated industries. Despite these misgivings, the plantation continued and production and yield grew over the years – by the mid-1920s a new factory at the plantation could

138 S.P. Ham, 'De proefaanplantingen van caoutchouc-en getah-pertja leverende hourtsoorten te Tjipetir. Onder de leiding van het Boschwezen', *Tijdschrift voor Nijverheid in Nederlandsch-Indië*, 1900, Volume LXI, pp. 64–77.

139 *Indisch Genootschap*, 'De Gouvernements getah pertja-kultuur', *Indisch Genootschap Algemeene Vergadering* van 20 November 1900, p. 231.

140 *Indisch Genootschap Algemeene Vergadering*, pp. 203–235.

141 Terry, 1907, p. 273.

142 ST Singapore, 14 October 1909, p. 7.

143 H. van Lennep, *The Netherland Indies Government Gutta Percha Estate Tjipetir (Java)*, Buitenzorg, 1922, p. 8.

PLATE 6 *Gutta percha from the Tjipetir planation in the early twentieth century.*
Note: Gutta percha produced at the Tjipetir estate. The size of the tablets dates them to the period near to the International Rubber and Allied Trades Exhibition held in London in 1911. Telcon made purchases from the estate after the company saw samples at the Exhibition.
Source: Collection Nationaal Museum van Wereldculturen, Coll. No. TM-10012863, with permission.

produce up to 400,000 pounds of gutta percha in a year.[144] Still, at that rate the factory would have worked for almost two years at full production to achieve the product required, for example, just for the 1866 Atlantic cable link that required 740,800 pounds of gutta percha.[145] In later years, the factory mainly produced 'white' gutta percha used for golf balls.[146]

Despite Britain's predominance as the largest market for gutta percha and cable manufacture, it was not until 1915 that the cable company Telcon finally decided that 'to avoid complete dependence upon the native collection'

144 J.W. Vander Laan, *Production of Gutta-Percha, Balata, Chicle and Allied Gums*, Trade Promotion Series No. 42, Crude Rubber Series, Washington, Department of Commerce, Bureau of Foreign and Domestic Commerce, Government Printing Office, 1927, p. 13.

145 See Obach, 1898, p. 103.

146 SFPMA 18 June 1931, p. 19.

it would establish a plantation in Malaya to be run in a similar fashion to a rubber estate.[147] A plantation of 6,000 acres was established in Pahang, named *Selborne* after Lord Selborne, the then Director of the Company. Evidence from the Telcon archives suggests that this decision in fact took place after Telcon's cable producing arm, the Gutta Percha Co, bought gutta percha that had been produced at the 'primitive' factory then operating at Tjipetir. Examples of the production from that factory had been seen at the International Rubber and Allied Trades Exhibition held in London in 1911. Sample orders placed by the Gutta Percha Co in 1912 and 1913 were made on the condition that they were in accordance with the product shown at the Exhibition.[148] That is, once the technology and associated planting had been proven through the scientific endeavors and investment of others, then the Gutta Percha Co, after almost seventy years in the business of cable manufacture, began to consider both its need to ensure the future supplies of gutta percha and its position in the market place.

Plantation growing of itself would not necessarily have been a practical solution to the vexed question of guaranteeing supply since, unless the trees could regularly supply gutta without the trees being destroyed, the viability of plantations was by no means guaranteed. Tapping was not as feasible as expected, despite decades of criticism of indigenous collectors for not doing so, in fact European experts could not find a means of how this could be done.[149] Eventually a solution was found that overcame the problems – this was achieved by the refinement of processes to obtain gutta percha from leaves, twigs and bark. Both the Tjipetir and Selborne plantations functioned successfully only because of this new technique. Early efforts to achieve this had mixed success. As an indication of how much growth was required to obtain gutta percha in this manner, in 1948 the Selborne plantation factory took in 12 tons of material from the trees *per day* for 300 days to achieve an annual output of just 110 tons of gutta-percha.[150] While Tjipetir and Selborne were eventually productive, they were not large scale operations: at least not large enough to have sustained a continuing global demand for submarine telegraph cables insulated with gutta percha had it remained at the rates seen in the nineteenth century. Many more plantations would have been required had telegraphy continued

147 J.N. Dean, 'A visit to Selborne Plantation', *Zodiac House Magazine*, Telcon Incorporating Staff Association Social and Athletic Club and Shop Stewards Committee, (March) 1949, p. 2.

148 Telcon (TCM) *Telcon Archives*, National Maritime Museum, Greenwich, London, TCM 21/52.

149 IRGPETJ 20 June 1904, p. 604.

150 Dean, *Zodiac House Magazine*, p. 3.

to have been reliant upon gutta percha insulation, or had the existing world network of submarine cables needed to be replaced by cables insulated in the same fashion. This was not done. A combination of good fortune and technology development meant that global communications networks had averted a very considerable risk to their continued viability.

The European quest for understanding, and more importantly, for sources of supply of gutta percha, harnessed both science and commerce as counter-balancing forces. Gutta percha 'adventurers' like Sérullas, Schlechter or Combanaire might be among those described in the English newspaper *The Spectator* in 1892 in an article entitled 'The Incentives to Exploration':

> We send out the same explorers intent on geographical discovery, scientific 'exploitation' and a mighty dividend...Science, too, of a purer kind, though it will not urge so many men as the thirst for adventure and novelty, or the lust for wealth, will still do much to force on the completion of exploration.[151]

While there is no doubt that research and expeditions added to the knowledge required to develop the trade, personal disagreements and lack of funding also held back the development of European engagement in the quest for gutta percha. The expeditions, studies and personal enterprises outlined above led to an improvement in scientific knowledge and in a commercial sense traders like William Lingard profited from the trade but European manufacturers, traders and governments did not establish their own commercial operations. Building on the knowledge obtained from many expeditions – and seedlings collected over time – there were some plantations established in the long hunt for gutta percha. Yet they were never to reach commercial levels of supply required for cables had the technology continued its existing course; and in fact, plantation supplies played only a modest role in submarine telegraphy in the waning years of gutta percha insulation (see Appendix 1.2).

In 1848, one of the earliest reports about the rise of the gutta percha trade and its future challenges proved to be the source of a fortuitously, yet largely accurate summary of the future of the trade:

> Although the Dutch have many able botanists amongst them, they too, it would appear, have been obliged to await the slow indigenous process by which useful knowledge is diffused to the Indian Archipelago[152]

151 *The Spectator*, London, 28 May 1892, p. 12.

152 J.R. Logan, *Journal of the Indian Archipelago and Eastern Asia*, Vol. 11, 1848, p. 531.

Between 1843 and 1903 – from the first commercial export of gutta percha to the laying of the Pacific telegraph cables – a total of 1,856,847 pikuls (around 110,500 tons) of raw gutta percha was exported from Singapore and additional amounts from elsewhere in the region.[153] That gutta percha was collected in the forests by indigenous collectors and traded and processed by Chinese traders and merchants. How the trade developed and how the needs of submarine telegraphy were met is considered in the following chapter.

153 See Appendix 2, Table 1.

CHAPTER 4

Factory to Forest: Opportunity at the Periphery

> The Singapore papers are filled with the description of an article called gutta percha[1]
>
> *SYDNEY CHRONICLE*, *4 November 1847*

∴

While cable manufacturers sought a continuing supply of quality gutta percha, most of the time the hunt for this valued resource was undertaken not by them, but by local participants in Southeast Asia. Simply put, collectors and traders did the hunting for gutta percha in a series of commercial transactions. One study of the benefits of commodity chain analysis concluded that amongst other things it provides a process for identifying and understanding how, and by whom, the chain is created.[2] Using that methodology, it becomes clear that global submarine telegraphy was dominated by a few influential companies and individuals – or chain drivers – whose influence will be explored further in this chapter.[3] It is also clear that much of what shaped the gutta percha commodity chain, what influenced these 'chain drivers', was not only profit, but technologically driven aspects – the cost and risk associated with a new technology had an impact on how cable producers depended upon gutta percha. The reluctance to risk using another type of insulation meant that the gutta percha suppliers of Southeast Asia were looked to for continued involvement in the trade.

Studying the commodity chain reveals the resulting interdependence between cable manufacturers, intermediary Chinese and indigenous trade networks in Southeast Asia. This is set out in this chapter, in part through correlation of evidence from official and commercial archives and trade figures. As will be shown, the commerce of cable telegraphy meant that this integration took profitable but sometimes unpredictable turns. The gutta percha commodity chain developed an unusual set of trade arrangements based in

1 *Sydney Chronicle*, 4 November 1847, p. 2.

2 J. Bair, 'Global Capitalism and Commodity Chains: Looking Back, Going Forward', *Competition and Change*, Vol.9, No.2, June 2005, pp. 153–180, especially p. 157.

3 Bair, *Competition and Change*, p. 157.

 | DOI 10.1163/9789004357280_006

part upon local political, commercial, social and cultural conditions. Patterns of European colonial expansion and local trade and knowledge networks were important aspects in the gutta percha trade. The networks were a key feature in dealing with the geographic, economic and political challenges facing the trade. Also revealed is the speed with which commercial decisions by cable manufacturers affected the market.

Singapore and the Development of the Gutta Percha Trade

In 1843, Christopher Nickels, later a principal of the Gutta Percha Co, attended a meeting of the Joint Committee of the Royal Society that discussed samples of gutta percha sent to it from Singapore by a Dr William Montgomerie. Following that meeting, Nickels contacted Jewesbury and Wilkinson, a company of Colonial Agents (later Jewesbury and Co) and asked them to obtain samples of this previously unknown product. Jewesbury's agents in Singapore, Ker Rawson and Co, arranged for what became the first shipment of gutta percha as 'an article of commerce' and as a result in 1845 the Company imported 202 cwt.[4] It was used to produce the variety of industrial and consumer products described earlier. The demands for gutta percha grew rapidly and the entrepreneurial and processing capacities of Singapore were geared to respond to new commodity trades like this. Indeed, the market grew so quickly that in 1848 exports reached a reported 14,670 cwt (12,326 pikuls) and this was before the demands of submarine telegraphy began in 1850. Newspaper 'Market Reports' were already noticing this new commodity, commenting that rates were going up due to great demand and the number of 'useful purposes' to which it could be applied.[5]

It was perhaps fortunate that the samples of gutta percha had been found in Singapore because it possessed the ideal trading infrastructure for the exploitation of a new commodity found in the remote hinterland.

Singapore was an important entrepot for Asian trade, a centre to which goods were shipped for re-export. Such centres often function without the imposition of customs duties or other fees and possess port and storage facilities. Other services may be offered at an entrepot, such as sorting or repackaging of goods, access to commercial information, and efficient distribution networks

4 British figures were quoted in imperial measures – H.W. Jewesbury, *A Short History of the Introduction of Gutta Percha into Europe*, London, H.W. Jewesbury, 1883, p. 12.

5 *Straits Times and Singapore Journal of Commerce* (*hereafter* STSJC), 29 March 1848, p. 262.

facilitated by business contacts or through an arrangement of transport hubs.[6] An entrepot may include the 'presence of specialized agents who match buyers and sellers'.[7] An entrepot can also be an important venue for the resolution of cross-cultural differences between trade participants (including, but not limited to, language differences). All of these characteristics applied to the gutta percha trade and the entrepot of Singapore that facilitated that trade. Singapore had natural and geographic endowments that formed the basis of its economic prosperity, including a superb harbour around which were built excellent port facilities. An island on the edge of the Malay Peninsula, its other good fortune was a location that enabled it to take advantage of the shipping routes between the Indian Ocean and the South China Sea – busy shipping lanes between Britain, Europe and the U.S. and the major export centres of China and India. It was better placed than many other ports to attract shipping from regional trading points in between. Singapore was 'the grand central station of the East – the halfway house between Calcutta and China ... the products of all its neighbours are poured into its lap'.[8]

Commodities could be handled easily for sorting, re-processing and packaging, and then re-loading onto small ships for transport back to the larger ocean-going vessels at the harbour. The location of Singapore's wharves and transport facilities meant that both large and smaller vessels could be easily and quickly loaded or unloaded via an arrangement of roads and warehouses (godowns), with river and harbour transport provided by lighters.[9] Chinese-run shops and godowns were located along the riverbank for handling the arrival of the so-called *native craft* or *prahu* and other trading vessels – 'the extraordinary variety of nondescript native craft'[10] run by Asian traders. They were gradually superseded by coastal and larger square-rigged and then steam vessels run by

6 See R.C. Feenstra and G.H. Hanson, 'Intermediaries in entrepot trade: re-exports of Chinese goods', *Journal of Economics and Management Strategy*, Vol. 13, Iss. 1., March 2004, p. 5.

7 P. Moustakerski and W. Shang-Jin, *Outsourcing Tariff Evasion: A New Explanation for Entrepot Trade, IMF Working Paper* WP/05/102, Washington, International Monetary Fund, 2005, p. 3.

8 G.F. Train, *An American Merchant in Europe, Asia and Australia: A series of Letters from Java, Singapor, China... etc*, Broadway, G.P Putnam & Co, 1857, pp. 68,75.

9 See W.G. Huff, *The Economic Growth of Singapore: Trade and Development in the Twentieth Century*, Cambridge University Press, Cambridge, 1995, p. 137.

10 J. Cameron, *Our Tropical Possessions in Malayan India*, London, Smith, Elder and Co., 1865, Reprinted Oxford University Press, 1965, p. 39; see L.K. Wong, 'The Trade of Singapore 1819–1869'. *Journal of the Malayan Branch Royal Asiatic Society*, Vol.33, no.4, 1960, p. 71.

either European or Chinese traders.[11] The produce arriving in Singapore was obtained by Chinese traders who bartered imported manufactured goods as well as rice and cloth. The goods traded were often obtained from European or sometimes Indian merchants or in bazaars on *credit*; in return, they sold produce for *cash* to the European-owned agency houses.

The practical arrangements for trade were matched by bureaucratic and financial conveniences. The status of Singapore as a free port was important: traders and merchants were not subject to the payment of customs duties and other government charges, and they were protected against any chance that the British Government might change that in future – its mercantile community had 'prevented the adoption of so suicidal a policy'.[12] The trading infrastructure of Singapore was more elaborate than just a physical arrangement of port, godowns and facilities for shipping of various capacities and nationalities. Huff argues that Singapore provided specialized services: entrepreneurial/managerial, financial, processing and marketing.[13] These were the sorts of institutional arrangements that Wallerstein describes as deepening a region's engagement in the world-economy.[14] The port linked hinterland and metropolitan centres with specialized services and provided economies of scale in both services and facilities.[15] Agency houses of Singapore provided the interface between the metropolitan core and the many producers and traders; agency houses were at the hub of two widely differing worlds and capable of meeting those demands. They were the contact point for European buyers and sellers: agency houses, many of British origin, could arrange almost anything, and frequently acted as agents for insurance and shipping companies. Agency houses were an important factor in establishing and consolidating supply chains for the produce trade of the region; they provided a trusted and efficient means of operation.[16]

Agency houses did not directly purchase from suppliers or process much of the natural produce themselves; rather, they managed orders and established

11 Wong, *Journal of the Malayan Branch Royal Asiatic Society* p. 59; L.K. Wong, 'Singapore: Its growth as an entrepot port, 1819–1941', *Journal of Southeast Asian Studies*, 9, (1), 1978, p. 60.

12 G.M. Reith, *Handbook to Singapore*, Singapore, The Singapore and Straits Printing Office, 1907, p. 6.

13 W.G. Huff, pp. 16–17.

14 I. Wallerstein, *The Modern World-System III. The Second Era of Great Expansion of the Capitalist World Economy, 1730–1840s*, San Diego, Academic Press, 1989, p. 130.

15 Huff, p. 22.

16 See J.H. Drabble and P.J. Drake, 'The British agency houses in Malaysia: Survival in a changing world', *Journal of Southeast Asian Studies*, 12 (2), (September) 1981, pp. 297–328.

contacts to fill them, usually by purchasing the produce from Chinese merchants who had obtained it either from traders who had travelled to Singapore, or through a network of other Chinese traders who travelled throughout the region trading and bartering goods for produce. The other source of produce arriving in Singapore was through the European trader like William Lingard or John Ross, and other regional traders such as the Malay, Arab or Bugis traders who used Singapore as a hub. This process, by which most produce entered the global market through the system of traders, merchants and agency houses, provided a means by which cultural differences and transport and handling problems were managed. Importantly, the separation of roles underpinned the system of credit and bartering upon which the whole entrepot of Singapore depended.

All the countries of Southeast Asia were drawn into the global economy through various commodity trades. Yet colonial administrations and settler populations in these peripheral locations were not able to fill the breadth of roles that would be required to link closely with communities at the periphery. The organization of local political systems in most regional states was fragmented and this made it difficult for colonial administrators to engage in direct political relationships and an over-arching control of states.[17] As a result, most colonial powers relied instead on relationships with intermediaries, collaborators or other indigenous hosts to fulfil that role. European traders also relied upon intermediaries to provide the necessary links with widespread and remote indigenous trading networks. In Southeast Asia, those roles were fulfilled largely by Chinese traders or indigenous leaders; they were also appointed to share power or revenue collecting tasks that were assigned to them by European administrators.

For many years, the Chinese merchants in Singapore had obtained produce by bartering with traders who had come to Singapore. As coasting vessels gradually took over more of the trade, from around the 1870s onwards, this practice varied so that the Singapore-based Chinese merchants were linked to a network of Chinese traders who travelled into the region from Singapore, taking with them goods for barter in exchange for produce. These traders took with them cotton 'piece goods' or 'Manchester goods' as they were also known, along with other items that might appeal to produce collectors. These agents were effectively a network of traders who travelled, often at personal risk, 'on a clan, kith, and kind basis…The extra-insular credit structure was too difficult

17 See R.E. Elson, 'International commerce, the state and society, economic and social change', in *The Cambridge History of Southeast Asia, Volume Three – from c.1800–1930s*, ed. N. Tarling, Cambridge, Cambridge University Press, 1999, pp. 127–191.

to replace, while the Straits produce trade was a Chinese specialty, based on centuries of experience and traditional links'.[18] Such credit networks, combined with the practice of barter, found their way up rivers in Sarawak. The Chinese traders moved throughout the region, bringing with them the process of social and economic change – and, for good or ill, an increasing interdependence with the global economy.

Wong concluded that the main economic significance of Singapore lay not so much in its own economic achievements, but rather in generating an expansion of regional trade as a whole: and Singapore trade in the region did take on an increasingly regional focus.[19] Singapore's trade in 1833 was 26 percent with the region and 49 percent with Europe and India; by 1860, 45 percent was within the region and 36 percent went to Europe and India;[20] and there was an increase in per capita GDP from 1870.[21] In these years, cotton piece goods were the most valuable commodity exported from Singapore to the region.[22] The *Weekly Market Report* reflects this – with comments such as '…the demand for cotton manufactures [is]…chiefly in barter for produce',[23] and rice was the most important commodity in the intra-region trade of Singapore between 1870 and 1915 according to Chiang.[24] Yet some of this intra-regional trade grew from the use of these products as a means of exchange and so, while the figures of increased intra-region trade have sometimes been taken as a sign of improved standards of living, they are also an indication of changing regional trade mechanisms, and the growing linkages of the region into the capitalist 'world-system' as it was being developed at this time. They were in part, too, a reflection of the importance of Chinese trade networks in the region that had been established during previous centuries of a Chinese 'world-system' of trade in this part of the world. Huff's analysis of Singapore's international rice trade concludes that Singapore and its Chinese trade network 'promoted regional economic development by providing finance through a barter system largely based on the rice trade' and cites a Straits Settlement government report from 1920 that went so far as to describe rice as 'the pecunia of the

18 Wong, *Journal of Southeast Asian Studies*, p. 83.

19 Wong, *Journal of the Malayan Branch Royal Asiatic Society*, p. 85.

20 C.D. Cowan, *Nineteenth Century Malaya, The Origins of British Political Control*, London, Oxford University Press, 1961, p. 21.

21 A. Booth, 'The economic development of Southeast Asia in the colonial era: c.1870–1942', *History Compass*, 6/1, 2008, p. 27.

22 H.D. Chiang, *A History of Straits Settlements Foreign Trade 1879–1915*, Singapore National Museum, 1978, p. 117.

23 STSJC 29 March 1848, p. 262.

24 Chiang, pp. 118–119.

east'.[25] The two commodities, rice and cloth, were two of the most significant import items in Singapore trade in the years when gutta percha exports from there were particularly high, suggesting that the trades could well have had regional connections for several decades.[26] The interconnection of these three trades in Sarawak is examined in a later chapter. Rice barter was a significant element in the trade between Singapore and the Netherlands Indies[27] and, with that region being a key supplier of gutta percha to the Singapore market, there is likely fertile ground for research into the connection between rice and gutta percha exports during a period like that surveyed here.

As the trade of Singapore grew, it became more specialized; certain merchants concentrated on particular types of produce – with their capacity to collect, modify or store produce to be consolidated for shipment westwards to Europe acting as a magnet for further entrepot trade. Conversely, on arrival in Singapore, European manufactured goods could be broken into smaller more marketable shipments for barter, for example, eastward to Borneo. Chinese merchants often concentrated on the trade and processing of particularly classes of staple goods – for example, gambier and pepper (as they were usually grown together), or gutta percha and rubber and other resins. In the case of gutta percha, this went beyond just collection from traders and sorting and assembling into transportable loads; evidence examined below suggests that Chinese entrepreneurs in Singapore were 'creating' the gutta percha of commerce. These various arrangements resonate with the proposal from Wallerstein that a typical division of a world-economy involves a hierarchy of occupational tasks, requiring a higher level of skills or capitalization in 'higher-ranking' areas, where the marketplace tends to reinforce the existing distribution of reward.[28] The Chinese might be replaced by other Chinese but their network would also be difficult to penetrate: 'their agents...reside in Sumatra, Borneo...collecting produce by barter with the natives, to whom they are not infrequently related by social, as well as by commercial ties'.[29] The provision of credit provided an additional link: either a financial obligation or a bond of trust was established which further served to keep out new entrants. While the

25 W.G. Huff, 'Bookeeping barter: Money, credit, and Singapore's international rice trade, 1870–1939', *Explorations in Economic History*, Vol. 26, 1989, pp. 185, 171.

26 See H.L. Godfrey, 2011, Appendix 2.5.

27 Huff, *Explorations in Economic History*, 1989.

28 I. Wallerstein, *The Modern World-System I. Capitalist Agriculture and the Origins of the European World-Economy in the Sixteenth Century*, New York, Academic Press, 1974, p. 350.

29 J. Thomson, *The Straits of Malacca Indo-China and China*, London, Sampson Low, Marston, Low & Searle, 1875, p. 12.

role of European agency houses was further consolidated by a shared network of roles and relationships, the same could be said for that played by the Chinese and their extension of credit to produce collectors. Increasingly, as the century progressed, agency houses and Chinese traders developed networks and specialties that targeted particular markets and clients. These networks were soon to take advantage of, and make possible, the new trade in gutta percha.

There was, then, an established network and arrangement of trading facilities with which the new trade in gutta percha could be developed.

As fate would have it, there were other conditions that facilitated the early arrangements for the trade. The agency house that had arranged for the first commercial shipment of gutta percha – Ker Rawson and Co – was also agent for the Temenggong of Johor, Daing Ibrahim. Johor was a state in the southern part of the Malay Peninsula, immediately adjacent to the island of Singapore. Temenggong Ibrahim lived in Singapore where he was a friend and neighbour of William Ker, a principal of Ker, Rawson and Co.[30]

The Temenggong, no doubt readily able to learn and profit from Ker Rawson's superior market connections, declared gutta percha a royal monopoly in his territory.[31] He also restricted the rights of outsiders to trade for gutta percha in Johor and was said to have shifted a number of *Oran Sabimba*, a group of his serfs, from the island of Battam to Johor to collect gutta percha.[32] One British source, reflecting an excited view of the extent of the trade at the time, wrote that that the response in the region was such that these 'organized gum-hunters ... spread like a cloud of locusts' through Johor and nearby regions.[33] All these actions were reported on as early as 1847.

Daing Ibrahim had for some years been a focus of attention from British authorities because of his alleged engagement in maritime raids on regional sea-traders. Johor itself was also an area of political concern to the British;

30 C.M. Turnbull, *The Straits Settlements, 1826–67: Indian Presidency to Crown Colony*, London, Athlone Press, 1972, p. 280.

31 J.R. Logan, 'Range of the gutta taban collectors, and present amount of imports into Singapore', *Journal of the Indian Archipelago and Eastern Asia*, Vol. 11, 1848, p. 529; T. Braddell, 'The Sultan of Johore', *The Journal of the Indian Archipelago and Eastern Asia*, New Series, Vol. 11, 1858, p. 64.

32 J.R. Logan, 'The Orang Binua of Johore', *The Journal of the Indian Archipelago and East Asia*, Vol.1, 1847, p. 286; J.R. Logan, 'The Orang Sabimba of the extremity of the Malay Peninsula', *The Journal of the Indian Archipelago and East Asia*, Vol.1, 1847, p. 298.

33 H. Cleghorn, J.F. Royle, R.B. Smith and R. Strachey, 'Report of the Committee appointed by the British Association to consider the probable effects in an economical and physical point of view of the destruction of tropical forests', *Report of the British Association for the Advancement of Science*, London, John Murray, 1852, p. 96.

problems there, including piracy, contributed to the increasing British involvement in the Malay Peninsula.[34] Piracy had a vexed history in the region for colonizing powers although their own, not dissimilar, actions were such that they could not be accorded any morally superior position. Ibrahim's activities may in part have been altered by the increased activities of the British and French navies in the region, but more importantly by the fact that the Temenggong began exploiting the new gutta percha trade. Turnbull's analysis of British policy in the Straits Settlements attributes the Temenggong's changed behaviour to his discovery of this new legal, alternative way of making money, from the gutta percha trade; but the Temenggong is also widely recognized as very aware of the economic potential of Johor and its need for commercial development.[35] Whatever the nuances of his motivations, the Temenggong began supporting British attempts to control regional piracy.

Yet in 1848, the Singapore Chamber of Commerce had written to the British Resident Councillor asking for urgent action to be taken against the Temenggong who they alleged had positioned armed *prahus* 'to intercept all prahus with gutta destined for Singapore'.[36] He was, they said, either taking over the shipments or demanding they be sold to him at ridiculously low prices. The Temenggong controlled ninety percent of the Singapore gutta percha market, estimated at between 10 and 12,000 pikuls and valued between $150,000 and $200,000, said the *Singapore Free Press* when it reported on the issue, which it regarded as 'so serious a question' that it could not be tolerated for a moment.[37] While these reports come from British sources and how much the Temenggong actually controlled is unclear, the trade did grow and he gained support from much of the merchant community of Singapore. Another British report from 1847 had concluded that, while the Temenggong had declared the trade to be a monopoly, the price 'allowed by him was a fair one' and that it was to his 'energetic measures that Singapore has been indebted for a large part of the Gutta hitherto imported'.[38] Turnbull believes that the Temenggong's co-operation in

---

34 See C.M. Turnbull, *A History of Singapore 1819–1975*, Kuala Lumpur, Oxford University Press, 1977; and Cowan, *Nineteenth Century Malaya, The Origins of British Political Control.*

35 Turnbull, *The Straits Settlements, 1826–67: Indian Presidency to Crown Colony*, p. 277; see for example D. Hen and S. Aljunied, *Reframing Singapore. Memory, Identity, Trans-Regionalism*, Amsterdam, Amsterdam University Press, 2009, Ch. 3 re the Temenggong's development policies.

36 C.B. Buckley, *An Anecdotal History of Old Times In Singapore* 1819–1867, Singapore, Fraser & Neave, Reprinted 1965, University of Malaya Press, 1902, p. 482.

37 Buckley, p. 483.

38 J.R. Logan, 'Range of the gutta taban collectors, and present amount of imports into Singapore', *Journal of the Indian Archipelago and Eastern Asia*, Vol. 11, 1848, p. 529.

suppressing piracy was so 'precious' that Governor Butterworth was in fact not inclined to thwart his monopolistic gutta percha activities.[39]

Further complicating the political issues associated with the gutta percha trade was that fact that, for reasons with a complex history, there were two potential rulers of Johor – there was both a *Temenggong* (Daing Ibrahim) and a *Sultan* of Johor, and each claimed the right to political control and a share in the revenue of the country, and each courted British support for their claims. Because of his role in the suppression of piracy, the Temenggong not only won support for his alleged monopolistic approach to trade, he consolidated his claims to Johor at the expense of the Sultan of Johor. After Butterworth was replaced as Governor, the Johor question (whether the Temenggong or the Sultan should gain control) became increasingly complex. Singapore's European merchant houses alternately took sides either for or against the Temenggong (Ker Rawson's supported him, of course) or took sides depending upon whom the respective governor appeared to support. The Temenggong in the meantime opened Johor up to Chinese (especially Teochiu) gambier and pepper growers and an interconnection between the gutta percha, gambier and pepper trades was established. Those connections were to remain largely unbroken for decades.

The long history of the political and commercial intrigue associated with Temenggong Daing Ibrahim is beyond the scope of further review here.[40] As will be seen in Chapter 8, the role of the gutta percha trade in shaping the interplay between political and commercial networks was seen again two decades later when the trade was instrumental in other political machinations in Brunei and Sarawak that provided an added impetus for regime change. Wallerstein proposed the concept of 'Ministructures' in the process of globalization; these included such measures as the establishment of political structures with power, institutional infrastructure, security and appropriate currency arrangements.[41] The history of gutta percha trading emphasizes how commodity chains did not require economic dominance in a region to have an impact upon the political or social landscape. All are part of an inter-related set of 'ministructures'. In turn, the elements established as part of colonial

39 Turnbull, *The Straits Settlements, 1826–67: Indian Presidency to Crown Colony*, p. 278.

40 See Turnbull *The Straits Settlements, 1826–67: Indian Presidency to Crown Colony*, for a detailed study of its significance for the changing political landscape of the Straits Settlements; also, Swettenham, 1955, Ch. v.

41 I. Wallerstein, *The Modern World-System III. The Second Era of Great Expansion of the Capitalist World Economy, 1730–1840s*. San Diego, Academic Press, 1989, pp. 130–131.

'ministructures' facilitated patterns of regional development and commercial networks which further enhanced the power of local elites.

Most of what the Temenggong controlled, and whatever other gutta percha entered the market in Singapore, was purchased by the Gutta Percha Co: Jewesbury's claimed to have taken the *complete* exports of gutta percha out of Singapore from 1844 to 1855 on behalf of the company.[42]

The Temenggong of Johor's actions may have done little for the promotion of free trade if some British accounts were correct, but his actions had a twofold effect. Firstly, they helped to establish a consolidated commodity chain that ensured that gutta percha could be supplied readily and not dismissed as some obscure tropical product, the continued use of which could not be guaranteed and was subject to the vagaries of an unstable supply network. As one agent put it at the time, it proved not to be one of 'these new things that gave a great deal of trouble, and seldom came to anything'.[43] Secondly, it enabled the Gutta Percha Co to become established as the major gutta percha manufacturer. The Gutta Percha Co and its subsequent partner Telcon had access to scarce assets: gutta percha, and even capital – it was said to have had the enormous sum of £500,000 of capital available to it in 1846.[44] Later access to technology that included the patented techniques to purify gutta percha praised by the British Joint Committee, and finally the good fortune to have been connected with the Ker Rawson and Temenggong's control of the trade, all facilitated the Gutta Percha Co in achieving its subsequent domination of the global gutta percha trade. This in turn facilitated its dominant role in the global production of submarine telegraph cables depicted in Map 3; that is, it dominated both the supply and demand aspects of the trade. Werner Siemens wrote that the Gutta Percha Co controlled the market for gutta percha and restricted the supply of the purified form that was recommended by the Joint Committee even as late as 1869.[45] The advantage enjoyed by the Gutta Percha Co, and the difficulty facing new entrants in breaking existing relationships between agency houses and the marketplace in Singapore, are exemplified by the relationship between the Gutta Percha Co (Telcon), Jewesbury and Ker Rawson (or as it became, Paterson Simon) which continued for decades. It is unclear when the arrangements ceased; however, Jewesbury and the Gutta Percha Co (Telcon) were still

42 Jewesbury, p. 12.

43 Jewesbury, p. 10.

44 T.B. Armstrong, 'Gutta Percha', *Transactions of the American Institute, Sixth Annual Report of the American Institute of the City of New York*, 1848, p. 244.

45 W. Siemens, (Trans, W.C. Coupland), *Personal Recollections*, London, Asher and Co, 1893, p. 344.

obtaining gutta percha from Sarawak, for example, well into the twentieth century, as seen in the documentation still extant in the Telcon Archives.[46]

Based on trade figures up to 1848, the combination of the Temenggong's influence, the amount of the Gutta Percha Co's capital, and the activities of collectors, traders and agents, the gutta percha trade had gone from a non-existent international trade to a fully operational and profitable commodity chain in just a few years. To do so required an historical connection that built on old regional trade patterns, even including that developed by the Temenggong, and a trading infrastructure in Singapore that linked the region to the manufacturing core in Europe. That is, these were in part the ministructures of the world-system and an artefact of existing structures (a Southeast Asian-system one might argue) at the semi-periphery.

Singapore Chinese Gutta Percha Merchants

European-owned agency houses did not themselves deal with the arduous, tedious, dirty and financially risky aspect of buying and handling produce – especially forest produce – particularly given that such produce could come from geographically diverse sources and unknown suppliers. Many of those produce traders might have been considered poor credit risks, and would have been dealing in relatively small amounts of produce. Even then, much of the business of the agency houses was handled by Chinese employees, partly because there were simply not enough Europeans to handle these tasks and because it made sense to utilize Chinese connections that possessed local language skills.[47]

So it was with gutta percha where, on arrival in Singapore, it was traded to 'Chinese speculators' according to various listings published in *The Weekly Market Reports*.[48] Because it was essentially a Chinese organized trade, barter (especially for cotton goods) was a common feature of the gutta percha trade, and barter was often more common than cash sales. The barter – credit system obviated problems that existed for many years in the region – the continuing shortage of currency and fluctuations in currency valuations generally.[49] By the use of credit and barter, traders could ensure a continuing obligation for future supply. In addition, bartered goods offered under credit were often

46 For example, 1916 invoices for Sarawak gutta percha, supplied by Jewesbury and Co to the Gutta Percha Co; TCM 21/50/1.

47 See Chiang, *A History of Straits Settlements Foreign Trade 1879–1915*, 1978, p. 49.

48 Such as STSJC 21 March 1849, p. 410.

49 See Huff, p. 59.

marked up for considerable profit. Such systems and networks were ideally suited for exploitation when a new trade – that of gutta percha – was being established. Logan reckoned in 1847 that gutta percha and other forest produce were obtained by Malay traders from the Binua people of Johor in return for Singapore goods at profits; allegedly of between 100 and 400 percent profit mark up on the barter goods, and with the gutta percha in turn obtained at a reduction of 100 to 400 percent on the market price. Although, as Logan noted, the gutta itself was considered worthless by the Binua who had no great need for it.[50] Chinese traders went up-river in Borneo bartering goods, on credit, to hinterland forest produce collectors. They and other traders took advantage of downstream and coastal markets that grew up to meet the increasing demand for forest and coastal produce. Again, from the *Weekly Market Reports* it is common to see gutta percha being bartered; the barter price was higher than the cash price and cotton shirting (cloth) is the most common item of barter.[51] These reports are consistent with Chiang's comments about the importance of cotton goods as an item of barter for produce.[52] Later years' Market Reports continue spasmodically with similar reports of gutta percha trading – and do not mention gutta percha being taken by any *other* than the Chinese traders.

As the size of the gutta percha market grew, so too did the scope for specialist Chinese gutta percha traders. Specialized traders represent a further sophistication in the marketplace and, like agency houses, they gained the advantage of reputation, knowledge and supplier connections that limited entry by less well-connected entrants. The earliest specializations may well have been established via Chinese connections in Johor where Teochiu Chinese firms were heavily involved in gambier and pepper plantations, which the Temenggong had promoted. Records suggest that specialist Chinese gutta percha merchants in Singapore were also Teochiu, and the largest of these was a Teochiu who was also a gambier merchant by the name of Low How Kim. The Teochiu used the trade, financial, credit and clan links established in Johor, where gutta percha was readily and profitably available because of the Temenggong's activities, and then established bases in Singapore. The Temenggong needed Chinese traders to handle the machinery of the trade over which he had control, and produce was directed out of the nearest and most attractive entrepot, that is, Singapore.

Low How Kim appears in the *Singapore Directory* of 1877 as an assistant at the Singapore Opium Farm and Song says he was a 'Chinese holding a good position in the Colony' and held Power of Attorney for Low Ah Jit (a gutta

50 J.R. Logan, *The Journal of the Indian Archipelago and East Asia*, Vol.1, 1847, pp. 261, 287.

51 Such as STSJC 31 January 1849, p. 403 and March 21, p. 410.

52 Chiang, pp.117–118.

percha merchant, see below) and several of his firms until 1880.[53] With Tan Yeok Nee, he formed Low How Kim & Co, *Gambier and Pepper Merchants*. On his death, his son Low Long Teng became sole proprietor dealing extensively in gutta and rubber as well as piece goods; he became 'one of the leading men of the Teochew [sic] community'.[54] Low How Kim & Co was described as 'one of the largest gutta-percha dealers in Singapore' – in 1901 when Penoyer Sherman was sent out as Special Agent for the Philippines Forestry Bureau to investigate gutta percha in the Straits Settlements.[55] The company was still trading as Gutta Percha Merchants and appears in subsequent editions of the *Straits Settlements Directory* as Gutta Merchants, along with another firm, Teng Ho Chiang, in 1904.[56] In a 1908 publication, another important Teochiu businessman, Tan Swi Phiau was noted as being a shareholder in 'two important Chinese companies trading in pepper, gambier and gutta percha' so the linkages between those three commodity trades and Teochiu groups were very firmly established and continuing. He also held important positions in business and charitable societies.[57] Sherman says:

> The entire gutta percha trade is practically in the hands of the Chinese … and they guard the secrets of boiling, working over, mixing, adulterating, and colouring the gutta percha for European markets most zealously. All who have tried to investigate their methods agree that there is no connection between the various grades and the different tree species, and that pure gutta percha … is no longer found on the market unmixed with inferior grades.[58]

Sherman considered that botanical and geographical origins of the different grades were irrelevant to the commercial names which were purely chosen for marketing convenience.[59] He commented that 'the natives guard the secret of the different kinds of gutta-percha trees and their locality, so do the Chinese

53 *Singapore Directory for the Straits Settlements*, Singapore, Straits Times Office (hereafter SDSS), 1877, p. 102; Song, p. 191.

54 O.S. Song, *One Hundred Years History of the Chinese in Singapore*, London, J. Murray, 1923, p. 190.

55 Sherman, *Report of the Special Agent of the Forestry Bureau*, p. 339.

56 *Singapore and Straits Directory*, Singapore, various publishers (hereafter SSD), 1904, p. 188 E–F.

57 A. Wright and H.A. Cartwright, *Twentieth Century Impressions of British Malaya*, London, Lloyds Greater Britain Publishing Company Ltd., 1908, p. 146.

58 Sherman, *The Gutta Percha and Rubber of the Philippine Islands*, p. 16.

59 Sherman, *The Gutta Percha and Rubber of the Philippine Islands*, p. 18.

hide their methods of preparing gutta-percha for foreign markets'.[60] Sherman thought that supplies were coming from different countries and trees, combined or adulterated in different ways, with 25 different varieties on sale in Singapore by 1901 but that there was 'no connection whatever between the different kinds and the species of the trees'. He says that the Chinese were 'very skilful in the work', though they went solely by 'smell, color, toughness and the softening and hardening test in hot, then cold, water'.[61] Sherman refers to Low How Kim and he may have visited Low How Kim's facilities.

According to Song, Low Ah Jit & Co dealt in gutta percha and other produce.[62] Low Ah Jit appears in the listing of 'Principal Chinese Firms' in Singapore specifically as a 'Gutta Merchant' in the *Straits Settlements Directory*.[63] The connections of the Low family extended widely into both the gutta percha trade and the commercial and social world of Singapore. Low Ah Jit had two sons – Low Cheang Yee and Low Koon Yee – with two shops dealing in gutta percha. Cheang Yee was a member of the Chinese Advisory Board.[64] Low Cheang Yee was elected one of the founding committee members of the Straits Chinese British Association in 1900, and was a prominent 'Teochew' and much respected by European merchants.[65] The firm was still in business at its original premises at No 53, Boat Quay at the time of Song's writing in 1923. The Low family, its Teochiu heritage and its connections with gambier and pepper dealings, reflect the early connections established in Johor. As will be seen in Chapter 6, the linkage between Teochiu Chinese gambier and pepper dealers and gutta percha was replicated in Sarawak. Low How Kim's early involvement with the opium farm is also of interest; opium and gambier and pepper farming tended to be associated and also tended to be linked into the Teochiu communities.[66] The identification of other Chinese traders dealing in gutta percha in Singapore is not possible; however, the fact that the Low family linkages could continue so successfully for several decades is an indication of the importance and profitability of the trade.

Chiang's statistical snapshot of Singapore gives an outline of the importance of gutta percha exports from Singapore between 1870 and 1915: it was

---

60 Sherman, *Report of the Special Agent of the Forestry Bureau*, p. 339.

61 Sherman, *The Gutta Percha and Rubber of the Philippine Islands*, p. 19.

62 Song, p. 142.

63 For example, SSD, 1887, p. 163.

64 See J.F. Warren, *Ah Ku and Karayuki-san: Prostitution in Singapore 1870–1940*, Singapore, Singapore University Press, 2003, p. 336.

65 Song, p. 142.

66 See C.A. Trocki, 'Opium and the beginnings of Chinese capitalism in Southeast Asia', *Journal of Southeast Asian Studies*, 33, (2), 2002, pp. 297–314.

one of the top six commodities exported from the region throughout that period, and the third most valuable export in 1900,[67] when the market for gutta percha was nearing its peak in response to demand for production of the Pacific cables. However, as will be seen in the next section, other market fluctuations in Singapore and Sarawak were the result of changes in the operating environment of the telegraph cable manufacturers themselves.

The Trade Comes to Sarawak via Singapore

Sarawak was part of the near region that was linked into the global trade network with Singapore as its hub. Sarawak was part of the domain of the Sultan of Brunei until it was granted to Englishman James Brooke in 1841 when he became Rajah of Sarawak as a 'reward' for his assistance to the Sultan in suppression of resistance from local Malays and the indigenous people known as Dayaks. In later decades, as the Brunei Sultanate found itself increasingly unable to deal with regional 'piracy' and insecurity elsewhere in Brunei (and encouraged by the benefits of annual payments from Brooke), the Sultan agreed to – or was cajoled into agreeing to – further territorial annexations by Brooke so that Sarawak gradually spread north eastwards, a march continued under Brooke's successor, his nephew Charles, who was Rajah from 1868 to 1917. By the nineteenth century, antimony and gold (mined by Chinese workers) and trade in forest produce, including edible birds' nests and rattan, attracted Chinese and Malay traders who bartered products like rice, cloth, salt, beads and iron. The Brunei Sultanate and officials (*pangeran*) benefited from these developments by controlling access to trading ports and taking a share of the produce or taxing it. Under the early decades of the Brooke Raj, the sago and gutta percha trades were the two single most important exports of Sarawak. Aside from indigenous, Malay and Chinese traders and produce workers, the other economic force in the Sarawak economy was the Borneo Company Limited (BCL). The company was formed in 1856 and was given monopoly rights to access many of the country's resources, especially mineral resources, and it was also involved in the sago and later in other industries. As will be shown below, it played an early role in the exploitation of Sarawak's gutta percha trade, and in turn the trade itself was to change the strategic direction of BCL in a manner revealed here for the first time. The company paid the Rajahs royalties in return for their access to Sarawak's resources and developed relationships

67 Chiang, p. 192.

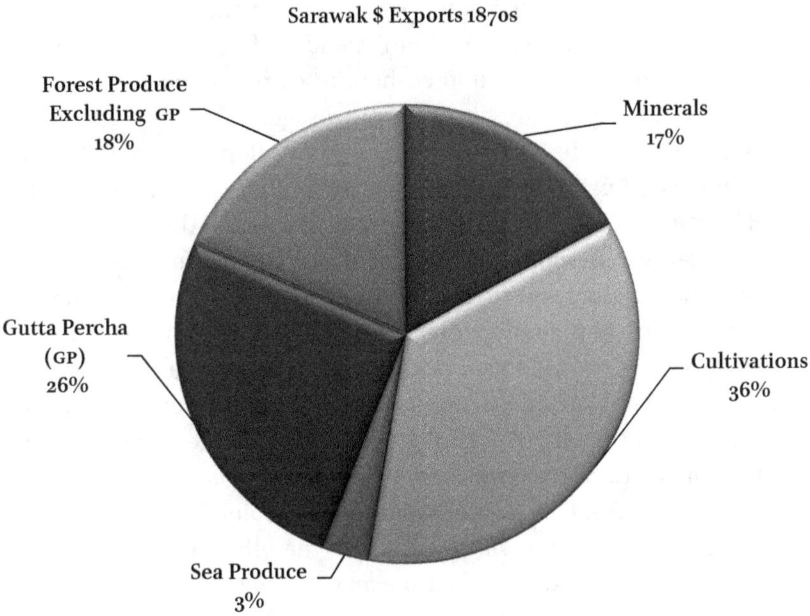

FIGURE 2 *Sarawak exports in the 1870s.*
Source: Sarawak Gazette various issues 1870–1904, and Consular Report in British Sessional Paper (BSP) 1871 [C.429], p. 809. See Appendix 2, Table 2 for details.

with many Chinese *towkays*. Its regional headquarters were in Singapore and in later years it expanded through much of the Southeast Asian region.

Various natural and mineral resources underpinned Sarawak's involvement in the world economy and in so doing brought the forces of international commerce to the hinterland population to an extent far more significant than ever experienced in previous centuries of interaction with Chinese traders. Hanson's survey of nations participating in world trade 1820–1900 concludes that by 1880 Sarawak was linked to the global economy.[68] A survey of its exports (excluding re-exports) for the decade leading up to that linkage shown at Figure 2, illustrates the importance of forest produce – and the significance of gutta percha in that process of connecting with the external economy.

Sarawak's trade during this period might, for our purposes, be divided into five main categories. Firstly, *Forest Produce, Gutta Percha* and *Sea Produce*. These were the items from which indigenous peoples profited most directly.

68 J.R. Hanson, *Trade in Transition. Exports from the Third World, 1840–1900*, New York, Academic Press, 1980, p. 19.

The Treasurer's Reports included gutta percha data in the category of *Forest Produce* (called *Jungle Produce* at the time); the identification of gutta percha as a separate resource sector is adopted here to clarify its place in Sarawak trade. The calculations given are based on the disaggregation of gutta percha statistics extracted from the totals in the Treasurer's Reports. Secondly, were the *Cultivations* and *Minerals* sectors, the focus of Chinese labor and capital investment. The mining sector profited the Borneo Co Ltd and the Brooke Rajahs in particular because of monopoly rights and royalty payments set aside for the company and the Rajahs respectively. Importantly, Forest Produce operated on a commercial basis in the marketplace, without monopoly restrictions and with no specific royalty payments to the Rajahs or to BCL, other than an export levy which varied over time and in application. The figures reveal that, while Sarawak had entered the global economy, trade was in large measure dependent upon forest/sea resources collected by indigenous people (47 percent as shown in Figure 2 above). That is, the traditional and colonial economies were interlinked – the latter did not simply overtake the other with the demands of European trade and the advent of European colonization and intermediary traders. Cleary interprets this pattern, which he sees as applying elsewhere in British Borneo, as evidence that the metropolitan core could not at will 'simply crank the resource-rich periphery into life'.[69] A key factor in the capacities of global trade networks arose from the fact that the metropolitan centres were not fully able to control colonized communities. These aspects are examined in this study through the prism of the gutta percha trade; what they point to is what Trocki describes as an illusion of total European control – that it is not accurate to see Southeast Asians as merely reacting to European initiatives, and that indigenous peoples 'took advantage of the opportunities that presented themselves'.[70] Such opportunities are evidenced by the activities of the gutta percha collectors/traders examined in this study and set out in Diagram 1. One way to envisage this process is to conceptualize the commodity chain itself. Broadly speaking, gutta percha was collected in forests and traded through one or more Chinese traders until eventually the supply would arrive in (most commonly) Singapore where it would be handled, usually by Chinese traders who prepared and packaged gutta percha for on-sale to European agency houses.

While it is difficult to reconstruct detailed trade histories from this era, by taking the combination of official reports, trade data, commercial and other documentary information on gutta percha in the Singapore and Sarawak markets,

69 M.C. Cleary, 'From hornbills to oil? Patterns of indigenous and European trade in colonial Borneo', *Journal of Historical Geography*, Vol. 23, No. 1, 1997, p. 44.

70 C.A. Trocki, *Opium, Empire and the Global Political Economy: A Study of the Asian Opium Trade, 1750–1950*, London, Routledge, 1999, p. 77.

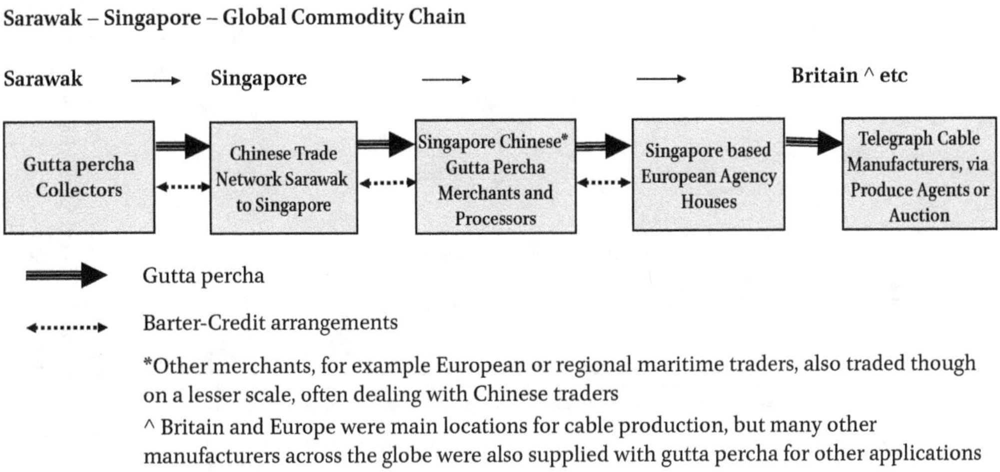

DIAGRAM 1 *Sarawak – Singapore – Global telegraphic gutta percha commodity chain.*

and matching all this with records related to submarine cable manufacture, it is possible to draw together a sense of how this commodity chain functioned. For some years, the gutta percha export figures are missing but the extant figures, if matched against other forms of evidence, enable conclusions to be drawn. All trade figures of the era must be regarded with caution, as reporting requirements and monitoring were not to be judged by today's standards. That said, patterns across several years, when used in connection with other evidence, can be harnessed. At Figure 3 is represented the exports from Singapore in pikuls of gutta percha from 1850 – the year of the first submarine cable attempt – to 1903 following completion of the Pacific cables (exports in Straits Dollars were not available for several years). At Figure 4 are the exports from Sarawak from 1863, the year from which accurate trade data commences (in Sarawak Dollars – pikul amounts were not available for some years). Overlaid on those graphs are the estimated lengths of submarine telegraph cables constructed for the given years, following on from the graph given at Figure 1 in Chapter 2.

The graphs reveal how the demand for gutta percha was related to the manufacture of cables, including increases leading up to cable construction or because of re-stocking after completion of large ventures. The variations in the export figures that are presented here provide a valuable opportunity to examine both western business history and its inter-linkages with the trade (in forest produce) of the periphery in the nineteenth century. The material also illustrates how actors in global commodity chains at this time were remarkably sensitive to changes in demand and to European commercial decision making.

Some of the causes of the changes in both markets can be explained by the choices being made by the major users of gutta percha. This is set out in the

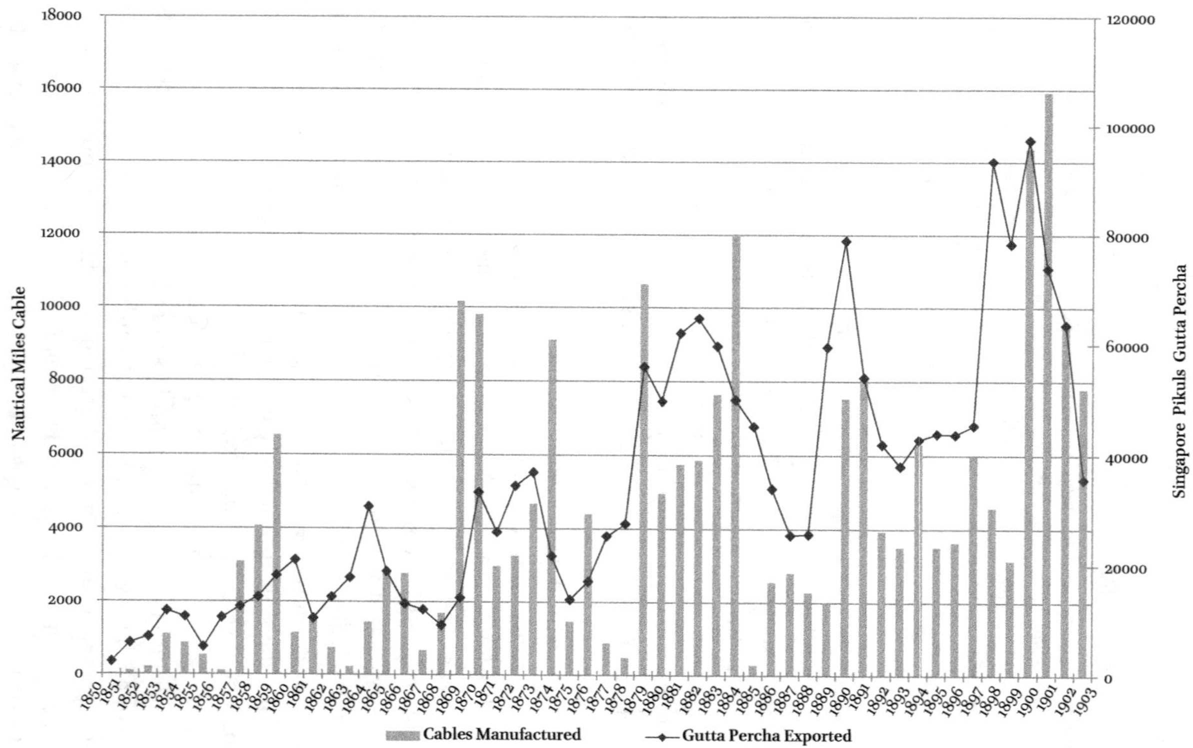

FIGURE 3 *Singapore gutta percha exports and estimated world submarine cable production 1850–1903.*
Source: H.W. Jewesbury, *A Short History of the Introduction of Gutta Percha into Europe*, London, Jewesbury, 1883, p. 12. E.F.A. Obach, 'Cantor Lectures on Gutta Percha', *Journal of the Royal Society of Arts*, London, 1898, p. 96. A.M. Burn-Murdoch, 'Some facts about gutta percha', *Agricultural Bulletin of the Straits and Federated Malay States*, Vol IV, No. 2 (February) 1905, p. 40. See Appendix 2, Table 1 for details

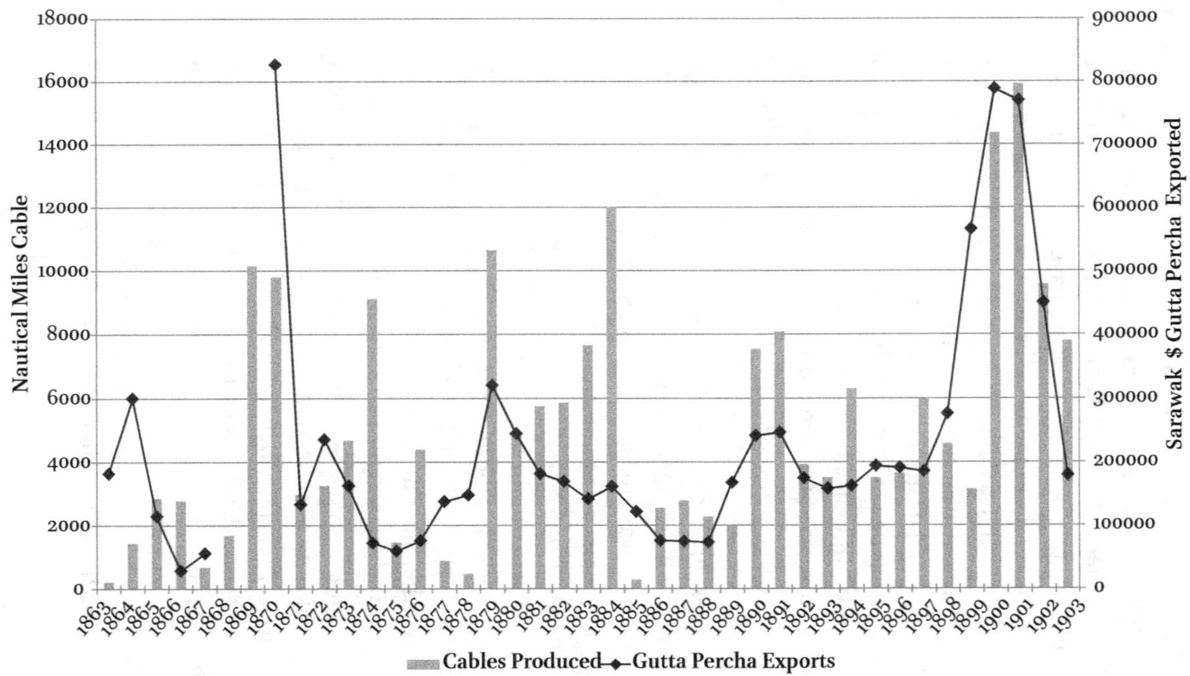

FIGURE 4 *Sarawak gutta percha exports and estimated world submarine cable production 1863–1903.*
Source: For Sarawak GP (Figure 4): Treasurer's Reports reprinted in
Sarawak Gazettes, various issues 1870–1904; and Consular Reports in British Sessional Papers (BSP): 1865 [3478], p. 28; 1866 [3582], p. 805; 1866 [3669], p. 266 for the years 1863–65 respectively. Figures for 1868–69 are not available.
For N.M. Miles Cable (Est'd), both Figures: *Pall Mall Gazette*, 'The West Indian Telegraph Expedition', 1870, July: n.p.; United States Hydrographic Office, *Submarine Cables*, Washington, Dept of Navy, Bureau of Navigation, 1892, pp. 41–59; *India Rubber World*, New York, 1 February 1902, p. 136; Telcon, *The Telegraph Construction and Maintenance Co. Ltd on the Occasion of the Visit of the Delegates to the International Telegraph Conference*, London, Telegraph Construction and Maintenance Co. Ltd, 1903, Appended; C. Bright, *The Life Story of Sir Charles Tilston Bright Civil Engineer, with Which is Incorporated the Story of the Atlantic Cable and the First Telegraph to India and the Colonies*, London, Constable & Company, 1908, p. 313 and W. Glover, *Cable Timeline 1845–1900* and *Cable Timeline 1901–1950*; 2017/Continuing: http://www.atlantic-cable.com. See Appendix 2 Table 1 for data.

following section. Some of the evidence illustrates how swiftly and far-reaching a boardroom decision, even management failings, in London, could, in the 1870s, extend a global reach to those in the forests and trading stations of remote rivers in Borneo. In other years, the trade from Sarawak, for example, is shown to have suffered from other factors that were irrelevant to the Singapore and metropolitan markets – the causes related to activities and conditions unique to Sarawak that are examined in this and subsequent chapters.

There is important documentary evidence for the close linkages between the metropolitan and peripheral venues of the trade based in part upon Minutes of the Gutta Percha Co (that is, the manufacturing arm of Telcon). These Minutes focus on the needs, supplies and general management of the company, including its gutta percha requirements, albeit not at a detailed level. Such commercial information is unusual for forest produce and, by extension, for forest economies of this time. Given the peculiarly dominant position Telcon (and its cable producing arm the Gutta Perch Co) had, as the world's major customer for gutta percha, this means that any decision the Company took in relation to the commodity influenced both the overall market and most of those connected with it. Also used are Directors' Minutes from the Borneo Company Limited (BCL), and the India-Rubber, Gutta Percha and Telegraph Works Company (Silvertown). Although BCL was founded as a joint stock company in 1856 to exploit the resources of Sarawak, it also operated as an agency house in Singapore from 1857 with trading interests in the region. It is to be expected that BCL's records are relevant to Sarawak (and Singapore); however, in considering the records for this research, a previously unknown and unexpected connection between BCL and Silvertown was discovered and that connection was based in part on a common interest in seeking gutta percha, the effects of which were felt in both Singapore and Sarawak. Again, corporate decision making in London will be shown to have had an impact on the life of forest communities on the other side the world.

For the era from 1870 onwards, evidence for Sarawak's changing involvement in the trade, and the impact of the trade becomes quite detailed as from that year the trade and social conditions in Sarawak were sufficiently extensive and optimistic for the publication of a weekly newspaper, the *Sarawak Gazette*. Indeed 1870 had seen a considerable increase in trade owing to the rush for gutta percha which had run up to 'fabulous prices'.[71] The Gazette had the two-fold objectives of informing European inhabitants of official business and matters of public interest, and providing information on 'natives and trading interests' of Sarawak for circulation to other countries and settlements so its

71 Sarawak Gazette (hereafter SG), 28 February 1871.

focus is especially useful for the purposes of this research.[72] This material helps to make the connection between the impact of the trade at the periphery and corporate decision making and the demands for global communication. Like earlier material from the period, its cultural perspective needs to be treated with caution, and the perspectives reflect a European rather than a necessarily indigenous or Chinese perspective.

London to Sarawak via Singapore

The first peak in exports of gutta percha from Singapore was before the first submarine cable was produced. This occurred in 1848 and was said to be due to the demand for the manufacture of police truncheons to control riots during the Chartist uprising in England, as written up in the Market Reports in 1848.[73] Prices then fell until demand rose for gutta percha after the first attempted submarine cable was laid in 1850. Thereafter, exports started to rise again, with the next peak in 1853. In that year, several cables were manufactured and the price of gutta percha in Singapore rose from $35 to $75 per pikul.[74] Exports again fell, then steadily rose through the 1850s as the demand for cables was met, including the (unsuccessful) Atlantic and Suez cables, peaking in 1860, possibly after re-stocking upon completion of those cables. Then a serious correction is seen. Detailed figures for Sarawak become available only for 1863 onwards, but there is a report of a surge in gutta percha trading in Sarawak occurring in the year 1853. This coincided with a surge in cable-laying following the success of the 1851 Channel crossing; with multiple domestic and international cable connections between Britain and the Netherlands, Belgium, and Ireland being laid. The response in Sarawak was that least 3,500 pikuls were exported to Singapore alone, with the market recording high prices and 'stimulated exertion' even though the trade was in its infancy and in spite of the fact that 'but a fraction of the population is as yet acquainted with the process of collecting the gutta'.[75] As the history examined later will show, the population soon learned.

In 1864, there was a larger peak in exports from Singapore that was also evident in Sarawak. This can be linked to the demand for gutta percha to manufacture the (heavily insulated) Atlantic cables to be laid in 1865 and 1866. After the initial failures, Atlantic cables were always the most heavily insulated with gutta percha because of their deep-sea position and the rough laying

72 SG 26 August 1870.

73 STSJC 1 August 1848, p. 18.

74 Buckley, 1902, p. 579.

75 Sydney Morning Herald, 3 October 1854, p. 4.

conditions; both features demanded heavy insulation. Exports from Sarawak were very large. Also, it was in 1864 that the Gutta Percha Co and Glass Elliot and Co, which had previously been separate cable manufacturing companies, formally merged into the Telegraph Construction and Maintenance Co (Telcon), under Sir John Pender. The result was a revitalized management and access to influential and very wealthy investors, not least of them Pender himself, who was well able to invest in risky cable ventures and the expensive commodity that was gutta percha.

There was, however, an additional cause for the level of demand seen in 1864, particularly in Sarawak. Minutes from both the Borneo Company (BCL) and from the cable manufacturer Silvertown show a previously unknown connection: the two companies had decided to form an alliance to build on their mutual strengths (access to gutta percha, and telegraphy respectively) and solving shared weaknesses (problems in accessing capital). Prior to that date, Silvertown had focused on manufacturing industrial products with india rubber, now it would have access to gutta percha and could be involved in the telegraph cable industry. In fact, it was *after* the two entered this partnership that the company known as *The India Rubber, Gutta Percha and Telegraph Works Company Limited* came into being and it was decided that:

> ...the manufacture of Gutta Percha, as well for general purposes as also specially for purposes of submarine telegraphy, shall be immediately commenced and vigorously prosecuted by the company ...[76]

The agreement noted that BCL was engaged in the importation and purchase of gutta percha and that the new Company would use BCL 'for the raw material required for its works upon terms to be arranged for ... mutual advantage'. The BCL had a dominant position in Sarawak and its new arrangements would have been influential in the market there. Silvertown produced its first cable in 1865, so the effect of the alliance was felt in the market soon after its formation. Although BCL operated as an agency house in Singapore, and so could purchase gutta percha there, its position in Sarawak also gave it opportunities to exploit Sarawak resources and to buy at more favorable prices. That is, BCL's position would have advantaged Sarawak's exports in particular. The Consular Report for Sarawak for 1864 notes that almost a third of the gutta percha exported from Sarawak was exported *directly* to England (and not through Singapore, the more usual practice for export of Sarawak produce), probably under BCL's auspices.[77] By amount (pikuls) exported, 1864 represented the largest

76 BCL Board Minutes, 21 June 1864.

77 Consular Report on the Trade of Sarawak for 1864, BSP 1866[3582], p. 805.

gutta percha exports ever achieved from Sarawak – 13,168 pikuls; only the lower price of gutta percha at that time kept the value of exports from exceeding that of the record level to be achieved in 1870 (see Appendix 2, Table 1).

While Sarawak exports benefited from these corporate arrangements, the trade could also suffer because of corporate decision making; indeed, the trade in the whole region was influenced by corporate activities. So it was that, despite the great boom in Sarawak in 1864, exports of gutta percha diminished considerably in the next few years. There was a fall in exports of gutta percha caused by a 'falling off in demand' according to the Consular Report for trade of 1865.[78] Singapore export figures also show an apparent crash in the market between 1865–1869. British Customs figures add further confirmation of the trend suggested by the company records. In 1864, 1865, 1868 and 1869, British gutta percha imports from the Straits Settlements fell – while at the same time imports from the Netherlands rose.[79] The explanation for these falls emphasizes the fragility and global nature of commodity chains and the fact that changes could occur for quite prosaic reasons. Company records show why the market may have suffered. The largest player in the market, the Gutta Percha Co, was now buying much of its gutta percha closer to home: from the Netherlands, rather than Singapore. The Netherlands was the second most important market for gutta percha because it received gutta percha from Dutch colonial territory in the Netherlands Indies. Those territories collected great quantities of the product and, while much of it was channeled through the profitable Singapore market and thence to Britain, there were public auctions held in the Netherlands.

The Gutta Percha Co Directors' Minutes provide evidence to explain how quickly the impact of decisions in London could affect collectors and traders in Sarawak and Singapore. In November 1868, the Directors authorized a large purchase from the Netherlands for £15,000.[80] The significance of the Gutta Percha Co's dominance of the gutta percha markets – the major 'chain driver' in the trade – is indicated by the fact that, in September 1868 alone, the Managing Director of the Gutta Percha Co was authorized by the Telcon Board to make gutta percha purchases 'to the extent of £60,000'. This represented around two thirds of total UK imports for the year (which were £91,830 in total for that year).[81] In other words, a purchase made by the Gutta Percha Co could, in just a single decision, equal the bulk of annual UK imports. The authorization does serve to emphasize both the importance of the Company's share of the

78 Consular Report on the Trade of Sarawak for 1865, BSP 1866[3669], p. 268.

79 Obach, *Journal of the Society of the Arts*, 1898, p. 97; the Straits Settlements included Singapore and the lesser colonies, in terms of amount of exports, of Penang and Malacca.

80 Gutta Percha Co, BICC I/18/04 161, 10 November 1868.

81 *Statistical Abstract for the United Kingdom*, 1863–77, BSP 1878 [C.2073], p. 34.

market and in turn how significant an effect the Company's decisions could have in both Singapore and in the markets that supplied it – such as those in Sarawak or in the Dutch colonies. Exactly why the Gutta Percha Co chose to purchase from the Netherlands is unclear. Certainly, the cost of shipping from there was significantly less than out of Singapore. Alternatively, it may be that the gutta percha was needed in a hurry and the faster delivery from the Netherlands would have been preferred to that from Singapore. Perhaps they could avoid paying agent's commission to Jewesbury or Ker Rawson in Singapore? The Minutes do not make it clear; either way, what happened would invariably result in a serious dampening of demand in both Singapore and Sarawak.

While gutta percha exports bottomed in 1868–9 for Singapore (and, as noted, figures are not available for Sarawak in these years), there were some positive signs of change. In 1868, the British Government introduced legislation to nationalize the domestic British overland telegraph network and part of the British international submarine telegraph network (cables that crossed the English Channel and Irish Sea). In 1869, the Government introduced a money Bill into Parliament to make the funds available to buy out the telegraph operators and as a result huge sums of money became available, some of which was released for investment into international (submarine) telegraphic ventures.[82] Also in 1869, another Atlantic cable was commissioned, this time by the French Government, to connect it with North America. In the same year, the cable was produced for a connection for Suez-Aden-Bombay to provide the British link to India discussed earlier. Other, smaller ventures were also prepared.

In addition to these developments, there were further activities on the part of BCL and Silvertown in connection with a new venture, the West India and Panama Cable Co, with BCL undertaking to provide gutta percha in exchange for £12,000 worth of shares in the new W.I. & P.C. Co as it was known.[83] The network was, at that time, the largest telegraph cable venture ever undertaken. So much cable was required that Charles Bright, the engineer in charge of the expedition, had the ship that carried most of the cable cut in half and extended by forty feet to carry the great load.[84] The ship and its cargo were insured for the enormous sum of £300,000.[85] So, by 1869 a significant amount of gutta percha had either been used or was being sought to produce cables and some of those ventures resulted in specific increases in exports from Sarawak.

82 See J. Kieve, *The Electric Telegraph: a Social and Economic History*, Newton Abbot, David & Charles, 1973.

83 BCL Board Minutes, 22 August 1869.

84 C. Bright, *The Life Story of Sir Charles Tilston Bright Civil Engineer*, London, Constable & Company, 1908, p. 310.

85 C. Bright, p. 315.

As a result, there was some confidence returning to the gutta percha market. In 1869 in Sarawak, demand

> ...sprang up in the middle of the year for gutta percha and...had gone on steadily increasing... [this] has infused great activity into the whole country[86]

The British Consular Report commented on rising competition for gutta percha in Sarawak; reporting that demand had already been feeding into the market in Sarawak, where prices per pikul for quality gutta percha rose from $25 in April 1869 to $115 later that year.[87] The Report went on to record that whereas there had previously been 'one firm being the large purchaser' (BCL), there were now agents from other Singapore firms (that is, Chinese traders) in the Sarawak market and this combined with increasing scarcity of supply, had caused prices to soar. Chinese traders became the linchpin in the Sarawak gutta percha market from around this date, and continued so from that time. More details of the role of Chinese traders in the Sarawak trade follows in later chapters.

The following year, 1870, telegraph cables were produced for connections between India and Singapore and between China and Singapore – they were lengthy and required a considerable amount of gutta percha. It is not surprising that in 1870 there was a significant rise in the export of gutta percha from Singapore – increasing around 150 percent from that of 1869, and the exports from Sarawak were so enormous that Acting Vice-Consul St John was correct when he observed of the gutta percha exports 'I believe that the trade ...will hardly [ever] reach the amount shown in 1870'. The price of gutta percha in Sarawak equaled $135 per pikul.[88] With the sudden surge in expected demand, entrepreneurs saw potential in Sarawak. The entry of the new Singapore Chinese agents onto the scene also signaled the arrival of Sarawak as an increasingly important gutta percha trading port and increasing linkages of Sarawak into the global economy. In the same Report, St John noted that the very large influx of silver coin consequent on the gutta percha export 'caused an unusual stimulus throughout the commerce of Sarawak'. The impact of the 1870 'stimulus' is examined in Part 2 of this book.

There was, however, another factor in the 1870 rise in demand for gutta percha, and another likely cause for Sarawak enjoying a particularly good year. Company records again reveal the linkages between the long arm of commerce in London and collectors and traders in Sarawak: BCL and Silvertown

86 *Consular Report on the Trade of Sarawak* for 1869, BSP 1871[C.343], p. 7.

87 *Consular Report on the Trade of Sarawak* for 1869, BSP 1871[C.343], pp. 7–8.

88 *Consular Report on the Trade of Sarawak* for 1870, BSP 1871[C.429], p. 806.

were yet again exchanging shares for gutta percha – for £21,000 worth of shares in Silvertown itself (rather than for a specific cable venture).[89] For comparison, it might be noted that the official value of the whole of the UK imports of gutta percha from the Straits Settlements in 1869 had been worth only some £52,000.[90] Not unexpectedly, Sarawak benefited from the connection between BCL and Silvertown. Another reason to believe that BCL and its arrangements with Silvertown were a significant factor in the 'boom' of 1870 Sarawak trade, was that the exports from Sarawak (including the amount sent directly to the UK), were equivalent to around one third of the total amount exported from Singapore in that year – a proportion that Sarawak was never again to achieve. The gutta percha boom of 1870, however, proved to be short-lived.

In 1871 export levels from Singapore again fell – by around 25 percent, and by around two thirds out of Sarawak. Such a dramatic change in the market must have left some evidence for its cause? Was the rise in 1870 in part due to speculation that future demand for gutta percha would continue because of the capital coming available following the nationalization of the British telegraph companies? Perhaps there were expectations of a mini-boom in telegraphy along the lines of fortunes made (and lost) in the railway boom of earlier decades. Some of the fall was due to a reduced demand for cables. However, commodity chains may be merely fickle: the fall in trade was caused by something far removed from the demands of the cable market.

Addressing a shareholders' meeting in 1871, the new Chairman of Telcon, Sir Daniel Gooch reminded those present of the problems he had raised at their meeting in 1870 when he had told them they could not carry out all their contracts 'for want of gutta-percha'.[91] Gooch said he was confident that now, in 1871, they did have sufficient for any work they had, as at any one time they carried up to 500 tons of gutta percha in stock (approximately 8,400 pikuls). In other words, by 1871 the fall in global demand for gutta percha was because the major customer for gutta percha had filled its order books and could reduce the amount of gutta percha it was seeking. Decisions taken by the managers in Telcon (for its supplier the Gutta Percha Co) had a rapid impact in Singapore and Sarawak – they generated part of the rise in 1870 and caused most of the subsequent fall in 1871.

In fact, the evidence of Telcon's activities goes even further, both in explaining the market changes in 1870 and 1871, and in illustrating the extent to which

89 BCL Board Minutes, 22 August 1870.

90 E.F.A. Obach, *Cantor Lectures on Gutta Percha* [*Lecture Reprint with addition of Appendices*, London, Society for the Encouragement of Arts, Manufactures and Commerce, 1898, p. 97.

91 *The Money Market Review*, 4 March 1871.

global trade can be altered by the apparent minutiae of management decision making – or, indeed, mismanagement. There was more to the story than Gooch was admitting to his shareholders. The Gutta Percha Co Minutes confirm that, by late 1870, not only did the company have sufficient stocks to reduce its purchases in Singapore, but by October 1870 it decided to have Jewesbury and Co, its agent, telegraph Singapore to '*cease purchases altogether*' (original emphasis) until further notice. Given that Telcon, through the product supplied to it by the Gutta Percha Co, was by far the world's greatest consumer of gutta percha, this was a very significant decision to make – causing serious disruption in the commodity chain – having a disturbing financial impact on collectors, traders and agents. Furthermore, officers of the Gutta Percha Co were ordered to report on the excess gutta percha stock they were holding which was 'beyond authorized levels'. A subsequent Director's meeting concluded there was mismanagement by managers in the Gutta Percha Co. In December 1870, Jewesbury and Co, still acting as agents, wrote to the Managing Director of the Gutta Percha Co about the decision to cease purchases, and advised on 'the importance of keeping up the machinery at present existing at Singapore for its collection' and recommending the continuation of regular monthly orders, even if they were at a 'minimum' levels. The Gutta Percha Co postponed consideration of the question, although Jewesbury's suggestion was finally agreed to in late January 1871. However, the fact remained that six months had passed between when the Company began very significant reductions in levels of gutta percha purchases until it finally agreed to commit to regular though significantly reduced purchases out of Singapore. Its agents there in the meantime must have been most reluctant themselves to continue committing to purchases whilst the situation remained in limbo. Thereafter, the Gutta Percha Co had to submit monthly 'Gutta Percha Stock Returns' for approval of the Telcon Board to ensure that such stock management problems did not recur. Although there are no records of it, the response in Singapore to the likely withdrawal of the Gutta Percha Co from all gutta percha purchases would have been dramatic and it is easy to see why Jewesbury would have recommended maintaining some purchases to 'keep up the machinery' in place; in effect, they were trying to keep the commodity chain in some form of working order.[92] Just as the Temenggong of Johor had helped to establish, and then altered the patterns of the trade by controlling it, so Jewesbury were now seeking to protect the patterns of the trade to maintain confidence in it, as the purchasing decisions of its most significant customer could seriously jeopardize its operation.

So, the plunge in 1871 – felt so dramatically in both Singapore and more noticeably in Sarawak (because the rise there in 1870 had been greater) – seems to

92 Summarised from Gutta Percha Co, BICC, I/18/04, Minutes for October 1870–January 1871.

have arisen not just because Telcon/Gutta Percha Co had sufficiently stocked up in 1870 as Sir Daniel Gooch had concluded – but because staff at the Gutta Percha Co mismanaged purchases and their mismanagement led to a serious correction in 1871 – a correction that was swiftly felt thousands of miles away. The vulnerability of the commodity chain – what Jewesbury recognized as the 'existing machinery' – emphasizes the global nature of the gutta percha trade and the rapidity with which a tug from the core could result in a rapid stagnation in the market. The effect in Sarawak can be seen by a glance at the export figures for 1871 as illustrated by the graph at Figure 4.

The figures show an increase in 1872, and the Minutes of the Gutta Percha Co record very significant purchases in each of Singapore, Sarawak and the Netherlands; indeed, in August 1872, the Gutta Percha Co, bearing in mind its existing contracts and, taking into account its 'Brazilian and other contracts', decided to purchase 'all suitable gutta percha available in Holland' and to ask a client, the French Atlantic Cable Company, for a £50,000 advance to purchase more if required. An additional reason for 1872 being a better year in the Sarawak trade was that the Gutta Percha Co was purchasing not through the usual source of Jewesbury, but through BCL instead, including 270 tons to the value of £92,255. When BCL was the supplier of choice for cable manufacturers, then Sarawak was more likely to benefit than had another agency house been chosen, since much of BCL's gutta percha could readily be sourced in Sarawak. For BCL, its position was commercially ideal as it was supplying both Silvertown and the Gutta Percha Co.[93] That much of the BCL's product is likely also to have come from Sarawak is seen from the trade records of the time. Sibu, for example, on the Rejang river in Sarawak, exported almost $20,000 per month of forest produce in 1871.[94]

Cable manufacturing, being an expensive venture, appears to have followed relatively short 'just in time' supply channels and this can be seen reflected the demand for gutta percha and its effects in Sarawak and Singapore. In late 1873, the Gutta Percha Co decided again to reduce its purchases to ten tons per month, and kept them low in 1874. Falls were reflected in the market in both Singapore and Sarawak (and imports to the UK from the Netherlands). As a result, 1873 was 'unprecedentedly bad' for Sarawak trade, and this caused 'a stagnation in our bazaar...Gutta declined considerably in price, leaving large stocks in the country'.[95] What may have made things worse for Singapore and

93 Ibid.

94 SG 24 January 1871.

95 SG 2 March 1874.

Sarawak exports was the fact that there had been higher UK imports of gutta percha from the Netherlands again in 1873.

The relationships between the Sarawak exports and those of the Netherlands (which were really exports of gutta percha from Dutch colonies) emphasizes how the economies in the region, by participating in the gutta percha trade, had in effect become even more intricately linked to the world-system through the regional gutta percha trade system and its onward connections to Britain and the Netherlands. If one state or community benefited it might be to the detriment of another. Local, regional and global economic networks became increasingly interlinked, entangled even. When Sarawak benefited from the organization of the trade in 1865 – when BCL linked with Silvertown to export gutta percha from Sarawak directly to England – this was likely at the expense of Dutch territories in the region whose international exports suffered due to competition from its neighbor. This was possibly the first time that Sarawak had a significant and displacing role to play in the region's trade; it may have repeated this again in 1870 when its exports were enormous but those from the Netherlands fell despite the great cable manufacturing activity underway. In turn, Sarawak trade suffered and others in the region benefited when cable manufacturers bought their gutta percha in the markets at the Netherlands which sold gutta percha from the Dutch colonies.

Major new cable ventures, the nationalization of the British telegraph networks and consequent availability of capital for investment in new submarine cable ventures, the commercial ventures between BCL and Silvertown, and decisions by Telcon/Gutta Percha Co each appear to have had a direct and swift impact on the market for gutta percha in Singapore, Sarawak and the region. It took only a variation in corporate decision making for collectors, traders and agents in in Southeast Asia to be affected within weeks of decisions being made in London. The commercial evidence illustrates the global linkages between the various actors in the trade, and the amazing rapidity with which corporate decision making was reflected in the regional collecting and trading zones. These interlinkages are illustrated in the three graphs at the end of this chapter at Figures 5, 6 and 7.

After the dates discussed above, the Gutta Percha Co Minutes no longer feature information about their gutta percha purchases and this valuable source of information cannot be used to study changes in later years. However, the occurrences have been set out sufficiently to reveal the nature of the commodity chain – and the linkages between collectors and traders in Sarawak, traders in Singapore and the region and the commercial activities of the cable manufacturing centres in Britain. When Telcon appeared to be removing itself from the 'machinery for its collection' it clearly expected that the machinery of trade would be there when it called upon it again. In the meantime, BCL could use its position in Sarawak, as well as its operation as an agency house in Singapore, together with its commercial arrangements with Silvertown (and Telcon), to benefit as a supplier of gutta percha to two of the most successful cable manufacturers. By exchanging gutta percha for shares in telegraph ventures, BCL could invest in the commerce of communications in a way that it would not otherwise have had the capital to do. In turn, the trade of Sarawak benefited by the position that BCL occupied in these alliances.

In the 1890s exports began to rise again, especially in 1890 from Singapore, and this may have been due to buying for cables being constructed and to speculation relating to the anticipated construction of the British Pacific Cable. In 1894, the Canadian Government, which had been the principal proponent of a Pacific Cable scheme up to that point, advertised for tenders for the cable, with a requirement for it to be ready for laying in 1898.[96] Any potential tenderers would have considered buying gutta percha at that point if possible. The impact on the exports from Singapore, for example, can be clearly seen in the graph at Figure 3 – yet it was not merely the island and its exports that were part of that picture of boom. Much of the trade included in those statistics derived from the Malay peninsula which experienced such an active gutta percha trade in the last few years of the century that it has been described as a 'rampage', so vibrant as to have attracted collectors from Borneo.[97] In the event, due to a combination of very complicated political and commercial disagreements, the Pacific Cable was delayed until 1900 and Telcon was awarded the contract in December 1900. In 1899 a meeting of Silvertown shareholders was told that the company had been purchasing 'large quantities of gutta percha' in view of probable work, and that this had proven to be a wise precaution since 'owing to an extra increase in consumption…available stocks have seriously diminished'.[98] In the previous month, the New York Times noted a significant increase in the price of golf balls due to dramatic increases in the price of gutta percha.[99] The British Pacific cable was laid by 1902 (ahead of schedule). The French also contracted for trans-Atlantic cable connections and the U.S. Pacific Cable

96 See D.R. Winseck and R.M. Pike, *Communication and Empire: Media, Markets and Globalization 1860–1930*, Durham, Duke University Press, 2007, p. 160.

97 See Kathirithamby-Wells, *Nature and Nation Forests and Development in Peninsular Malaysia*, Singapore, Nias Press, 2005, p. 70ff.

98 IRGPETJ 23 December 1899, p. 498.

99 New York Times, 21 November 1899, p. 3.

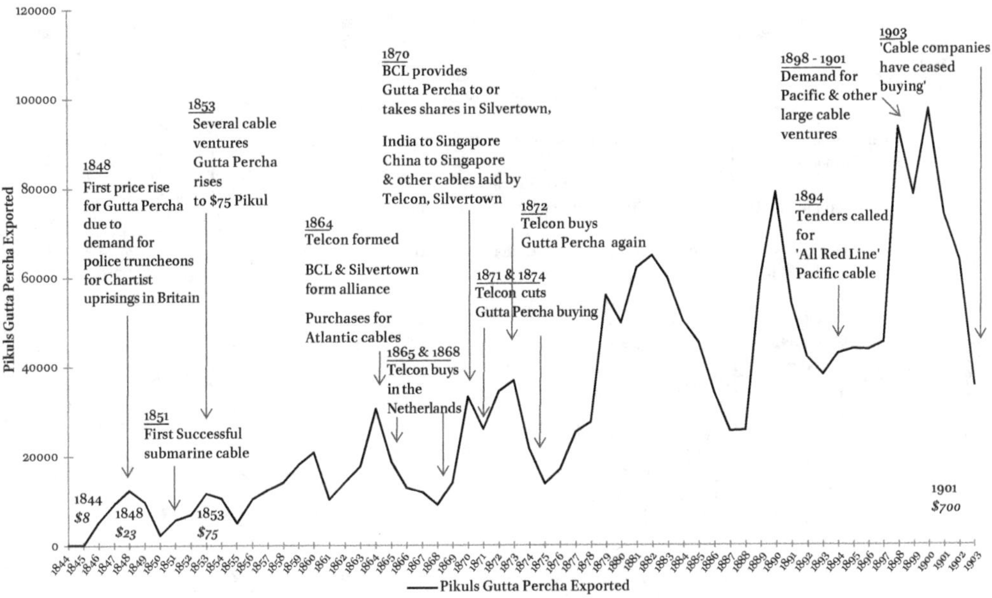

FIGURE 5 *Sensitivities in the Gutta Percha Commodity Chain Reflected in Singapore Trade. Source:* For Singapore GP Annual Export Figures: H.W. Jewesbury, *A Short History of the Introduction of Gutta Percha into Europe*, London, Jewesbury, 1883, p. 12. E.F.A. Obach, 'Cantor Lectures on Gutta Percha', London, *Journal of the Royal Society of Arts*, 1898, p. 96. A.M. Burn-Murdoch, 'Some facts about gutta percha', *Agricultural Bulletin of the Straits and Federated Malay States*, Vol IV. No. 2 (February) 1905, p. 40. See Appendix 2, Tables 1 and 6 for statistical details. GP prices: 1844 – $8; 1853 – $75; 1901 – $700 from C.B. Buckley, *An Anecdotal History of Old Times In Singapore 1819–1867*, Singapore, Fraser & Neave, 1902, pp. 404, 579. For 1848 – STSJC, May 31, Vol. 41 No 279. Prices in Straits Dollars per pikul. Quote for 1903 – Sarawak Gazette, 3 May 1904, p. 99.

connection that commenced manufacture in 1901. All combined to lead to the enormous demand for gutta percha mentioned previously. Both Singapore and Sarawak recorded steady patterns of exports in the 1890s – the peak being in 1898 to 1900 in Singapore and 1899 to 1901 in Sarawak. After the great demands for these cables were completed by 1902, there was a serious diminution commencing in both markets. In fact, the market for gutta percha never again reached such peaks.

What becomes apparent from this commercial evidence, matched with the trade data collected here, is that the demands and operations in the trade could change within months of commercial decisions being made. As will be seen in

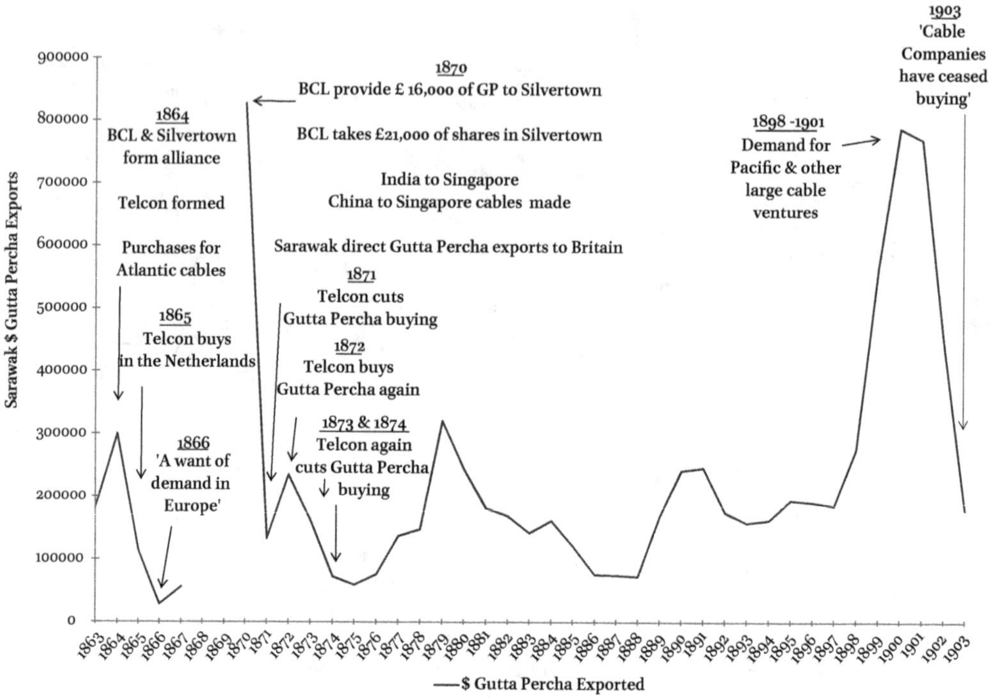

FIGURE 6 *Sensitivities in the gutta percha commodity chain reflected in Sarawak trade.*
Source: For Sarawak GP Figures: Treasurer's Reports reprinted in Sarawak Gazettes, (SG) various issues 1870–1904; and
Consular Reports in British Sessional Papers (BSP): 1865 [3478], p. 28; 1866 [3582], p. 805; 1866 [3669], p. 266
for the years 1863–65 respectively. Figures for 1868–69 are not available.
See Appendix 2, Table 1 for statistical details.
REFERENCES FOR QUOTATIONS 1866: BSP 1867–68 [3953-I-VIII], p. 355; FOR 1903: Sarawak Gazette 3 May 1904, p. 99.

subsequent chapters, collectors and traders in remote environments responded very quickly to these changes. The major causes and effects discussed above in relation to the trade are graphically summarised at Figures 5 for Singapore and at Figure 6 for Sarawak. At Figure 7 is the data for UK imports comparing Straits Settlements and the Netherlands as described above, to show regional inter-relationships (see Appendix 2, Table 1 for data).

The impact of the Gutta Percha Co and BCL/Silvertown's commercial arrangements on the regional gutta percha trade, as reflected in the UK import statistics described above, is seen in this last graph.

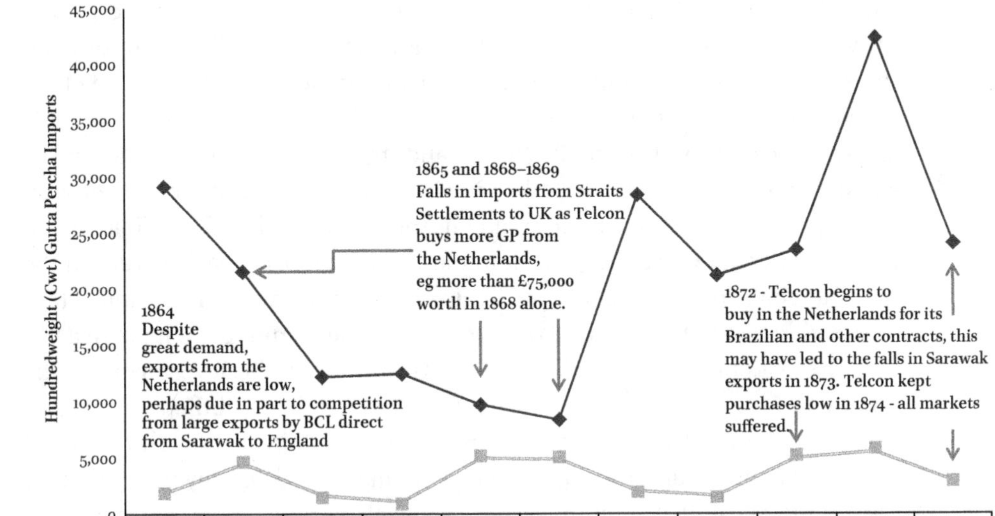

	1864	1865	1866	1867	1868	1869	1870	1871	1872	1873	1874
Strts. Sett.	29,151	21,585	12,211	12,487	9,727	8,387	28,410	21,227	23,487	42,340	24,085
Netherlands	1,779	4,682	1,485	899	5,199	5,073	1,945	1,498	5,202	5,825	2,941

FIGURE 7 *Sensitivities in the gutta percha commodity chain reflected in intra-regional gutta percha trade variations.*
Note: Measurements are in hundredweights – Imperial weights were used by British Customs. The Customs figures were listed under 'Straits Settlements', an aggregation that included mainly Singapore, but also Penang and Malacca statistics. The latter two were of much less value than those Singapore.
Source: E.F.A. Obach, '*Cantor Lectures on Gutta Percha*', London, *Journal of the Royal Society of Arts*, 1898, p. 97.

Of note is the low figure for 1864 for the Netherlands. As 1864 was an important year for demand for gutta percha for the Atlantic and other cables, the figures for the Netherlands should have been higher. The low levels may have been due in large part to the success of BCL in exporting large amounts of gutta percha from Sarawak direct to England. Because BCL was exporting large amounts directly to England, this appears to have 'crowded out' demand for gutta percha that could have been sourced from the Dutch colonies and sold in the auction houses in the Netherlands. Similarly, of note are the 1870 levels for the Netherlands, which might also have been expected to be higher given the level of cable manufacture but again, the enormous levels of gutta percha exports from Sarawak, reduced demand for supply from the Netherlands.

Dougherty concluded that it is the 'entwined' nature of the individual strands of the commodity chain that has an impact on those involved in the chain.[100] That is, the relationships and outcomes are more complex, more 'entwined' than simply commercial transactions of demand and supply, and transactions in one area of the chain have an impact elsewhere in the chain even if those nodes are not directly linked to each other. The corporate records link the role of imports from the Netherlands by the Gutta Percha Co as having an impact on the trade of Sarawak (and vice versa), so the connection between telegraphy and the gutta percha trade in the region is clear. With Telcon and the Gutta Percha Co as the most dominant of the global 'chain drivers' in the gutta percha commodity chain, such inter-connections were clearly important and readily highlighted in the extant records of the commodity chain. In addition, the evidence directly linking Telcon/Gutta Percha Co, Silvertown and the Borneo Company (BCL) to Sarawak's gutta percha exports is difficult to ignore – backed up by the trade figures shown. Sarawak's connection with the world gutta percha trade and the commerce of submarine telegraphy was forged by the unique role of BCL – for that reason Sarawak, perhaps more than any other country, was in some years both a significant contributor and most vulnerable to the market and commerce of submarine telegraphy and demand for global communications.

With such close dependencies, the ground was set for the gutta percha trade to be linked to the region, and the historic record illustrates such interdependencies. As Cleary notes in his analysis of the rise of Southeast Asian colonial economies, European traders were content to rely upon intermediaries to underpin trade.[101] In those years when cable communication was dependent upon gutta percha, it was the European colonial trading outposts in Southeast Asia where the trade was established with local and intermediary traders exploiting the forest sources of gutta percha in the enormous amounts required. In fact, the Temenggong of Johor was perhaps the first and one of the most dominant of the chain drivers in the gutta percha trade. Bair concluded that a benefit of commodity chain analysis is that it is an approach that allows for the mapping and understanding of spatially dispersed networks, and reveals how chains are developed within larger social and cultural, and political-economic environments.[102] This was seen when traders of the 'periphery' kept

100 M.L. Dougherty, 'Theorizing Theory: Origins and Orientations of Commodity Chain Analysis', *The Global Studies Journal*, Volume 1, Number 3, 2008, p. 29.

101 M.C. Cleary, 'Indigenous trade and European economic intervention in North-West Borneo c. 1860–1930', *Modern Asian Studies*, Vol.30, no.2, 1996, p. 305.

102 J. Bair, 'Global Capitalism and Commodity Chains: Looking Back, Going Forward', *Competition and Change*, Vol.9, No.2, June 2005, pp. 158, 168.

the trade functioning during this era; some, such as the Temenggong of Johor, actively shaped the operation of the trade. Others, such as the extensive network of Chinese traders, were at the core of the commercial transactions and production of the commodity itself.

Yet consumer demand can be shaped by unpredictable motivations; be they strategic alliances (such as between BCL and Silvertown); market domination, as shown when the Gutta Percha Co/Telcon acted, or mismanagement such as when the Gutta Percha Co over-purchased supplies of gutta percha and then severely reduced its buying. The gutta percha commodity chain was highly dependent upon the operation of a few corporate entities and their alliances. Management and mismanagement – and technological developments – each played a part in shaping the operation of the chain. Similarly, the nature of the commodity itself, and its technological applications, created a demand for quality and a fear of substitutes. These characteristics have been highlighted by analysing the commodity chain to this point. The entry and withdrawal from the market of these companies had speedy and noticeable impact on the trade. When the demands for gutta percha for the metropolitan market fluctuated – and were on occasion extreme and company specific – networks could respond with considerable speed to market needs. Towards the end of the nineteenth century, it was possible for the Gutta Percha Co/Telcon to manufacture, ship and lay a cable within five months from the date a contract was signed.[103] The evidence seen here reveals how quickly collectors and traders were affected by consumer decisions to withdraw from the market – forcing them to identify other commodities to exploit, all while cable contractors expected that they would be there when they needed them next.

Such market fluctuations were noticed in small economies like that of Sarawak. Nonetheless the gutta percha trade provided important opportunities to government, the indigenous community and Chinese traders. This recalls Marichal et al. in their view that globalization results from processes in which the local is embedded in the international and vice versa.[104] Sverrisson's proposal

103 H.D. Wilkinson, *Submarine Cable Laying and Repairing*, 2nd edn, London, 'The Electrician' Printing and Publishing Company, 1908, p. 88.

104 C. Marichal, S. Topik and Z. Frank, 'Conclusion: Commodity chains and globalization in historical perspective' in *From Silver to Cocaine. Latin American Commodity Chains and the Building of the World Economy, 1500–2000*, eds S. Topik, C. Marichal and Z. Frank, Durham, Duke University Press, 2006, p. 360.

to view linkages from global to local suggests following the commodity chain itself: an alternative way of understanding the global economy, his approach would 'emphasize the scope for local entrepreneurship at all levels and focus research on that'.[105] The following chapters continue that process.

---

105 A. Sverrisson, 'Local and global commodity chains' in *Linking Local and Global Economies: The Ties That Bind*, eds C. Pietrobelli and A. Sverrisson, London, Routledge, 2003, p. 30.

PART 2

Power, Profit and the Periphery

> The whole year shows a marked improvement, principally attributed to a very large demand for gutta which arose at that time for the purpose of telegraphic cables. The supply at such short notice being hardly equal to this excessive demand…the influx of silver coin consequent on this caused an unusual stimulus throughout the commerce of Sarawak.

Acting Vice-Consul St. John on the Trade of Sarawak 1870
BSP 1870 [C.429], p. 806

CHAPTER 5

The Gutta Percha Trade of Sarawak

> The increase of trade is principally owing to a rush after gutta percha which has run up to fabulous prices
>
> SARAWAK GAZETTE, 28 February 1871

The local conditions that shaped the gutta percha trade are set out in this chapter, which in effect considers the factors shaping interaction between the peripheral and external economies in global trade relations. Using Sarawak as a case study, the chapter sets out factors shaping the trade environment: the geography, people, and unusual system of government under the Englishmen James and later Charles Brooke, the so-called 'White Rajahs' of Sarawak. With few resources that could be exploited, the gutta percha trade became a significant feature of the Sarawak economy, and a key factor in linking the country to the global economy. Indigenous groups were the only people capable of collecting this forest product, but Chinese intermediary traders were the key to enabling its export to the outside world.

Sarawak, Its People and the Brooke Administrations

Borneo is in a region that historically had formed part of the trade route to China – part of an 'Asian World Economy'.[1] Foreign trade fed through the coastal regions of Borneo where it was organized and taxed by Malay elite. Knapen proposes that the interior population of Borneo, even in isolated and out-of-the-way areas, incorporated changes that affected settlement patterns, agriculture, trade and population growth because of changing political conditions and contact with foreign trade from as early as 1600 (and prior to European

1 C. Dixon, *South East Asia in the World-economy*, Cambridge, Cambridge University Press, 1991, p. 61.

 | DOI 10.1163/9789004357280_007

arrival).[2] Building upon that history, the expansion of the global economy in the nineteenth century and Sarawak's proximity to the entrepot of Singapore gave it a great potential to link into the benefits of world trade and with British colonial networks that were outlined in the previous chapter.

The British Consular Report on Sarawak for 1863 noted the country's various natural advantages for trade: the river system, proximity to the regional entrepot of Singapore, and its location on the steam shipping route between England and China, and being especially suited to traffic in both monsoon seasons.[3] The Report praised Sarawak's free ports and light taxation that compared well with neighbouring provinces. Kuching, the capital of Sarawak, became a small entrepot for the near region as it garnered some of the trade from Brunei, Labuan and parts of the Dutch territory (Netherlands Indies) in Borneo. James Brooke aimed to provide peace and stability in a territory that had suffered serious instability, misgovernment and economic backwardness. His earliest laws claimed to introduce free trade, equitable taxation, reliable weights, measures and currency for trade, and to abolish slavery and debt slavery and provide protection for all – Malay, Chinese or indigenous.[4]

The trade of Sarawak was particularly shaped by its geography – its extensive river system and steep inland terrain falling to gradually undulating hills then to coastal plains – and its monsoon weather patterns and heavy rains. Harsh geography made road construction difficult. The system of rivers and streams allowed for river transport that was relatively extensive and cheap, if not sophisticated. James Brooke highlighted 'friendly intercourse with the natives, and the existence of navigable rivers' as the two key factors in penetrating the interior of Sarawak and expanding the prospects of trade.[5] In 1864 the Rejang river, for example, was said to be navigable for 120 miles by ships of 1,000 tons and the Baram river was navigable for considerable distances,[6] but many rivers were not navigable fully upstream or during the monsoon seasons. The harsh terrain of Sarawak meant that, as Chew notes, it was more appropriate for distances in Sarawak to be expressed in the number of day's

2 H. Knapen, *Forest of Fortune? The Environmental History of Southeast Borneo, 1600–1880*, Leiden, KITLV, 2001, p. 6.

3 Consular Report on the Trade of Sarawak for 1863, BSP 1865 [3478], p. 31.

4 See R. Payne, *The White Rajahs of Sarawak*. Robert Hale, 1960, Reprinted Singapore, Oxford University Press, 1992, p. 47.

5 J.C. Templer, (ed.), *Private Letters of Sir James Brooke K.C.B. Rajah of Sarawak, Narrating the Events of His Life*, 3 Vols, London, Richard Bentley, 1853, I, pp. 23, 25; see also pp. 96, 135.

6 Consular Report on the Trade of Sarawak for 1863, BSP 1865 [3478], p. 27.

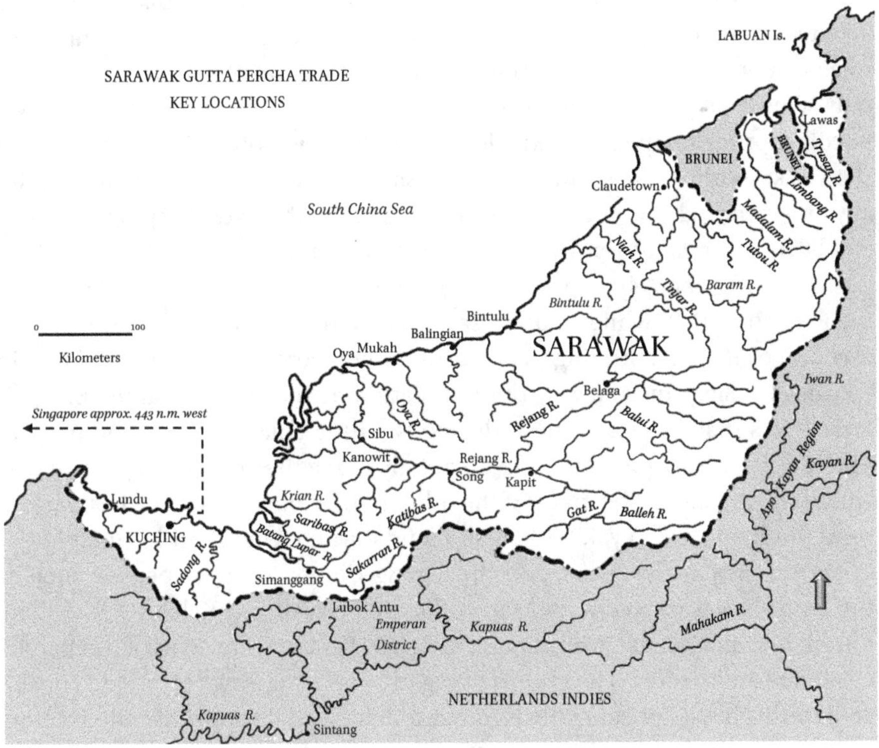

MAP 5 *Historic map of Sarawak and environs.*
Note: Historic map of Sarawak: names are those used in the nineteenth century and adopted here for consistency with the archival records.

journey required, rather than in the distances calculated in absolute terms.[7] Hose writes about the skills with which boat crews could negotiate the river systems, but it could take three days to travel up a river that could take just four hours to come down.[8] Map 5 shows key Sarawak rivers and trade points.

Social and trade networks formed around river-based groups, defining political and social relationships.[9] Aside from their significance to indigenous populations, these river networks were crucial to outsiders, particularly Chinese

7 D. Chew, *Chinese Pioneers on the Sarawak Frontier*, Singapore, Oxford University Press, 1990, p. 5.

8 C. Hose, 'In the heart of Borneo', *Geographical Journal*, Vol.16, no.1, (July), 1900, p. 42.

9 See B.W. Andaya and L.Y. Andaya, *A History of Malaysia*, Basingstoke, Macmillan, 2001 and A. Kaur, *Economic Change in East Malaysia. Sabah and Sarawak since 1850*, Basingstoke, Macmillan Press Ltd, 1998, Ch.2.

traders for whom rivers became the only comprehensible means of accessing an environment both physically and culturally impenetrable.[10] River transport was at the core of the forest trade networks of colonial Sarawak, particularly that of gutta percha. The Chinese traders who relied on river transport to travel to collectors or to up-river trade depots would inevitably have become lost in the forests without these 'highways'. As Sarawak developed, small and gradually larger coastal shipping and steamers added to the river shipping networks and became a major force in the Sarawak trade landscape.

Borneo was sparsely populated – the naturalist Alfred Russel Wallace commented upon this, while Knapen says that Borneo's low population density was one of its most striking peculiarities when compared with elsewhere in Southeast Asia.[11] In Sarawak, the population at the time of James Brooke's arrival was mainly Malay and indigenous with a small Chinese component. The Muslim Malay population lived mainly along the coastal fringe, together with the indigenous Melanau. Of the other indigenous groups, the best known were the Iban ('sea dayak'), Bidayuh ('land dayak'), who lived in the middle/upper reaches of the rivers, and the 'Orang Ulu' or 'up-river people' such as the Kenyah, Kayan, Kedayan, Murut, Punan, Bisayah and Kelabit.[12] The Iban, Kayan, Kenyan, Kelabit, Murut and Punan (or Penan) were active gutta percha collectors and traders. The groups living in Sarawak during the period considered in this book are generally referred to here as they appear in the primary sources. Today, these names are often overtaken by nomenclature reflecting more detailed group identifiers; however, the titles used here were those used in nineteenth-century writings (they are still recognized by most groups today) and are maintained here for reasons of consistency with the historical record and to avoid continual cross-referencing between modern and historical titles. An exception is the use of Iban for the historic 'sea dayak' for simplicity sake and because 'dayak' is used in various contexts in the historic record.[13]

Along the coastal fringe, Malays generally practiced wet rice cultivation while the Melanau population lived off the produce of the sea and swamp

10 See W.M. Crocker, 'Notes on Sarawak and Northern Borneo', *Proceedings of the Royal Geographical Society and Monthly Record of Geography*, New Monthly Series, Vol.3, No.4, (April), 1881, p. 194.

11 A.R. Wallace, *The Malay Archipelago*, 10th edn, London, MacMillan & Co, UK, 1869, Reprinted Singapore, Graham Brash Pty Ltd, 1987, p. 69; Knapen, *Forest of Fortune? The Environmental History of Southeast Borneo*, p. 105.

12 See A. Kaur, 'The babbling Brookes: Economic change in Sarawak 1841–1941', *Modern Asian Studies*, Vol. 29, No.1, 1995, p. 66 and H. Munan, *Beads of Borneo*, Singapore, Editions Didier Millet, 2005, pp. 16–20.

13 See R.M. Pringle and O.C Doering, Preface to *Rajah's Servant*, A.B. Ward, Cornell University Southeast Asia Program, Data Paper No.61, Ithaca, Cornell University Press, 1966, p. ix.

sago. Further inland, most other groups lived on some form of rice cultivation, swidden agriculture or shifting agriculture except in the steep inland areas where poor soil made such settlement untenable and hunting/gathering was also required. Swidden agriculture is characterized by the clearance of forest areas to provide land for cultivation. Clearance involves the removal of trees and other growth by burning off. Once the cultivated land becomes unproductive, the process begins again and another area of forest is cleared and cultivated.[14] In all areas, barter trade was used to supplement the way of life. In the deeper hinterland, the Punan were hunter-gatherers rather than cultivators, supplemented by trade in forest products.

The size of the population, and the numbers engaged in the gutta percha trade during the period of this study is not clear. Early population estimates are based upon fragmentary and often unreliable evidence. The following estimates appear in various sources.[15]

	Total*	Chinese	Malay	Melanau	Sea Dyak*	Land Dyak*	Other
1841	10,500 or 177–189,000	1,000–2,000					
1857	150,000	4,500					
1871	141,546	3,467					364
1877	222,000	7,000	60,000	30,000	90,000	35,000	
1883	250,000						
1898	300,000						
1904	500,000						
1909	416,000	45,000				3,000	

* Estimated total, largely indigenous, except where other figures shown. Blank spaces indicate figures were not listed in the source cited. The disparities shown above illustrate the problematic nature of grappling with estimates of population and population growth in Sarawak, for example, the two 1841 figures; the 1909 figures are included here to illustrate the difficulty of being certain of figures even in the twentieth century. The terms 'Sea Dyak' [Iban] and 'Land Dyak' were used for the population listings for 1877.

14 See for example A.A. Batabyal and H. Beladi, 'Swidden agriculture in developing countries', *Review of Development Economics*, Vol.8, No.2, May 2004, pp. 255–265 for an overview of swidden agriculture in developing countries.

15 Sources: for 1841 – Y.L. Lee, 'The Population of Sarawak', *The Geographical Journal*, Vol.131, No.3, Sep. 1965, p. 346; R.M. Pringle, *Rajahs and Rebels: The Ibans of Sarawak under Brooke Rule, 1841–1941*, Ithaca, Cornell University Press, 1970 pp. 15–16; 1857 – Pringle, *Rajahs and Rebels*, pp. 106–107; 1871 – *Sarawak Gazette* (hereafter SG), 15 December 1871; 1877 SG

The indigenous population was by far the largest, ninety percent at the turn of the twentieth century has been estimated by Lee, and the Chinese population grew over this period, with Europeans initially numbering perhaps dozens.[16] By 1883 Charles Brooke said that he had thirty European staff – and Europeans only numbered 691 even in the 1947 census, at which time the total population of Sarawak was still calculated at only 546,385.[17] For 'Other' in the listings above, these referred to 'Indian, Javanese, Bugis etc.' – see references. The Chinese population was based largely around the ports and urban centres such as Kuching.[18]

Lee writes that the 1841 (10,500) figure, and the 1871 and 1909 figures were mere guesses, and that large parts of the country did not come under official influence and were thus unlikely to have been counted.[19] The Sarawak Gazette observed that the 1871 census did 'not pretend to strict accuracy', and should be 'taken as an experiment'.[20] In 1870, official reports from Sarawak had said that the population for the Rejang district alone was between 70,000 and 80,000, and yet in early 1871 the *Gazette* said that the Rejang had a population of around 50,000 so this gives an added sense that the figures are not entirely reliable.[21] Noakes, in an Appendix to the 1947 Census of Sarawak, wrote that it would be difficult to conclude whether or not the indigenous population of Sarawak had in fact increased over the previous one hundred years.[22] The view has been put forward that overall the population did not significantly increase due to population decline arising from disease, warfare and migration.[23]

---

30 April 1880, p. 28; 1883 – C. Brooke, in *Pall Mall Gazette*, London, J.K. Sharpe, 19 November 1883, p. 1; 1898 – Consular Report 1899, Paper: 1900 [Cd.1][Cd.352] p.307; 1904 – Directory and Chronicle for China, Japan, Corea [sic], Indo-China, Straits Settlements, Malay States, Sian, Netherlands India, Borneo, the Philippines &c, Hong Kong, Hong Kong Daily Press, 1904, p. 784; 1909 – S. Baring-Gould and C.A. Bampfylde, *A History of Sarawak under its Two White Rajahs, 1839–1908*, London, Henry Southeran, 1909, p. 33.

16 Lee, *The Geographical Journal*, p. 345.

17 Brooke, 1883, *Pall Mall Gazette*, p. 2; Y.L. Lee, *The Geographical Journal*, p. 345.

18 Y.L. Lee, 'The Population of British Borneo', *Population Studies*, Vol. 15, No.3, March 1962, p. 238.

19 Lee, *Population Studies*, 1962, p. 227.

20 SG 15 December 1871.

21 *The Phoenix. A Monthly Magazine for China, Japan & Eastern Asia*, London, 1870, p. 20; *The Straits Times*, Singapore, June 11, 1870, p. 2; SG 24 January 1871.

22 J.L. Noakes, *A Report on the 1947 Census, Sarawak and Brunei*, Kuching, National Government, 1950, p. 8.

23 H. Brookfield, L.M. Potter and L. Byron (eds), *In Place of the Forest: Environmental Socio-Economic Transformation in Borneo and the Eastern Malay Peninsula*, Tokyo, United Nations University Press, 1995, especially Ch. 2

Estimating migration numbers, temporary and permanent, adds to the vexed question of demographics since at any given time population numbers were fluid. People moved across the border for purposes of trade, to avoid (or engage in) hostilities, avoid local taxation and so on. Large groups also moved both within Sarawak and across the border in search of 'old jungle' that was sought for padi growing.[24] Godfrey listed just a few extracts from the Sarawak Gazette that reveal how thousands of people moved into the Baram district alone in the last two decades of the nineteenth century, while the general tenor of the Resident's reports for the District over that period also suggests a continuous movement of groups.[25] Taking all these commentators views into account would suggest that the other nineteenth century figures must also be viewed with caution. Baring-Gould and Bamfylde's 1909 figures listed above, totaling 416,000, include 268,000 from groups potentially living in gutta percha collecting communities (that is, excluding Chinese, Malay, Melanau and 'others'). In summary, at best it can concluded that a large portion of the population of Sarawak was likely living in areas where gutta percha collecting could have been possible, and that the size of the population possibly grew, but not by a great deal and any estimate of this would be just that, an estimate.

The economic conditions in Sarawak at the commencement of Brooke rule were parlous, in part because of the depredations on trading activity brought on by lawlessness inland and in the coastal regions. Along the coastal periphery, 'piracy' was a serious disincentive to the operation and development of trade. Further inland, hostilities were a problem under both James and his successor, his nephew Charles Brooke. James Brooke made relatively limited efforts to expand the Sarawak economy. In April 1841, he wrote about Sarawak's greatly under-exploited produce.[26] In 1851, he spoke about his approach to governing Sarawak which could also describe his approach (or lack of one) to an economic policy: 'When I went there I had no fixed plan, no deep design. ...I had no design at all...'.[27] He professed himself to be a protector of 'native' interests and this became a hallmark of his and Charles' rhetoric. Other correspondence written by James in 1841 outlines broad strategies based around his obtaining the monopoly over Sarawak antimony resources, attracting Chinese labor and benefiting not just from their labor but by taxing them.

24 See for example, SG 1 October 1898, p. 196.

25 H.L. Godfrey, 'Trade and Transformation on the Borneo Frontier: towards a study of the Baram region of Sarawak in the late nineteenth century', *Borneo Research Bulletin,* 2014, Vol. 45, pp. 121–122.

26 In Templer, Vol. I, pp. 95–96.

27 J. Brooke, 'Speech of Sir James Brooke on Borneo', *The Leader,* Issue 2, No.88, 1851, p. 1131.

He mentioned the possible benefits from the 'small quantities of bees-wax, birds' nests, and the like', that is, forest produce (gutta percha not having at that stage been brought to European notice).[28] The Brookes' activities eventually enforced varying degrees of security in both coastal and up-river areas that enabled a trading environment to develop, albeit not always peacefully. While more settled conditions were a necessary pre-condition for economic development, forest produce generally, and gutta percha specifically, provided a foothold in the early decades of Sarawak's development as it required no investment and little additional labor or technology for its establishment. James Brooke had held monopoly rights over Sarawak's mineral resources. Later, Sarawak was open to outside investment but this was done largely under the aegis of the Borneo Company Limited (BCL), especially in mining. The Company in return was given preferential mineral royalty rights for its gold, antimony and cinnabar mines. The relationship between the Brookes and BCL was mutually beneficial and profitable, although export earnings from mining were less important to Sarawak than either forest produce or cultivation for many decades.

Poor soil conditions made agriculture difficult, even had the willingness been there to invest in it; perhaps 'forest produce and mining offered better prospects'.[29] Mining, however, required an investor, significant migrant labor, and time to get started whereas forest trade could exploit existing resources and historical trade patterns. The Sarawak Treasurer provided the Rajah with figures on the various export sectors, details of which were published in the Sarawak Gazette from 1870 onwards. Mining earnings were less significant than might be expected – around 15 percent of the total over that period, only slightly more than gutta percha itself which achieved 12 percent (see graph at Figure 9 later in this chapter). In theory, it was a deliberate policy of James and Charles Brooke to promote agriculture for both the long-term economic future of Sarawak and, as the Commercial Report of 1864 put it, to 'spread a civilizing influence over the natives'.[30] In reality, most agricultural development was undertaken by Chinese rather than indigenous groups. Charles Brooke was keen to promote any industries likely to benefit trade.[31] Experimental gardens were established under his instructions and officers were required to report on agricultural matters, but it required Chinese labor and entrepreneurialism (rather

28 Templer, Vol. I, p. 116.

29 J.T. Lindblad, 'Review [Of Free Trade and Native Interests: The Brooke and the Economic Development of Sarawak, 1841–1941 by Ooi Keat Gin]', *Bulletin of the School of Oriental and African Studies, University of London*, Vol. 62, No.1, 1999, p. 196.

30 Consular Report on the Trade of Sarawak for 1864, BSP 1866 [3582], p. 803.

31 SG 1 August 1894, p. 118.

than anything Brooke achieved) for success, for example, in the gambier and pepper trades.[32] Charles Brooke sought to promote the economic development of Sarawak by encouraging Chinese migration and investment, which he recognized had been a key factor in the development of Southeast Asian trade.

It was with the later arrival of migrant labor and investment of intermediaries – Chinese labor and traders – that profits from cultivation were to rise as seen in the graph below at Figure 8. Note that significant changes in the overall export economy began when cultivations began to increase, especially in the 1890s

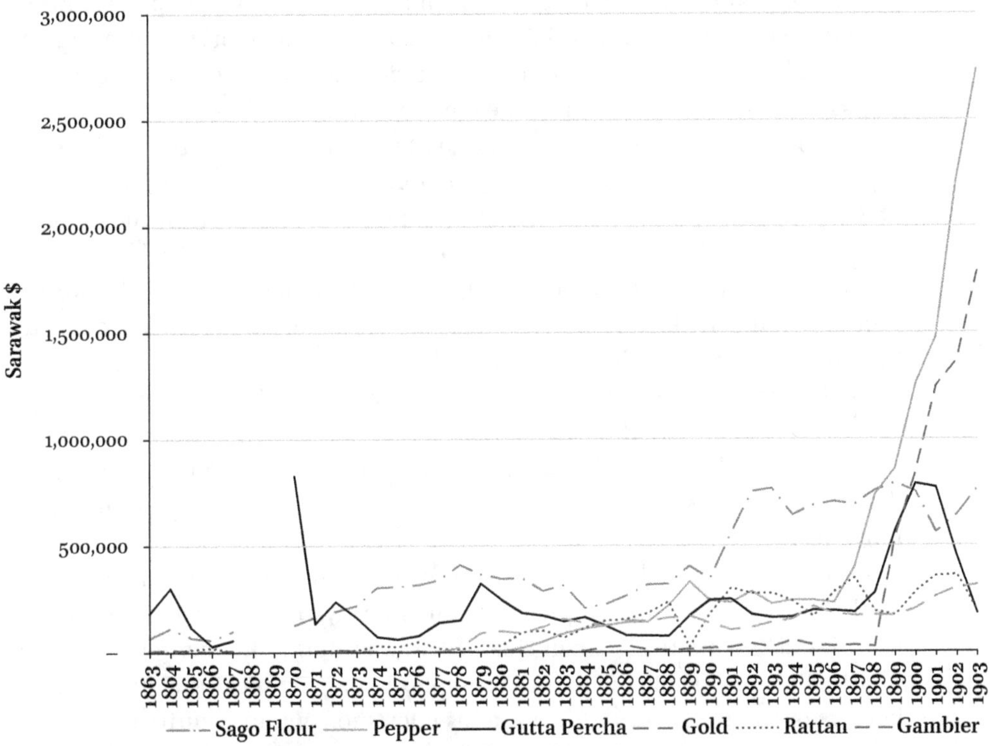

FIGURE 8 *Sarawak – Six major exports 1863–1903.*
Note: These commodities were the top six export trades based on total exports for the period; in descending order of value these were: Sago flour, Pepper, Gutta Percha, Gold (due mainly to exports from 1899 onwards), Rattan and Gambier.
Source: Treasurer's Reports reprinted in Sarawak Gazettes (SG), various issues 1870–1904; and Consular Reports in British Sessional Papers: BSP 1865 [3478], p. 28; 1866 [3582], p. 805; 1866 [3669], p. 266 for the years 1863–65 respectively. Figures for 1868–69 are not available. See Appendix 2, Table 3 for more statistical details.

32 See K.G. Ooi, *Of Free Trade and Native Interests, The Brookes and the Economic Development of Sarawak, 1841–1941*, Kuala Lumpur, Oxford University Press, 1997, pp. 40–41.

with the gradual increase in sago flour (largely exploited by the Chinese and BCL), but mainly from pepper, especially from 1900.

Also of note is the increase in gold exports at the end of the century. Prior to 1899, gold exports were steady if modest, but after that date BCL took over most of Sarawak's gold mines; and, more importantly, began using the newly developed cyanidation extraction technique which, as seen in the graph, radically raised the level of gold exports.[33] Gutta percha and gold both lifted noticeably in 1899, pepper in 1900 – but only gutta fell by the noticeable level seen in 1903 (for the reasons set out in Chapter 4). Both mining and cultivation depended very largely upon Chinese labor, and to a certain extent upon Chinese capital. Sago production, especially processed sago (sago flour) was also largely controlled by the Chinese and BCL. This meant that the indigenous population's role was largely confined to the forest and sea produce sectors of the Sarawak economy – gutta percha and rattan in the trades shown above. Between 1870 and 1903, forest produce and sea produce combined made up one third of Sarawak's exports – down from 47 percent at the beginning of the period, in the 1870s (see Figure 2 earlier). The changing role of forest produce in the economy reflected Sarawak's increasing inter-connectedness with the world economy – it also illustrates the extent to which indigenous societies could become economically 'disenfranchised' by the changing demands of global trade. The changes reflected the strength of the Chinese in the economy, both in economic terms and their ability to allocate labor and capital to agriculture. What Dixon called 'the Asian world economy' had a pre-existing foothold in the region and was positioning itself in the European world-system that was moving into the region.

In terms of regional and administrative measures taken to establish Sarawak, James Brooke had succeeded in many ways because of his personal charm and the relatively small size of Sarawak during his period as Rajah. Charles Brooke – dealing with more complex issues and seeking to expand the borders of Sarawak – was ruthless in his efforts to contain security problems. He sought to do this while maintaining an economic and administrative system on a tight budget. Administratively, Sarawak was separated into various 'Divisions', each under the responsibility of a European Resident, with a subordinate Assistant Resident or Cadet. They were limited in number – as few as fifty Europeans administered the whole country.[34] The system gave the Rajah

33 See Ooi, pp. 135–136.

34 N.S Talib, *Administrators and Their Service. The Sarawak Administrative Service under the Brooke Rajahs and British Colonial Rule*, Selangor Darul Ehsan, Oxford University Press, 1999, p. 51.

administrative oversight and made his rule visible, albeit modestly. He chose to use a network of existing Malay elite, and sometimes indigenous or Chinese leaders, as an administrative and ruling force, in a bid to both control them and to win their support. While each Division was under the administrative control of a European with a very small number of European staff, it was the Malay, indigenous or Chinese political figures who were responsible for collecting taxation and maintaining traditional law and order, subject, in theory, to oversight from the European staff. These local leaders usually took a commission on the taxation or fines they collected rather than being paid salaries, and 'whether he was as Iban *penghulu*, a Malay *tuah kamphong*, or a Chinese *kangchew*, the community headman managed local affairs in accordance with the customs of his people'.[35] The use of these intermediaries was consistent with colonial policies exercised elsewhere in the region by other European powers. As far as was feasible, traditional laws were to be respected except in so far as this interfered with the concept of western ideas of justice or the security of Sarawak. For these latter, the European administrators would be responsible; the two systems were seen to be complementary. Security, where required, was provided by Iban under official organization. In part Charles Brooke's decision to use Iban forces was an economic one; he could not have afforded an army even had he been prone to use one (or to pay for it). The aim was to both maximize, and control, what was seen as the respective strengths of the three sections of society, and to keep them separate.

Charles Brooke dealt with widespread security issues that expanded as Sarawak expanded. Some Iban, for example, were hostile to the Brooke regime or were adversaries of other groups in Sarawak or in the neighbouring Dutch territory of Netherlands Indies. Other groups such as the Kayan were hostile to the Iban, and to the Brooke administration. However, not all Iban or all Kayan were necessarily hostile to either the Brookes or other Iban, so the security situation was fluid and complex – and there were many other groups who had different targets of disaffection and different loyalties. Very small incidents could at times inflame old enmities and feuds could last for many years, and revenge attacks were an important feature of social relations that made settled conditions even more difficult to achieve. Of course, these conditions made trade difficult as well. Charles Brooke, over the many decades of his rule, undertook a series of expeditions, some only arguably necessary, some punitive, aimed at enforcing peace and at reinforcing his position as Rajah. In keeping with his policy, and the necessity of keeping costs at a minimum, Charles looked to the most capable force he had – the Iban – to control hostile groups. The risk was

35 Pringle, *Rajahs and Rebels*, p. 144.

that, by allowing Iban forces to fight, and to 'take heads' whilst on government expeditions, it would be difficult, if not impossible, to stop those same forces from taking heads at other times.

Charles' attention was often taken up with dealing with the security problems of Sarawak, with hostilities arising between Sarawak residents and with those from the neighbouring Dutch held Netherlands Indies. At times difficulties arose when Sarawak inhabitants moved into Dutch territory as national boundaries were largely a construct of European administrations and they held only a notional meaning for indigenous inhabitants whose predecessors had for generations traversed much of Borneo. Another of the factors that caused hostilities was the push made by several groups of Iban to move into areas further northwards in Sarawak territory, which had not traditionally been Iban territory. Much of the Iban urge to move has been regarded as the need to seek out new land for the purposes of swidden agriculture; however, these new lands were also very valuable sources of forest produce such as gutta percha and this added greatly to the urge to explore new areas. Charles Brooke had two broad approaches for dealing with the Iban pressure for expansion. On the one hand were aggressive attacks referred to above and the other means was to satisfy Iban needs for migration by granting them permission to move into new areas. The risk in both policies was that Iban, even with official permission, might move into areas already inhabited by others who would not appreciate their land or resources being shared by a group who were very successful swidden cultivators and collectors of forest produce. Charles' deliberate disregard for traditional concepts of social groupings and resource ownership was not unique in colonial history, but this did not make it any less problematic either to those whom it affected or to the Brooke administration itself. The impact of Charles' policies in this regard had ramifications for the stability of Sarawak and especially for the gutta percha trade which will be considered in Chapter 8.

Both James and Charles Brooke also dealt with the risks of hostilities from outside Sarawak. The Brunei Sultanate, including the Sultans themselves, as well as various *pangeran* and hostile groups in Brunei, were naturally concerned to protect their own interests which did not necessarily match those of the Brookes. British interests in the area were rather complex and divided over the years of Brooke rule, and have been studied elsewhere.[36] British responses were based on the need to protect their interests in their colony of Labuan, an

36 See G.W. Irwin, *Nineteenth Century Borneo: A Study in Diplomatic Rivalry*. Singapore, Donald Moore, 1967; L.R. Wright, *The Origins of British Borneo*, Hong Kong, Hong Kong University Press, 1970 and N. Tarling, *Britain, the Brookes and Brunei*.

island off Borneo which was dependent upon trade connections with Brunei. British interests were also concerned about being saddled with the burdens of Sarawak should the Brookes or their descendants fail, or worse still if Sarawak should fall to foreign interests. Further complicating British policies were disagreements between the Foreign Office and the Colonial Office about how to treat Sarawak and Borneo, and even how the Brookes should be regarded – were they British subjects or were they foreign heads of state? Even the personalities of officials holding the position of Governor of Labuan, and then the establishment in an area north of Sarawak of a Chartered Company in British North Borneo (later Sabah), made matters more complicated. In short, the British attitude to Borneo and Sarawak reflected its need to determine its overall colonial objectives. These complexities were to create many distractions for both James and Charles Brooke because they were to have an impact on the relations among Sarawak, Brunei and their populations – and the trade connections between Sarawak and Brunei. Gutta percha trading became entangled in all these aspects. As will be seen in later discussion, the trade became an issue that had an impact on the relationship between people and rulers of Sarawak and Brunei and the British administration.

Significance and Growth of the Gutta Percha Trade in Sarawak

Brooke policies have been described as an 'ideology of gradualism'.[37] In an economic sense this was true – this was one of the reasons that the gutta percha trade was so important to Sarawak in the nineteenth century.

The Sarawak Gazette described the country's economic development in one of it early issues:

> ...the demands of traders have been for commodities which entail little or no attempt at cultivation. The produce of the jungles procured with ease and little labor by the Dyaks, would be the first articles offered by them for sale or barter and this natural instinct has been strengthened and encouraged by the course which European and Chinese traders have taken in the country[38]

37 R.H.W. Reece, 'Economic development under the Brookes' in *Development in Sarawak*, eds R.A. Cramb and R.H.W. Reece, Centre of Southeast Asian Studies, Melbourne, Monash University, 1988, p. 33.

38 SG 16 July 1872.

In these few short words the Gazette had, unwittingly, described the foundation of Sarawak's entry into the fold of the modern global economic network. There were many reasons for the gutta percha trade being taken up so quickly and for its influence on both the economic and social fabrics of Sarawak for many decades. The forest produce of Sarawak had been traded for centuries and local trade networks were far-reaching and international and pre-dated European contact, in particular because of Chinese traders interested in highly valued produce such as edible birds' nests that were collected there. Forest collectors and traders were the descendants of many generations of collectors and traders who had established networks and practices of trade. The value of these networks and experiences were important as they established a firm basis upon which to build future trade in other forest produce such as gutta percha; that is, one commodity chain could be partially re-invented to suit a new commodity.

An overview of the trade returns for Sarawak up to the end of the nineteenth century illustrates the dependence upon forest produce, the indigenous economy and its associated trade networks – providing what has been described as a 'lifeline' for governments in British Borneo.[39] After analyzing the nature of gutta percha itself and the conditions in Sarawak in the period under review, it is clear that there were conditions which made gutta percha a commodity that suited both the peculiarities of the physical and human environment, and the entrepreneurial conservatism and political style of the Brooke administrations. It suited the limited options available to the indigenous population and suited the historical experience and cultural preferences of these communities. Gutta percha was profitable, relatively easy to collect, required no capital investment – and, as Dove notes, an important differentiating point was the right indigenous peoples had to exploit forest resources.[40] Collectors in Sarawak were relatively free to operate this part of the commodity chain – and, importantly, the trees supplying gutta percha could be cut down without detriment to any local indigenous need. It had no food value, any cultural or subsistence trade requirements, and practical use could continue to be met as far traditional society was concerned.

39 M.C. Cleary, 'Indigenous trade and European economic intervention in North-West Borneo c. 1860–1930', *Modern Asian Studies*, Vol.30, No.2, 1996, p. 314.

40 M.R. Dove, 'The life-cycle of indigenous knowledge, and the case of natural rubber production' in *Indigenous environmental knowledge and its Transformation. Critical Anthropological Perspectives. Studies in Anthropology*. A. Bicker, R. Ellen and P. Parkes eds, Amsterdam, Harwood Academic Publishers, 2000, p. 236.

Probably the single most important characteristic was that it did not require capital (it was naturally occurring in forests, awaiting collection) when compared with some other commodities, particularly those grown on plantations where capital investment was essential and expensive. Plantations would have put gutta percha out of the range of indigenous collectors in Sarawak. Yet, they would also likely have put the trade beyond the scope of the Brookes – either because they lacked the will to spend or the finances to do so, or were cautious about allowing in foreign companies. Gutta percha could be a low risk 'cash cow' waiting to be exploited. It becomes clear below that the opportunities it provided to collectors and the conditions in which it grew were beneficial to indigenous collectors; it was not affected by resources they did not have (capital) but suited those they did have (labor and knowledge of the forest).

Had gutta percha been grown on plantations, collectors would have been required to work for employers, possibly European, Malay or Chinese – therefore totally changing the role they held in the trade and imposing an obligation to remain in one place to carry out the work. Such subservient roles and tied responsibilities were alien to most of the population. Most of the population preferred to maintain their existing traditional lifestyle and locale rather than to seek out the regularly remunerative but restrictive burdens of paid employment. By contrast, in the twentieth century, when para rubber became an important part of the Sarawak economy, it changed the lives of many families who were able to plant even small plots in their gardens; yet, as one Iban has said '*Our feet were stuck by the rubber*' because they had to remain where their rubber was planted if they were to benefit from it and, as a result, traditional Iban migrations were constrained.[41] In a sense, gutta percha collecting was as much about not choosing as it was about choosing – collectors could be engaged in the trade but chose to do so without necessarily needing to become dis-engaged from links to their traditional way of life. Gutta percha collectors could opt in or out of the trade according their own perceived interests. Indirectly, of course, once they traded gutta percha for barter or were linked into credit arrangements, they began a process of social and cultural as well as economic change but collecting gutta percha did not fix collectors in one geographic location or to one means of subsistence. This is a crucial feature: gutta percha gave rise to a commodity chain that involved indigenous collectors in a globalized trade that did not require them directly to sacrifice their social, cultural or economic traditions to exploit a new opportunity; they did

41 V.H. Sutlive, *Tun Jugah of Sarawak: Colonialism and Iban Response*, Kuala Lumpur, Sarawak Literary Society, 1992, p. 31.

not need to settle down and establish plantations, plant invasive crops, move to new grounds or to settle for long terms in one spot.

Just as importantly – in those years when the price for gutta percha fell dramatically or when there were other reasons for being engaged in other activities – collectors were able simply to ignore gutta percha trading completely. In part, this was an economic choice. Had they been tied to plantations or to crops that required regular maintenance or some form of commitment, this would again have left them tied to their investment or to a European or Chinese employer. Collecting forest produce was also consistent with swidden agriculture since, for some groups like the Iban, this could require the burning of old growth forest and it was reasonable to collect forest produce from areas which would in any case at some point need to be cleared. In fact, swidden agriculture by its very nature required only seasonal labor obligation from the male population, leaving them free to undertake more extensive forest produce collecting expeditions like that required for collecting gutta percha. Sago-dependent populations, such as those along the coastal periphery in Sarawak, were less free to do this because that commodity did not have the same seasonal 'down-time'.

It is unclear when the first shipment of gutta percha was sent from Sarawak, although the Sarawak Gazette records that 50 pikuls were exported in 1848, and that 90,000 pikuls were exported between 1854 and 1874.[42] About 3,500 pikuls were reported to have been exported to Singapore in 1853.[43] Figures for the period up to 1870 are incomplete, although they were found for the years 1863–67 for this survey, but apart from the 1848 and 1853 figures above, earlier figures are missing. McDougall describes gutta percha in 'The Productions of Borneo' in a letter dated July 1851, so the trade was well underway by that date.[44] The operation of the trade over the next few years is not recorded in any detail, but the Consular Report for the trade year 1865 noted that reduced exports were said to have been caused not by a failure of supply or production, but a falling off in demand due to the late commercial crisis in England and Singapore, as discussed previously. If there were a lack of supply in future years, it was said, this might arise due to a lack of labor arising from 'scarcity of the native population and their unformed habits of regular labor' or when the 'jangal [sic] productions on the main water routes have been collected and people are compelled to journey farther into the interior to obtain the desired

42 SG 1 May 1875.

43 *Sydney Morning Herald*, Sydney, Charles Kemp and John Fairfax, 3 October 1854, p. 4.

44 H. McDougall, *Letters from Sarawak: Addressed to a Child*, London, Thomas Priest, 1854, pp. 53–54.

supplies – the prices of such commodities will then of course rise'. The decline in exports was paralleled by a decline in imports in cloth and brassware (that is, in prestige items made of brass, discussed in later sections).[45] The connection between BCL and Sarawak has been noted in the previous chapter, and 1864 was the year in which BCL and Silvertown formed their alliance. In 1865 and 1866 there were direct exports of gutta percha to both Singapore *and* the UK, with gutta percha being the chief export in both those years – the UK exports were most likely those contracted by BCL. In later years, as more of the trade was handled by Chinese traders, it was almost exclusively exported to Singapore – thereby reflecting developments in the commodity chain.

As noted in Chapter 4, in 1869 BCL gave an undertaking to provide gutta percha in exchange for shares in the West India and Panama Cable Co.[46] As seen from the export figures at Figure 8, 1870 was the most significant year for Sarawak gutta percha exports; much of the West India and Panama Cable was probably made with Sarawak gutta percha. The $135 price per pikul of gutta percha reported in 1870 was a significant increase from the $10 reported in 1866.[47] The 1870s were important for the gutta percha trade and this contributed directly to the expansion of bazaars and trade generally in Sarawak. For example, the important station of Sibu, established in 1862 on the Rejang river, was a major export point for gutta percha, beeswax and other forest produce. By 1871 the Gazette was reporting that around $20,000 of trade was sent from there each month by Chinese traders who went there especially looking for gutta percha.[48] The opening of bazaars along the Rejang river and the quest by Chinese traders for the now valuable gutta percha was a physical manifestation of the spread of global communications and the commodity chain linked to it – the global had moved to the local even at distant, interior forests, in a process facilitated by Chinese intermediary trade networks. See Plate 7 below.

Early in 1871, a report in the Sarawak Gazette illustrated the flow-on effect to the wider economy of the good prices fetched by gutta percha. Interestingly, security problems continued but the good prices being achieved for gutta percha seemed to have been sufficient to have overcome even those problems:

> The year 1870 saw a considerable increase both in the trade and revenue returns ... In the early months of the year the Dyaks of Rejang caused some disturbances and considerable anxiety ... These disturbances had

45 Consular Report on the Trade of Sarawak for 1865, BSP 1866 [3669], p. 268.

46 BCL Minutes, 22 August 1869.

47 Consular Report on the Trade of Sarawak for 1870, BSP 1871 [C.429], p. 805.

48 SG 24 January 1871.

PLATE 7 *Examining the quality of gutta percha at a Sarawak Chinese trader's warehouse.*
Note: Despite its significance to global communications and the detailed analysis of gutta percha that was undertaken by scientists, the product as it arrived in the local market place had modest beginnings. The photograph illustrates the importance of Chinese traders and trade linkages and labor in the trade, notwithstanding the dependence upon indigenous collectors and traders. Here, samples are being examined by a representative of Norddeutsche Seekabelwerke (NSW), the German cable company, in the early twentieth century.
Source: Archive NSW GmbH, Germany, with permission.

> but little effect on the trade, as gutta worked by the enemy found its way to the Chinese market.[49]

Demand for gutta percha continued to grow but the Sarawak Gazette recorded that 1873 trade compared 'unfavourably' with 1872 with a marked falling off in most areas of import and export. It was noted that almost all outstation trade was stopped due to an outbreak of cholera that deterred people from working forest produce and drove people into the hinterland.[50] Trade for 1874 continued to fall; the Treasurer's report to the Rajah concluded that, while ever Sarawak's export trade continued to depend chiefly upon 'jungle produce', its

49 SG 28 February 1871.

50 SG 2 March 1874.

trade returns would continue to fluctuate according to the condition of the inhabitants and the security situation. The Sibu station report for 1874 noted that cholera and disturbances of the population were particularly detrimental to the gutta percha and jungle produce export trades.[51] That, combined with any falling prices, would cause a material decline in the trade figures.[52] In 1875, the trade of Rejang river was 'completely paralyzed' by smallpox and it was expected to be sometime before confidence would be restored.[53] The annual report from the Treasurer for 1875 re-confirmed his findings for the previous year: small-pox and cholera and the disturbed condition of the country had again had a great and damaging impact on the collection of forest products and a consequent fall off in trade.[54] Even after the outbreak of cholera or smallpox itself had passed, the disruption continued as people fled further into the forest.[55] In August 1877, forest produce awaiting transport to Kuching was delayed as shippers were afraid to go to Kuching because of the cholera problem. Trade in the up-river areas was depressed as the epidemic had caused people to flee and it would take a year or two for a steadier trade to return.[56] The Gazette concluded that the most serious event that marked the year 1877 was the cholera epidemic that traversed the whole country.[57] Chew notes that reduction of trade in forest produce was most apparent when cholera struck.[58] There was another outbreak in 1888, reducing exports, which had a severe impact on the livelihood of shippers.[59] Knapen suggests that it was a response to smallpox (or rumour of it) that led the Punan to practice 'silent barter' (trading exchange without face-to-face contact). He notes 'economic life could be completely disrupted by the presence of smallpox or cholera epidemic'.[60]

By 1877, the Sarawak Gazette thought that recent rises in the price of gutta percha would have a beneficial effect on the security situation because it would 'produce wants' on the part of the Dyaks by inducing them to trade.[61]

51 SG 16 March 1875.

52 SG 16 March 1875.

53 SG 4 December 1875.

54 SG 20 May 1876.

55 S. St. John, *Life in the Forests of the Far East*, London, Smith Elder and Co., 1862, Vol. II, p. 23.

56 SG 27 September 1877, p. 70.

57 SG 26 February 1878, p. 8; it was also commented upon by H. De Windt, *On the Equator*, London, Cassell, Petter, Galpin & Co, 1882, p. 40.

58 Chew, p. 102.

59 SG 2 September 1889, p. 131.

60 Knapen, pp. 146, 186.

61 SG 16 January 1877.

But lest readers became too excited, the paper observed a few months later that it was these high prices of raw produce and the general prosperity of the country that had caused a considerable increase in the importation of rice and padi (cheaper, unhusked rice). Imports of rice rose tenfold from $12,011 in 1865 to $124,424 in 1876 the paper informed with some disapproval, because collecting raw produce had drawn the populations away from their agricultural activities.[62] Put in a theoretical sense, being part of the world-system, when considered at the local level, manifests itself as both a strength and a weakness. Exports of gutta percha continued to improve in 1878 as well; in fact, it was mainly gutta percha and sago that showed improvement, most other 'jungle produce' having decreased.[63]

Exports improved during the 1880s and by 1898 the market demand for gutta percha was at last beginning to move again; exports from Singapore doubled in that year, and increased significantly from Sarawak. The Rejang river continued to be an important source of gutta percha. The Resident of the Upper Rejang reported that Chinese traders were keen to take advantage of the very high prices of gutta percha and that as a result they were willing to take the risk of entrusting valuable cargoes of it onto small vessels, the Chinese 'Bandongs' – with two vessels leaving in September taking shipments worth $4420 and $10,710.[64] The shift in 1898 foreshadowed a dramatic turn in the gutta percha market brought about by the Pacific cable ventures. With the dramatic rise in demand and price of gutta percha around the turn of the century came a dramatic rise in gutta percha exports from Sarawak.

In January 1901, the Sarawak Gazette reported that 'The value of Gutta exported has been phenomenal ... more in the first nine months of 1900 than in the whole of 1899' and the year 1899 had itself been a good year for gutta percha exports.[65] Once the needs of the Pacific and other cables were met, it was inevitable that demand and prices would fall – and in 1902 exports from Sarawak reflected that. However, as had occurred in other years, there was another, local, factor that returned to reduce exports – cholera. The outstation reports note that the working of gutta percha and other jungle produce had been interfered with by the outbreak, export levels being less than half that achieved in the peak of 1899/1900 – this contrasted with the relatively marked but more steady decline in exports from Singapore. In 1902 there was a fall in

62 SG 16 April 1877, p. 26.

63 SG 24 June 1878, p. 42.

64 SG 1 November 1898, p. 208.

65 SG 2 January 1901, p. 1.

the number of people working gutta in Sarawak.[66] In some areas, the impact was severe, with one area recording no trade returns at all. These two factors coincided with a fall in the price and demand for gutta percha.

Over the years, the Sarawak Gazette and the Consular Reports at times suggested that the reasons for the fall in Sarawak's gutta percha exports were a scarcity of supplies, or that the gutta had been 'worked out' due to the non-sustainable way in which it was collected and that 'no provision is made for the future by planting'.[67] The trees appear to have been worked out in certain areas and the means of collecting was destructive; however, the idea that supplies as a whole were 'worked out' in Sarawak was not supported by the considerable rise in gutta percha exports that occurred between 1899 and 1903. Such levels were very significantly higher than those that had occurred for many decades, highest in fact since the peak years of 1870 and 1864, which does not suggest that overall the resource had been destroyed by the turn of the century. Some of Sarawak's gutta percha exports were supplied by sources in neighbouring Dutch territory in Netherlands Indies as traders brought produce into Sarawak (see following chapter), and the Residents Reports in the Gazettes continue to mention local supplies coming to the market.

In 1903 exports of gutta percha continued to fall, and to fall dramatically, due to a fall in market demand. The great days of telegraph cable manufacture using gutta percha insulation were easing (although they had not ceased altogether). The Superintendent of Customs, in his report to the Treasurer for the trading year 1903, noted 'cable companies have apparently ceased buying'.[68] Gutta jelutong and other forms of 'lesser' gutta became alternative exports, but they were never as significant as was gutta percha either in economic or in social terms. The demand and price never again returned to the peaks seen in the previous half century. Gutta percha remained an item of trade for many decades to come, but it never again became the 'phenomenal' export it once was.

The contribution of gutta percha to the Sarawak economy, compared with other sectors of the economy for the years 1870–1903, when the great era of gutta percha and the submarine telegraph peak, is depicted in Figure 9

The very scale of the gutta percha export figures suggests the likely hundreds of thousands of trees that annually needed to be located and felled – and the product then collected and boiled, carried overland, traded, shipped to outports and then to Kuching and Singapore – revealing a degree of labor that engaged much of the population. Given the relatively low population of the

66 SG 3 March 1903, p. 54.

67 Consular Report on the Trade of Sarawak for 1885, BSP 1887 [C.4915] [C.4923], p. 193.

68 SG 3 May 1904, p. 99.

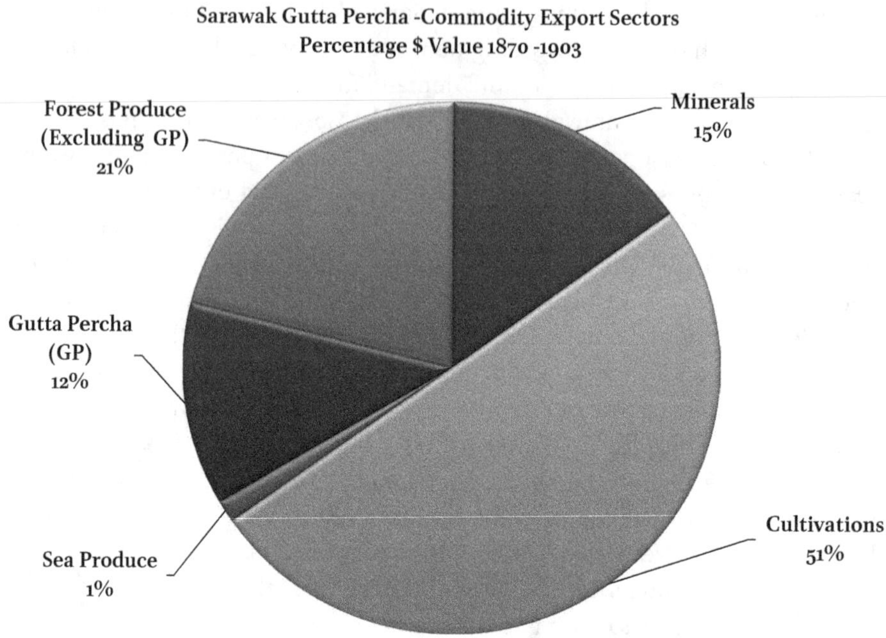

FIGURE 9 *Gutta Percha as a proportion of Sarawak trade 1870–1903 compared with the key export sectors.*
Note: The gutta percha trade on its own almost equalled the entire value of Sarawak's mineral exports, which the Rajahs and the Borneo Company (BCL) had valued.
Source: Sarawak Gazette (SG), various issues 1870 – 1904, and Consular Report in British Sessional Paper (BSP) 1871 [C.429], p. 809. See Appendix 2, Table 2 for statistical details.

country, this would suggest that at almost any location or period, the gutta percha trade was a feature of Sarawak life during the period studied here. Whatever exploitative intentions might have come from the external market, local actors responded to the opportunities for trade in ways that were unique to them, but that were to have both material and social effects which are considered in subsequent chapters. Or, as Howe concluded, colonized cultures were capable of richly adaptive patterns of response that cannot be reduced to a 'simple dialectic of domination and resistance'.[69]

Sarawak was a far less important entrepot than Singapore, with its own peculiar limitations and, arguably, its own strengths in the export trade. It was

69 S. Howe, 'David Fieldhouse and "Imperialism": Some historiographical revisions', in *Managing the Business of Empire: Essays in Honour of David Fieldhouse*, eds P. Burroughs and A.J. Stockwell, London, Frank Cass, 1998, p. 223.

rapidly drawn into the global market by the demand for gutta percha and the fluctuations in the market occasioned by the commercial activities of the telegraph cable companies and of BCL – and its pattern of exports to some extent tracked the patterns seen out of Singapore. Yet, because of the interests of BCL, Sarawak was able to benefit from the submarine telegraph cable industry. Indeed, the role of BCL in Sarawak and its connection with the cable producer, Silvertown, meant that in the 1860s and 1870s at least, Sarawak was, of all the region's countries, probably the most influenced by the changing demand for global communications and the commerce of submarine telegraphy. In some years, however, the conditions in Sarawak itself meant that its gutta percha exports suffered. There were other peculiarities about Sarawak that both energized and impeded the gutta percha trade – these factors are considered in subsequent sections.

The following chapter explores the key stages of production by setting out a survey of each of the groups engaged in those processes – in this context more accurately described as 'collection' and distribution. The chapter then considers the various strategies and responses of those at the Sarawak sections of the gutta percha commodity chain.

CHAPTER 6

Operation of the Gutta Percha Trade in Sarawak

> Everyone capable of doing so had engaged in the trade; and so long as the present high rates are given, the native dealers and collectors will continue to bring it to market.[1]
>
> CONSULAR REPORT ON THE TRADE OF SARAWAK FOR 1869

This chapter sets out the key stages of collection and distribution of gutta percha from its original sources in the forest through to the links with the outside economy: the groups, network relationships and mechanisms that were developed to cope with the peculiar challenges and opportunities that existed in the Sarawak trade environment. The operation of the trade was in part dependent upon the use of local, ecological knowledge by indigenous traders and their connections with Chinese bazaar and river traders who provided the essential links with the external economy through clan connections with Singapore traders. The impact of the geographic environment, overland trading paths and the collector's use of omen and the importance of social capital and risk taking is also examined in this chapter. Finally, the operation and significance of credit and barter in the trade is set out. In effect, the chapter suggests that the study of local factors is important in understanding the operation of the world-system.

The Trade Network

The survey begins with an overview of the key groups involved in the trade, based in part on a structure suggested in Dunn's widely recognized analysis of the subsistence economy of forest zones. In this model, Dunn concluded that inhabitants in these areas need an intimate, particularistic acquaintance with a vast range of information. The interactions between inhabitants and

1 Consular Report on the Trade of Sarawak for 1869, BSP 1871 [C.343], p. 7.

 | DOI 10.1163/9789004357280_008

their environments formed an 'ecological-economic system'.[2] That system, and the conditions in Sarawak in the nineteenth century, shared some common aspects: including the capacity of inhabitants as traders being able to identify forest species of economic value.[3]

In Dunn's model, the produce *Collector* has a close association with the forest and great knowledge of its resources, along with an ability to live and work there, but tends to be locked into a forest-dwelling, collecting life. Collectors are skilled traders since they also sell the produce they collect. To sell their produce, Dunn proposes the Collector has dual roles as both collector and trader.[4] To do this, the Collector/Trader needs to be informed about market demand and to establish a trading relationship with what Dunn calls the *Secondary Trader* to whom they transport and trade produce. The *Secondary Traders* were usually Chinese (and sometimes Malay) traders living in riverside towns or on river-based boats, who acted as middlemen between the Collectors and the other group of traders who deal with the external market, called *Tertiary Traders*. Secondary Traders have a general understanding of the ecology of the area (growing conditions, seasonal variations and so on) which they learn from local collectors, but are aware of market needs which they become aware of from their relationship with the Tertiary Trader to whom they trade produce. The *Tertiary Trader* is engaged in import/export, handles a variety of manufactured goods and raw materials, some of which are forest products. The Tertiary Trader has direct connections with regional trade.[5]

There were peculiarities in Sarawak that required an adjustment to the type of trade network described by Dunn. The inland geography of Sarawak, its steep hinterland, lack of road network and reliance upon river transport meant that there was a great dependence upon coastal and riverine shipping for all forms of commerce. Also, barter goods arriving from Singapore were an important item of trade because few goods were manufactured in Sarawak and most had to be imported and then shipped to river ports for upriver or overland trade. This meant that shipping was very important to Sarawak for import as well as export – and imports provided the goods for barter in exchange for produce. Although he did not propose it as such, Dunn's model is an important theoretical

2 F.L. Dunn, *Rain-forest Collectors and Traders: A Study of Resource Utlilization in Modern and Ancient Malaya,* Monographs of the Malaysian Branch, Kuala Lumpur, Royal Asiatic Society, No. 5, 1982, pp. 65, 99.

3 Dunn, pp. 55, 58.

4 Dunn calls Collectors in this latter capacity '*Primary Traders*'; that title is not adopted here as, while the separation of roles is clear, the dual nomenclature can become confusing.

5 Summarised from Dunn, pp. 99–103.

bridge between 'systems' issues highlighted by the world-system approach – the connections can be identified using commodity chain analysis – and the need to understand local responses. However, to achieve this bridge between the global and the local, it is necessary to identify the actors in the Sarawak network who performed the role of actually transporting goods for export and importing goods for barter. That is, to extend Dunn's network directly to the external network with the actors who performed this role. Therefore, an additional Sarawak 'node' or actor should be that of *'Shipper/Carrying Trader'*. In many cases, these shippers could act as both carriers and traders; that is, an addition to Dunn's model – in effect they become a link into intra-regional/international trade of the model proposed by Dunn. The *Shipper/Carrying Trader* was a key player in pulling Sarawak into the web of the world-system, an important actor on the local stage. *The Shipper/Carrying Trader* managed freight and trading activities, handled manufactured (imported) goods and raw materials (for export); managed credit operations; understood riverine, coastal or regional shipping patterns; travelled between Sarawak and regional ports; managed credit, trading and cross-cultural relations and sometimes engaged in international trade. Adding the *'Shipper/Carrying Trader'* to the model proposed by Dunn might be applicable to other settings or other studies in global history where transport and geographic challenges were a factor in economic and social development.

Dunn's actors performed roles akin to the 'node' in a commodity chain: this provides a useful model in understanding the stages of the collection and trading of gutta percha (and forest produce generally) in Sarawak. Gereffi et al. suggest that analyzing a commodity chain shows how production, distribution and consumption are shaped by social relations characterizing the sequential stages of the chain.[6] Social relationships and networks are needed to form the bonds of trade, in this case both between groups of collectors or between collectors and traders. Those relationships were built upon different forms of social capital. Putnam described both *bonding* social capital and *bridging* social capital. Bonding social capital refers to benefits arising from connections or sociological 'superglue'.[7] In this case, it could be between like groups (clans, ethnic enclaves, or in Sarawak it might be longhouse communities or groups such as Kayan or Iban, or Chinese speech groups such as Teochiu). Strong bonding capital is a conceptually useful way of understanding how traditional laws

6 G. Gereffi and M. Korzeniewicz (eds), *Commodity Chains and Global Capitalism*, Westport, Praeger, 1994, p. 2.

7 R.D. Putnam, *Bowling Alone*, New York, Simon & Schuster, 2000, p. 23.

(*adat*) enabled different groups to share or protect rights of access to forest resources, or to conceive the traditional enmities between groups who did not share a bond. A difficulty, conversely, might be the exclusion or even outright hostility to others from outside the group. Bridging capital on the other hand was the benefit gained from relationships with outsiders; that is, it 'bridged' the difference between 'us' and 'them'.[8] Bridging capital is a concept that helps to contextualize the nature of relationships between, for example, indigenous traders and Chinese traders; or between different indigenous groups who might join up in groups such as Iban and Punan searching for gutta but sharing little in the way of longer term bonds or social structure. In that sense, trade relations can both respond to – and promote – formation of social capital, particularly *bridging* social capital.

The role of each of these groups (Collectors, Secondary and Tertiary Traders and the Shipper/Carrying Traders) in the commodity chain and the importance of social capital within and between these groups is described below.

Gutta Percha Collectors

While often forgotten in distant metropolitan centres of government and commerce, in some ways these were the people at the core of global submarine telegraphy.

Understanding the forest and the likely locations of trees supplying gutta percha was one of the most important skills needed by gutta percha collectors – in Sarawak this was a skill that excluded outsiders from taking over the trade. Knowledge of where best to find sources of supply and being able to negotiate with local populations to find the best place to collect gutta percha was an essential part of the collecting expedition. Navigating through the forest and being able to subsist in the forest for months on end during expeditions required skill and endurance. Sparse populations in Sarawak made it all the more important to be able to cope for long periods on what the forest could provide in the way of food and cover since it was not always possible to rely upon locals to provide hospitality or supplies. As Dunn concludes, forest inhabitants regard themselves as part of the forest.[9] Again, with its long history of trade in forest produce, Sarawak's population was well geared to meet the demands of this new market. A useful concept to understand this knowledge, and the social

8 Ibid.

9 Dunn, p. 23.

framework within which this capability is embedded, is that of 'local knowledge' or as it is sometimes termed, 'traditional knowledge'. One definition of this concludes that it is:

> A cumulative body of knowledge, know-how, practices and representations maintained and developed by peoples with extended histories of interaction with the natural environment. These sophisticated sets of understandings, interpretations and meanings ... encompass language, naming and classification systems, resource use practices, ritual, spirituality and worldview[10]

In other words, the skills associated with trading and negotiating the natural and commercial environment were embedded in the spiritual and historic patterns of belief at the bedrock of the local collectors' world. While the Chinese traders moved into up-river and other trading points to deal with collectors, there is no evidence that they became directly involved in collecting gutta percha in Sarawak. Indeed, they usually moved into new areas after they had been opened up to some form of at least rudimentary protection or support by the Brooke administration. The density of the Borneo forest and the ease with which newcomers could become lost, along with the widespread fear of the 'headhunters', shaped the Sarawak commodity chain in a unique way. In Sarawak only those willing to take on the risk of the unknown forests and face up to the risks of hostilities over disputes about resources, were able to hold their position in the collecting 'nodes' of the commodity chain. The risks taken and knowledge required to manage that risk, as well as to locate sources of gutta percha, essentially combined to exclude all but indigenous collectors from the trade.

The indigenous populations of Sarawak collected and traded forest produce with the long-distance networks wherever possible and had long done so as a means of dealing with shortages of food or to trade for items such as salt, cloth, iron to make tools or weapons, or prestige goods. Despite this, most indigenous communities lived within a largely subsistence economy based upon the consumption of rice or sago; or, in the case of the Punan, on hunter-gathering activities in the hinterland. This did not provide for an easy existence. Even in the 1960 Sarawak census, the indigenous population was assessed as living a basically subsistence agriculture life, with rubber at times

10 International Council for Science, ICSU *Series on Science for Sustainable Development No. 4: Science. Traditional Knowledge and Sustainable Development,* Paris, 2002, p. 9.

being a cash crop.[11] Reinhardt concluded that Sarawak was largely a subsistence economy until well into the twentieth century when earnings from rubber and oil gained a foothold in the nation.[12] Freeman, in his very extensive study of Iban life and livelihood in the mid twentieth century, for example, found that 'it must constantly be borne in mind that we are dealing with what is primarily a subsistence economy' and that the 'subsistence economy of the Iban is one of scarcity rather than plenty ... there are few, if any, Iban families which have not, at some time or other, found themselves ... with insufficient padi for their barest needs.[13] In parallel with the need to supplement a harsh, subsistence living, a feature of Sarawak (and Borneo) indigenous economies was the fact that valued forest produce might also be bartered in return for prestige items – called *pusaka* – such as items made of brass ('Brassware' in the trade statistics), antique ceramic jars, certain types of beads or ornate cloth (see next chapter). That is, the subsistence and the prestige trades were both key parts of the local-external trade network and had been so for centuries (however paradoxical it may have seemed to outsiders).

While collecting some types of forest produce might be opportunistic activities, gutta percha was most often obtained on lengthy and distant, planned expeditions. As a result, knowledge of the forest and a capacity to live and work in it for long periods were essential when collecting gutta percha, as well as knowledge of where to find sources of supply or the capacity to negotiate with others for assistance in finding local sources. At times groups worked together (such as Kayan and the Punan), or combined to form larger expeditions.[14] Usually, collectors travelled in groups with various numbers being recorded (for example, a group of 45 Iban, and a later group of 200).[15] They were sometimes divided into small sub-groups involving fewer than ten men; Seeligmann et al. suggest 'gangs of three or four', but Bock mentions groups of twenty to thirty who preferred to be in these larger groups in order to obtain the benefit of mutual protection against 'attacks from dyaks'.[16] In a practical sense, most

11 Y.L. Lee, 'The Population of Sarawak', *The Geographical Journal*, Vol.131, No.3, Sep. 1965, p. 347.

12 J.M. Reinhardt, 'Administrative Policy and Practice in Sarawak: Continuity and Change Under the Brookes', *The Journal of Asian Studies*, Vol. 29, No. 4 (August), 1970, p. 858.

13 D. Freeman, *Report on the Iban*, University of London, London, The Athlone Press, 1970, pp. 264, 266.

14 Sarawak Gazette (hereafter SG), 3 January 1893, p. 31.

15 SG 1 July 1884, p. 75.

16 See for example SG 1 August 1873; SG 2 April 1883, p. 38; T. Seeligmann, L. Torrilhon and H. Falconnet, *Indiarubber and Gutta Percha. A Complete Practical Treatise*, 2nd English edn., London, Scott, Greenwood & Son, 1910, p. 328; C. Bock, *The Head-hunters of Borneo.*

expeditions were planned around other seasonal activities, especially padi planting. The noticeable exceptions to this were during the years when gutta percha prices were particularly high and the padi obligations were set aside or, alternatively, when padi harvests were poor, or padi prices were so high that people turned to gutta percha as a means of buying padi or imported rice.

Hornaday gives the following description of felling and preparing the gutta percha in the forest:

> Went with Gumbong to see how he collected gutta. A mile from the house he found a gutta tree, about ten inches in diameter and, after cutting it down, he ringed it neatly all the way along the stem, at intervals of a yard or less. Underneath each ring he put a calabash to catch the milk-white sap which slowly exuded. From this tree and another about the same size, he got about four quarts of sap, which, on being boiled that night for my special benefit, precipitated the gutta at the bottom in a mass like dough. The longer it was boiled, the harder the mass became, and at last it was taken out, placed upon a smooth board, kneaded vigorously with the hands, and afterward trodden with the bare feet of the operator. When it got almost too stiff to work, it was flattened out carefully, then, rolled up in a wedge-shaped mass, a hole was punched through the thin end to serve as a handle, and it was declared ready for the trade. I have seen the Dyaks roll up a good-sized wad of pounded bark in the centre of these wedges of crude gutta, in order to get even with the traders who cheat in weight, but I have also seen the sharp trader cut every lump of gutta in two before buying it. If he found bark, you may well believe he did not pay for it at the price of gutta. The crude gutta has mottled, or marbled, light-brown appearance, is heavy and hard, and smooth on the outside.[17]

His description is similar to that found in other sources.[18] One report said collectors could quickly identify species by the quality of the bark, and would

---

A Narrative of Travel up the Mahakkam and Down the Barito, 2nd ed., London, Sampson, Low, Marston, Searle & Rivington, 1881, p. 118.

17 W.T. Hornaday, *Two Years in the Jungle. The Experiences of a Hunter and Naturalist in India, Ceylon, the Malay Peninsula and Borneo*, 7th edn, New York, Charles Scribner's Sons, 1904, p. 433.

18 See for example, A. Combanaire, *Au Pays des Coupeurs de Têtes: À travers Bornéo*, Singapore, Editions Pagode, 1902, Reprint 1993, p. 202; E.H. Gomes, *Seventeen Years Among the Sea Dyaks of Borneo*, London, Seeley & Co. Limited, 1911, p. 235 or C. Hose and W. McDougall, *The Pagan Tribes of Borneo*, 2 Vols, London, Macmillan and Co, 1912, Vol. 1, p. 151.

bother felling trees only if they were around one metre circumference.[19] It was thought better to collect the gutta percha in the morning, when it flowed more freely, and that expeditions undertaken two months after the end of the wet season would obtain a better harvest.[20] Collection of forest produce was an essential practice in most of Sarawak's indigenous societies since agricultural cultivation was not a guaranteed source of food in most areas because of poor soil. Different groups were formed into different social structures and exhibited different patterns of forest trade. As a result, there were differences in the extent and manner of their involvement with intermediaries like Chinese traders, and that in turn influenced their involvement with global trade. In 1875, the Sarawak Gazette advised that Sea Dyaks (Iban), despite their practice of 'head-taking'

> ... are clever traders and as sharp at a bargain as any Chinaman. They work nearly all the jungle produce that is brought into the market, such as rattans, gutta percha, India-rubber &c, and no one can hold a candle to them in such work.[21]

The Iban were the largest indigenous group of Sarawak – they were famous in the nineteenth century as headhunters and for the habit of young Iban men absenting themselves for months or years on long distance journeys known as *bejalai*. Kedit describes this as 'the Iban custom ... that encourages the participants to leave their villages and earn money to buy material goods, which enhance their social standing on their successful return' and Freeman considers it one of the Iban's 'most cherished customs'.[22] Wadley's definition of *bejalai* as 'any journey away from home for prestige and/or wealth that includes the possibility of an eventual return' notes too that one of the failures of the studies in Iban *bejalai* is to connect it with wider studies of labor migration.[23] With the increase in commerce occasioned by greater contact with European trade, the impetus for *bejalai* for the purposes of collecting forest produce likewise increased. Gomes noted that for Iban boys, once they reached the age of around fifteen, it was a mark of pride when they could carve their own

19 Seeligmann, et al., 1910, p. 329.

20 Combanaire, p. 192.

21 SG 3 November 1875.

22 P.M. Kedit, *Iban Bejalai*, Kuala Lumpur, Sarawak Literary Society, 1993, p. 1; Freeman, *Report on the Iban*, p. 24.

23 R.L. Wadley, 'Environmental Challenges in South-East Asia. (Book Reviews)', *Borneo Research Bulletin* (Annual), Vol. 30, 1999, p. 168.

dug-out canoe or accompany others on expeditions for gutta percha.[24] In terms of their economic impact, the Iban journeys for the collection and trading of gutta percha and other forest produce were a form of 'labor migration' but, in terms of their sociological and even spiritual motivations, they fit accurately into the far more complex area of *bejalai*. Kedit says *bejalai* groups were usually small, but always had a leader and were aimed at purchasing material goods for heirlooms (*pusaka*), and focused around personality traits of aggressiveness, individualism and mobility.[25] Such characteristics were ideal for exploiting the forest trading environment and responding to the opportunities and risk arising with such expeditions.

Iban *bejalai* and the gutta percha trade had a symbiotic relationship – as these collectors and the commodity had peculiarities that combined to provide a forceful motivation to trade. Gutta percha was a valuable commodity, the collection of which was profitably rewarded by lengthy expeditions into distant locations, often into hostile territories. The rewards of such expeditions – wealth, the opportunity to prove bravery in hostile interactions with other collectors – or arguably, the possibility to take heads ('headhunting') – were achievements that were valorized in Iban society. In Borneo, or the region more widely, no other society sought to or could deal with, let alone reward such risky behaviour and not all were as attracted to the risks and rewards of serious gutta percha collecting:

> ... among the Dyaks the best head takers are in nearly all cases the best traders: possessing energy for the one work, they are always ready to turn it to account in the other[26]

While the quote does reflect a cultural bias, and by no means were all Iban headhunters, Wadley notes that the Iban communities place an emphasis on personal autonomy and achievement, and are highly competitive for wealth and recognition.[27] The interconnection between collecting (and trading)

24 Gomes, 1911, p. 103.

25 Kedit, p. 2.

26 SG 1 July 1871.

27 R.L. Wadley, 'Slashed and burned: war, environment, and resource insecurity in West Borneo during the late nineteenth and early twentieth centuries', *Journal of the Royal Anthropological Institute* (New Series), Vol.13, 2007, p. 112; see also C. Sather, '"All Threads are White" Iban egalitarianism reconsidered' in *Origins, Ancestry and Alliance: Explorations in Austronesian Ethnography*, eds C. Sather and J.J. Fox, Canberra, ANU E Press, 2006, pp. 73–112.

gutta percha, the pursuit of wealth, headhunting and *bejalai* in Iban history could be underlined in linguistic evidence. Richards, in his Iban-English dictionary, proposes that the Iban word *Jangkang* could mean both 'trees yielding gutta... (palaquium)' and a '... collection of head trophies hung over the hearth ...'.[28] According to Williams, 'jangkai' or 'jankar' varieties in Sarawak were a lesser quality gutta, used as an adulterant to redden other poor quality guttas to make it appear like the 'true gutta-percha' on the market.[29] It appears in the trade data for Sarawak from 1900 onwards. Bromeley however says that the more usual Iban term for a collection of heads is *tampun*, so the connection is not proven.[30] Some Iban collectors ventured beyond traditional collecting zones of Borneo – they took advantage of newly available and frequent regional shipping and travelled beyond Sarawak:

> Jungle produce has been somewhat scarce but the Dyaks failing to gather much here have travelled in large numbers to other countries in search of gutta, rotan, camphor and several have already returned here with the money obtained by the sale of their finds in other markets ... Mingat and Munan with 105 followers and $4,500 sailed by the 'Normandy' to Singapore.[31]

Elsewhere, Baring-Gould and Bamfylde wrote:

> ... their energies are almost entirely confined to working jungle-produce ... they go into the far interior, and this is often the cause of their getting into trouble with remote and wild tribes; they also go to North Borneo, Dutch Borneo, Sumatra, the Malay peninsula and even as far as Mindanau, in the Philippines.[32]

In 1877 Henry Murton, the Superintendent of the Botanic Gardens of Singapore, was called in to undertake a survey of the extensive 'dyak' exploitation of the

28 A.J.N. Richards, *An Iban-English Dictionary*, Oxford, Clarendon Press, 1981, p. 123.

29 L. Williams, 'Lactiferous plants of economic importance v. Resources of gutta-percha-*Palaquium* species (Sapotaceae)', *Economic Botany*, Vol.18, no.1 (January-March), 1964, p. 13.

30 P. Bromeley, Universiti Teknologi MARA (UiTM), Sarawak, personal communication re use of 'jankang' and 'tampun' in Iban language, 7 April 2010.

31 SG 1 January 1890, p. 2.

32 S. Baring-Gould and C.A. Bampfylde, *A History of Sarawak under its Two White Rajahs, 1839–1908*, London, Henry Southeran, 1909, p. 376.

gutta percha resources in the state of Perak.[33] At the turn of the nineteenth century, coinciding with the demand and high prices for gutta percha for the Pacific and Atlantic cables, 'dyaks' from Borneo were also reported to be active as far away as Kelantan and Terengganu on the Malay peninsula.[34] All of these locations effectively covered the area marked in Map 4 of the gutta percha supplying region as sought by submarine cable producers. The significance of Iban values and their involvement in the gutta percha trade is examined in greater detail in Chapter 8.

Another group, the Punan, unlike the Iban, were less likely to travel far beyond remote forest regions and did not practice swidden or other forms of agriculture. Although they had a reputation for living entirely as hunter-gatherers in the mountainous interior regions, they were in fact highly skilled collectors and traders of forest produce.[35] Some avoided contact with outsiders except for trade and, whilst they were highly mobile, living in temporary huts, some less commonly visited bazaars but instead often (although not exclusively) acted as guides for other collectors, or dealt with traders who travelled to them. Some Punan were said to be reluctant to trade in a face-to-face relationship with outsiders, and practiced 'silent barter'. Beech tells of a Malay trader who dealt regularly with a Punan group. The trader announced his arrival by tapping on the buttress of a tree – the Punan would then examine his trade goods. After making their selection, they then gave the trader a piece of rattan marked with notches signifying 'the number of days in which they could produce the requisite amount of gutta, or whatever jungle produce was expected', keeping a similarly notched *rotan* [rattan] for themselves. The trader said the Punan would turn up on the agreed date'.[36] The Punan had more limited contact with Chinese bazaar traders, dealing instead with Chinese boat traders, sometimes using silent barter,[37] or with other indigenous traders with whom they had traditional trading connections. Furness reports on a visit to a Punan village where all but two of the men were absent hunting for camphor or gutta percha. He noted that they 'are always sought by the Sibops, Kayas, Kenyahs, and Ibans

33 I.H. Burkill, 'Botanical Collectors, Collections and Collecting Places in the Malay Peninsula', *The Gardens Bulletin, Straits Settlements*, Vol. IV, Nos. 4 and 5, August 1927, pp. 153–154.

34 J. Kathirithamby-Wells, *Nature and Nation Forests and Development in Peninsular Malaysia*, Singapore, Nias Press, 2005, p. 70.

35 See, for example, M. Dove 1990 and C.L. Hoffman, *The Punan: Hunters and Gatherers of Borneo*, Ann Arbor, UMI Research Press, 1986; C.L. Hoffman, 'The "Wild Punan" of Borneo: A matter of economics' in *The real and Imagined Role of Culture in Development: Case Studies from Indonesia,* ed. M. Dove, Honolulu, University of Hawai'i Press, 1990, pp. 89–118.

36 M.W.H. Beech, 'Punans of Borneo', *Man*, 1911, Vol.11, pp. 17–18.

37 Combanaire, pp. 233–224.

as guides in expeditions after camphor, gutta-percha, and bees-wax'[38] and the Sarawak Gazette reported that the Punan 'knew the best gutta localities.[39]

When Charles Hose went on an expedition in 1891, his group travelled with the local 'Sibops' (a Kayan sub-group) who were approached by a group of Punan who had been working 'jungle produce' and who offered to assist in collecting specimens and had sought contact in order to trade.[40] See image from Hose at Plate 8. Punan acted as guides to other collectors as well as to

PLATE 8 *Group of Punan.*
Note: A group of Punan from the Baram district of Sarawak. The Punan people were widely regarded as the most skilled at identifying and collecting forest sources of gutta percha and their assistance and knowledge was widely sought by many groups in Borneo.
Sarawak: Four Punan tribesmen. Photograph c. 1896, Charles Hose, V0037444.
Source: Wellcome Library, London. With Permission.

---

38 W.H. Furness, *The Home-Life of Borneo Head-Hunters: Its Festivals and Folk-Lore*, Philadelphia, J.B. Lippincott Company, 1902, pp. 174–175.

39 SG 1 September 1882, p. 73.

40 C. Hose, 'A journey up the Baram river to Mount Dulit and the highlands of Borneo', *The Geographical Journal*, Vol.1, no.3 (March), 1893, p. 205.

Brooke administrators.[41] Eghenter suggests that in East Kalimantan (formerly part of the Netherlands Indies) the Punan collecting camps were situated near groups of Kenyah, and notes Brosius' suggestion that Kenyah competed with each other to control trade links with Punan who in turn would break off relationships and form relationships with other groups if they felt that they had been poorly treated.[42]

The skills of the Punan emphasize the significance of traditional knowledge as a resource in itself, in that the Punan traded their knowledge by providing a service to guide other collectors. In so doing, they were operating within traditional trade networks that existed among certain collecting groups. European trade networks; the entrepot trade of Singapore and the lesser ports such as Kuching; the Chinese traders, and the shipping that criss-crossed oceans and rivers, were all key in the development of Sarawak trade. However, at their foundation, regional economies often still required the relationships that functioned in the hinterlands. That is, the global, the regional and the subsistence trade networks were interlocking but to some extent independent networks.

Aside from the Iban and Punan, other groups such as the Kayan, Kenyah, Kelabit and Murut were collectors of gutta percha and these are also referred to in consideration of specific events set out hereafter. Throughout the narrative documentation from the Sarawak Gazette, for example, these groups were very active collectors – and sometimes hostile to other groups. The populations were smaller than Iban and not all as prone to such lengthy labor migration. They were not encouraged to migrate by Charles Brooke, as had been certain Iban, although expeditions were not unknown. These groups lived in Sarawak, or crossed from Brunei or Netherlands Indies territory. Lengthy expeditions that combined collecting and trading could take months depending upon the locations of the produce and the markets; in Kayan these were known as *paji pelalau.* [43] The Kenyah referred to trading journeys as *peselai.* These practices share a similarity with the Iban *bejalai* in that they involved an orchestrated venture of forest produce collecting/trading, although not with the same emphasis upon winning social standing as a result of the expeditions that was implicit in *bejalai*, or necessarily the warfare that *bejalai* might involve.

---

41 See SG 1 June 1882, p. 41.

42 C. Eghenter, 'Towards a causal history of a trade scenario in the interior of East Kalimantan, Indonesia, 1900–1999', *Bijdragen tot de Taal-, Land- en Volkenkunde,* 157, (2001), No.4, 2001, p. 751.

43 Ibid.

The attraction of trading in Sarawak for those from across the border in Dutch territory was reported in the Sarawak Gazette in 1899, when several Kayan crossed in a large and well-orchestrated operation involving 250 people using thirty boats to travel on rivers when required. In the same month, a Chinese trader with another party of Kayan arrived, selling gutta from across the border.[44] Another large group of 500, on their third such annual visit, was reported to have travelled with 'gutta of the best quality', to the total value of $14,000 on a journey that had taken five months.[45] If such numbers are accurate they must have made an imposing presence on the local communities. More groups brought gutta percha to trade as additional territory was ceded from Brunei to Sarawak. By the end of the century, for example, Muruts in northern Sarawak were bringing large amounts of gutta percha to bazaars, taking advantage of the good prices that were being achieved in 1900 and were selling the produce to buy buffaloes.[46] Murut gutta percha collectors living in Sarawak were very active and engaged in disagreements with Iban collectors who moved into these areas to collect produce (see Chapter 8). Collectors dealt in different ways with the demands of trade and credit, and the associated risks. Denison wrote that the Kayan in the Baram region of Brunei in 1872 were taken advantage of in matters of trade, although he said it was dangerous for traders to attempt to bully them.[47]

The hunt for gutta percha was not just a physical and commercial endeavor for indigenous collectors – it was carried out in the context of spiritual beliefs and practices. Traditionally, the collection of certain forest assets was associated with highly observed, ritualized and controlled methods of collection, especially for collection of camphor (crystals in the trunk of the *Dryobalanops aromatica* tree) and edible birds' nests. Those two assets were very valuable and difficult to find and, in the case of birds' nests, dangerous to collect. For both of those assets, local elite also controlled access and collection rights and levied taxes on them from collectors. In relation to camphor, a special language was used before and during collection, with restrictions on speaking it at all at other times. It is not surprising, then, that omen and religious grounding played a part in the gutta percha trade.

---

44 SG 1 June 1899, p. 121.

45 SG 2 April 1900, pp. 64, 61; also reported in O. Beccari, *Wanderings in the Great Forests of Borneo. Travels and Researches of a Naturalist in Sarawak*, London, Archibald Constable & Co, 1904, Reprinted Singapore, Oxford University Press, 1986, p. 362.

46 SG 2 April 1900, p. 75.

47 N. Denison, 'Journal when on a trip from Sarawak to Meri on the North-West Coast of Borneo in the Brunei Territory', *Journal of the Straits Branch of the Royal Asiatic Society*, Vol.10, (December) 1883, p. 178.

Haddon observed a connection between gutta percha collection and omen. Bird omens were especially important, he writes:

> A kind of thrush (*Cittocincla suavis* – the *nendak*) is particularly useful to the Iban when looking for gutta or other jungle produce. ... Before starting on a gutta expedition they would require to see something before *beragai* (*Harpactes duvauceli*), as this is a burong tampak, that is, an omen animal that is potent for hunting. What they like is: first to get *nendak*; then wait three days while they are "owning" it; finally, to get *beragai* on the right [hand side]. This combination signifies certain success; not only would they find gutta, but would obtain plenty of it, and no harm or sickness would befall them. If, however, they went for gutta on *beragai* alone, and that perhaps appeared on the left, they would obtain a fair amount of gutta, but they would stand a good chance of some misfortune happening to them, and one of their party might fall sick or even die.[48]

According to Haddon, the beragai bird was also important to the Punan, who called it *Pengiok*, and to the Kenyah, who called it *Pengolong Lomit.*[49] Hornaday records one collector, Le Tiac, who postponed a planned six-day gutta expedition after he heard the cry of an omen bird, obliging him to return and wait two or three days:

> if he should go on after hearing the bad omen he would have bad luck – either be taken sick, cut his hand or foot, or perhaps the gutta-percha trees would not run any sap when cut ... only once did he venture to go on after hearing the bad omen, and before he returned he accidentally cut his hand with his parang.[50]

The Sarawak Gazette reported on a group of forty Muruts who had brought a large amount of gutta percha to the bazaar; a larger group was to follow but 'the chief of their districts had started down ...but returned on the third day out owing to bad omens'.[51] Augury was important because of the perceived need to be forewarned about likely success or failure; it was a form of 'divine guidance'.[52]

48 A.C. Haddon, *Head-Hunters Black, White, and Brown*, London, Methuen & Co., 1901, pp. 389–390.

49 Haddon, p. 384.

50 Hornaday, p. 432.

51 SG 2 September 1889, p. 131.

52 See Kedit, Ch. II, and D. Freeman, 'Iban Augury', *Bijdragen tot de Taal-, Land- en Volkenkunde*, 117, no.1, 1961, pp. 141–167.

Metcalf remarks on the complexity of bird augury in Borneo including its role in economic activity, social organization and warfare.[53] Gomes, who worked as a Christian missionary in Sarawak for seventeen years from 1852, also wrote of the Iban use of omen and bird augury in preparations for gutta 'hunting expeditions'.[54] Gomes spent time in the forest with Iban groups and even though a Christian himself, he wrote that he sensed a cosmological interpretation of the forest, that he 'understood why it is that the Dyaks believe that the lone forests are inhabited by the spirits of the wind and the rivers, of the mountains and the trees'.[55] Collecting activities had to be observed and respected as part of the wider pattern of beliefs that guided society. In the case of Iban, rituals were observed before, during and after return; favorable dreams before departure were required. Kedit (himself an Iban) outlines the Iban ritual observations associated with such expeditions (this is not to suggest other groups did not undertake observances):

> Iban who went on *bejalai* to collect jungle products such as gutta percha ... have to observe ritual prohibitions so as not to offend the spirit of the forest (*antu rimba*), and follow strict customs in order to survive. A leader (*tuai bejalai*) who was known for his knowledge of bird augury, jungle-craft, and leadership qualities was usually selected from among the group of 10–20 men.

Kedit adds that bird auguries were taken before and during each stage of an expedition. For example, before the journey the 'keduran',[56] the noting of dreams, and listening for the omen bird, the 'Nendak', were all processes to be followed. Once the journey was underway, for example, during the process of clearing for forest camping sites, if the cries of the *embuas*, or the *kenyalang* omen birds were heard, then the group had to move and make another camp at a different site. If they heard a *sabot*, which 'sounds like a human crying', then again, the group should leave immediately to avoid the likely threat of illness or death within the group. Before commencing work each day, the group would again listen for various omen birds such as the *tutok, pangkas, embuas* or *beragai*. If any of these were heard, then the day's activities were cancelled

53 P. Metcalf, Birds and Deities in Borneo, *Bijdragen tot de Taal-, Land- en Volkenkunde*, 132, no. 1, 1976, pp. 99–123.

54 Gomes, pp. 234, 238.

55 Gomes, p. 149.

56 A ritual performed before *bejalai*.

as continuing might result in accident or even death. Rules associated with rituals and taboos to be observed were taught by elders.[57]

Physical evidence appears that connects the gutta percha trade with *antu* (spirits) and invocation and augury during collection expeditions. Animal figurines made of gutta percha were often placed in shipments of the produce such as seen in the sketch below at Plate 9. Hose noted that Muruts made effigies of crocodiles to mark and give thanks for successful expeditions[58] and there are other references in the primary sources to this practice, such as:

> The gutta, when brought to Muka, is usually twisted into fantastic shapes, and in every cargo one mass is found presenting a rude resemblance to an alligator ... doubtless intended as a sort of charm or propitiation to the spirits of the jungle.[59]

The figurines may also have been used to indicate ownership of gutta percha awaiting collection, with one trade report suggesting that the figurines were a means by which Chinese traders identified the group of collectors with whom they were dealing.[60] The use of figurines continued into the twentieth century; the New York offices of the *India Rubber World* also had several examples.[61] Edwards told me that Telcon was receiving them in gutta percha supplies in its

PLATE 9
Gutta percha figurines.
Source: H.L. Roth, *The Natives of Sarawak and British North Borneo*, 2 Vols, London, Truslove & Hanson, 1896, Vol. II, p. 243. See also E.F.A. Obach, 'Cantor lectures on gutta percha', *Journal of the Royal Society of the Arts*, Vol. 46, 1897, p. 124 and G.L. Lawford & L.R. Nicholson, *The Telcon Story*, London, The Telegraph Construction & Maintenance Co. Ltd, 1950, p. 23 which also depict similar findings.

57 Summarised from Kedit, p. 20 and pp. 33–34.

58 C. Hose, *Natural Man. A Record from Borneo*, London, Macmillan & Co, 1926, Reprinted Singapore, Oxford University Press, 1988, p. 232.

59 F. Boyle, *Adventures Among the Dyaks of Borneo*, London, Hurst and Blackett, 1865, pp. 123–124.

60 IRW 1 January 1902, p. 149.

61 Ibid.

London factory into the 1950s and a photograph of them also appears in the history of the company, the *Telcon Story*.[62]

Aside from that above, Kedit for example says that, to the Iban, living beings (birds, plants, insects etc) have souls, as do inanimate objects such as *pusaka* ceramic jars and other heirlooms, the trade in which is discussed in the next chapter. Such beliefs were shared across other groups in Sarawak, and so collecting gutta percha took place in the context of wider belief systems and the need to accord appropriate respect to the natural and supernatural: 'Life to the Iban is a continual process of interaction between all beings, whether natural or supernatural'.[63] Walker notes that most groups in Sarawak believed that they shared the material world with a range of spirits[64] – it is the world of *antu* (*hantu*) in which collectors searched for gutta percha, so it was not surprising that the protection and support of *antu* was a key element of the trade. Such practices would make sense given that all collecting occurs in the forest.

The scale of the gutta percha trade, and the destructive collection methods (however justified given the problem of collecting the gutta percha) meant that the idea of sharing the gutta percha resource with outsiders was difficult to follow in practice. Greater contact with collectors from other areas added further pressure to any thoughts of sharing or protecting the rights to local forest resources. Yet one understanding that did remain was a shared conception of the physical–metaphysical prescriptions of the environment like that outlined. Davis and Wali noted that indigenous cosmologies place great importance on connections between the natural and supernatural worlds.[65] This shared cosmology in Sarawak enabled groups of collectors to work more collaboratively, or at least more respectfully, in a manner that could not have occurred had outsiders also been attempting to obtain valued forest produce; for example, if one imagines groups of European traders making their way through the Borneo hinterland without due regard to local omen restrictions or protected sacred sites. The trade was, then, shaped by both the harshness and difficulties of the Sarawak environment and by long established cultural and spiritual beliefs. On the other hand, resource hostilities among collectors were

---

62 L. Edwards, Manager of the Telcon Balata [and Gutta Percha] Processing Plant, Greenwich; *Telcon and Gutta Percha, Discussion with Author*, 19 July 2007.

63 Kedit, pp. 17, 15.

64 J.H. Walker, 'Hikayat Panglima Nikosa and the Sarawak Gazette: Transforming texts in nineteenth century Sarawak', *Modern Asian Studies*, Vol. 39, No.2, 2005, p. 432.

65 S.H. Davis and A. Wali, 'Indigenous land tenure and tropical forest management in Latin America', *Ambio*, Vol. 23, No. 8 (December), 1994, p. 485.

also fostered by increased collecting activity that was promoted by demands from the external economy. This is examined in more detail in Chapter 8.

Gutta Percha Traders

> Hitherto the Chinese traders in Sessing have been scattered up and down river, and have built shops where they pleased. They have now selected a site ... where all the shopkeepers will collect and build a substantial bazaar.[66]

Moving along the gutta percha trade network, the next group was the *Secondary Traders* of Sarawak, who could be divided into two broad categories – the peripatetic boat traders and the small bazaar-based traders stationed along the rivers. In the early years of Sarawak these were Malay, but very soon they were overtaken by Chinese traders. In 1872 Denison described the operation of the gutta percha trade in directly contributing to the shift from Malay to Chinese domination of the trading sector:

> In the evening, an old Brunei Chief came on board and said it was all up with the Brunei people, as the Chinese had got into the Baram and were giving the Kayans $40 per pikul for gutta, while they had been paying a lower price. It is very evident that, if the Chinese obtain a footing in this river and come into direct contact with the Kayans, the Brunei dealers' occupation is gone.[67]

Denison says that Brunei traders along the Baram river advised him that because of increased competition, profits which had in the past sometimes reached as much as 400 percent had been reduced to 100 percent. Trade with the Kayan peoples involved extending credit for up to twelve months which Denison thought was a considerable risk at that lower rate of interest.[68] In turn, the traders borrowed their money in Brunei for around two to three percent per month. The Baram river was in what became north eastern Sarawak (see Map 5), but at the time of Denison's writing it had just been ceded from Brunei and was only just being opened to the growing Chinese trade networks with links to Kuching and thence to Singapore. He also wrote about another

66 SG 1 February 1890, p. 34.

67 Denison, *Journal of the Straits Branch of the Royal Asiatic Society*, p. 176.

68 Denison, p. 178.

bazaar where ten Chinese traders had settled and 'by competition and combination' had pushed out Brunei dealers.[69] Charles Hose, writing in 1891 suggests there were Malay traders in the Baram river area, even in the late part of the century but Hose wrote that by 1900, Chinese merchants were displacing the Malay in all avenues of trade.[70]

Chinese traders were widespread even 'in the most out of the way and unfrequented places', living a financially and physically risky existence and were dependent upon credit to survive financially.[71] Credit was provided to them by other Chinese traders or *towkays* (Tertiary Traders) who were linked to Kuching and Singapore traders. The secondary traders did not live off the forest as did the local indigenous groups, although there is evidence that some Chinese traders inter-married with the local population. They often competed, even physically fighting to obtain gutta percha from collectors who arrived at the bazaar.[72]

The isolation of the more distant collector groups provided an opportunity for secondary traders, but as Richards has pointed out, these traders were largely left to their own devices and as pioneers in difficult situations they filled an important role as providers of both products and market services, as well as sources of credit.[73] When collectors were willing or able to travel to Kuching or to the main outports, they obtained a better price for their gutta percha, or obtained goods at a more competitive price, but not all groups could or wanted to do this. In turn, traders could profit from that situation; whether it was exploitation or the provision of a service or simply the operation of commerce depended upon the customers' satisfaction. Hornaday, travelling up the Batang Lupar river in the 1870's, gave an example that illustrates the extent to which the market varied between Kuching and up-river:

> A Chinese trader came to the village to-day to trade rice for gutta-percha, wax etc. ... Instead of taking their gutta down to Kuching, where they could sell it at 60 cents per cattie (1 1/3 pound) and buy rice at ten to eleven gantongs for a dollar, they loaf around the village until a sharp

69 Denison, p. 179.

70 SG 1 February 1892, p. 35; C. Hose, 'In the heart of Borneo', *The Geographical Journal*, Vol. 16, no.1, (July), 1900, p. 55.

71 SG 1 December 1874; see SG 1 July 1871.

72 SG 1 September 1890, p. 121.

73 A.J.N., Richards, Review [Rajahs and Rebels: The Ibans of Sarawak under Brooke Rule, 1841–1941], *Modern Asian Studies*, Vol.8, No. 4, 1974, p. 556.

> Chinaman comes along and takes their gutta at 37 ½ cents per pound, in exchange for rice at five gantongs to the dollar.[74]

The dominance of Chinese traders in the gutta percha and other forms of trade was almost a cliché in Sarawak society:

> John Chinaman is of course to be found where ever a cent can be made ... Gutta is at present the object of their search.[75]

These secondary boat traders often travelled up river to longhouses and, as Chew notes, they played a major intermediary role between bazaar and longhouse: it was they who were responsible for local engagement with the external economy.[76] With this contact came change. The changes were in the long run not just economic, but social and cultural – and irreversible. The traders sometimes faced dangers from aggressive locals, and from the difficulties of traversing the natural hazards. Chew retells several examples of some of these difficulties, not only of traders being robbed, but also of attacks involving loss of life, relying on their wits, bravery and tenacity to survive.[77] In addition, they incurred years of obligation brought on by debt to trade creditors. Chew notes that boat traders were often obligated by debts arising from gambling or opium (or alcohol) to the prominent *towkay* and that many boat traders were engaged in what was at best a marginal occupation.[78] The boat traders' credit connections with the Chinese *towkay*s were based on two interconnected threads: *hui* or kinship groups and, equally important, the idea of *xinyong* or trust that 'underlies the Chinese economic dealings by verbal agreement'.[79] While such links may have been unwritten (although actual debts were recorded), they were based on generations of understanding and trade and the pressures to obey such obligations must have been greatly felt by many of the impoverished boat traders. Huff notes the importance of trust, evolving in part from kinship, as a key factor underpinning Singapore's regional network of trade and barter

---

74 Hornaday, p. 433.

75 SG 30 July 1872.

76 Chew, p. 77.

77 Chew, p. 79 and see Ch. 4 of that book for examples of attacks and tactics.

78 Chew, pp. 85, 90.

79 B.T. Chee, *Chinese Overseas Comparative Cultural Issues*, Hong Kong, Hong Kong University Press, 2004, p. 181.

in the Chinese trade network, and this pattern clearly extended throughout Sarawak.[80]

Often bazaars sprang up around the forts and administrative facilities established by the Brooke administration, but some Chinese traders ventured further to more remote areas. By 1872, some Chinese bazaars were so successful that one merchant even sold champagne and soda-water.[81] In 1876, the Saribas area (east of Kuching), went from being one barely safe for trade to one where there was a rich trade in gutta and other produce.[82] How quickly the residents of Sarawak benefited from the global commerce in forest produce like gutta percha was reflected immediately in their ability to purchase rice, tobacco and sugar: the Resident in Charge at Baram district observed that large profits were made by the Chinese shopkeepers whilst the price of gutta percha was so high in Singapore.[83] See Plate 10 for an example of trade negotiation.

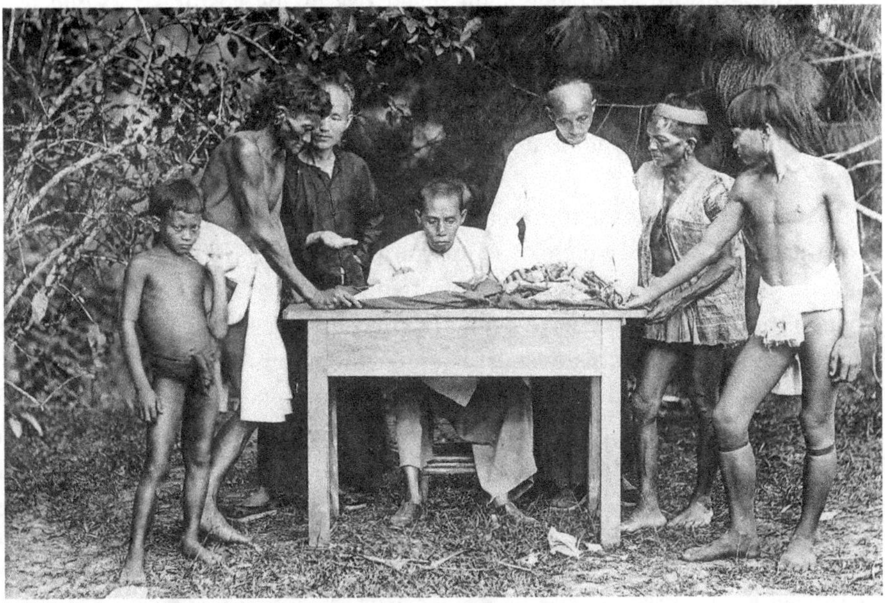

PLATE 10 *Kayan and Chinese produce trade negotiations.*
Source: W.H. Furness, *The Home-Life of Borneo Head-Hunters: Its Festivals and Folk-Lore*, Philadelphia, J.B. Lippincott Company, 1902, p. 168.

80 W.G. Huff, 'Bookeeping barter: Money, credit, and Singapore's international rice trade, 1870–1939', *Explorations in Economic History*, Vol. 26, 1989, pp. 173–174.

81 SG 1 April 1872.

82 SG 15 February 1877.

83 SG 1 March 1890, p. 38.

One reason for the success of Chinese traders was their practice of establishing 'social bonds' with the local population, including intermarriage, offering gifts, learning the local language, or forgiving small debts.[84] Managing relationships, building social capital with customers was a significant part of managing business. These traders tried to bridge the cultural divide between themselves and collectors, learning the local language – or at least enough of it for trade – and adapting their practices to meet the skills of others. Combanaire describes how Chinese boat traders in central Borneo adapted to the 'silent barter' preferences of local collectors. The trade took place in an alien environment – some Chinese thought some peoples in these areas were so primitive that they still, literally, had tails like primates.[85] While the general misconception now seems misguided, from the perspective of the actors involved, it underlines the great chasm of understanding that existed between the two groups – but for which trade demanded an accommodating arrangement. These informal arrangements, even risk-taking gestures by collectors and traders, were based on the need to deal with 'outsiders'. The problems of negotiation and interaction were not ones that could be solved by administrative or official intervention, but rather by behavioral change, risk taking, and the bonds of social capital, albeit bonds that arose largely from commerce rather than amity.

Yet the Brooke administration attempted to bring some form of regimentation into the system of weights and measures used by traders, and the Sarawak Gazette contains various examples of offenders being brought to justice. Although the system would have been difficult to police given the environment and limited administrative resources under the administration, nonetheless, collectors were attuned to the trading environment and often kept up to date with administrative changes. Collectors from Dutch territory brought gutta percha to Sarawak to trade because they had heard that the Rajah imposed 'more accurate scales' although in practice the scales were easy to manipulate without collectors necessarily being aware of it.[86] The commentary reveals the extent to which collectors tried to optimize their engagement with the external economy by keeping abreast of developments in the local trading environment. Traders did not have it all their own way – a report appeared in the Sarawak Gazette about 'inaccurate measures used by the Krian Dyaks' being the cause of constant complaints from Chinese and Malay traders in one District

84 See Chew, p. 221.

85 Combanaire, pp. 233–234.

86 Combanaire, p. 195.

which suggests that traders too had to be careful.[87] As with all the literature of the time, it is difficult to separate the extent of cultural bias, but the conclusion can be drawn that the opportunity for 'sharp' practices was available to all in the trade – the issue of trust and risk in the trade environment was never far from regular interaction.

Secondary traders – both the boat and small bazaar traders – operated on very small margins – better to have small but frequent profits, retain and grow markets and win customer loyalty (that is, to build up social capital), and to offer credit, than to have no customers. In any case, barter of goods offered scope for 'marking up' prices. By 1899, the British Consul's report on Sarawak trade concluded that:

> In the farthest reaches of the rivers, large communities of these people [Chinese traders] live amongst the dyaks, Kyans, and other inland tribes and grow rich by providing them with necessaries and luxuries in return for the valuable jungle produce they procure ... Shopkeepers are often financed by Chinese firms in Singapore.[88]

These shopkeepers were financed and supplied by *Tertiary Traders* established in Kuching (or the larger towns); that is, those ports with access to regional shipping and thus an ability to deal directly with the external economy. All were connected by clan and credit relationships with Singapore merchants; that is, they received credit benefits from Singapore. The tertiary trader's connections crossed national boundaries.

The *Tertiary Trader* was the direct link between Sarawak and Singapore for the gutta percha trade. Even after two decades of Brooke rule, by 1863 it was said that none of Sarawak's foreign trade was carried on with any port other than Kuching.[89] Yet after that, the demand for Sarawak's commodities in the network of global trade and its involvement with regional trading networks grew rapidly, such that by 1869, Sarawak's 'outports' had a role in international and regional trade.[90] In the case of major gutta percha traders, the links were with Singapore merchants specializing in gums and other resins or, in the case of smaller traders, with merchants dealing in forest produce. Tertiary traders operated in the larger trading towns and particularly in Kuching; most of the gutta percha was traded through to Singapore, with opportunistic

---

87 SG 1 June 1894, p. 87.

88 Consular Report on the Trade of Sarawak for 1899, BSP 1900[Cd.1] [Cd.352], pp. 307–308.

89 Consular Report on the Trade of Sarawak for 1863, BSP 1865 [3478], p. 27.

90 Consular Report on the Trade of Sarawak for 1869, BSP 1871 [C.343], p. 8.

export through Labuan, or direct to Britain in some cases, probably traded with BCL. The presence of tertiary traders, integral to the gutta percha trade, functioned because of, and promoted, new openings to trade. The importance of the tertiary traders is difficult to separate from that of other actors, all traders and bazaars being mutually dependent and reliant upon the capacity of the shipping trade. No one actor could function without the other; and no actor could function without the presence of bazaars or the credit network (see below).

Combanaire wrote of a Chinese trader he encountered in 1899 in central Borneo who had moved from Singapore to seek his fortune in the gutta percha trade, but who been forced into the hinterland of Borneo because the Sarawak trade was controlled by a 'dozen big Chinese dealers'.[91] As is the case with Singapore, it is difficult to identify many gutta percha traders in Kuching by name, but there are snippets of evidence. The earliest reference to a specific gutta percha trader appears in 1882, when Harry de Windt writes:

> The Chinese are ... the principal householders in the [Kuching] Bazaar, the richest among them being the Brothers Ken-Wat, a firm trading in gutta, gold-dust, and diamonds, with Singapore and China.[92]

De Windt was aide de camp to Charles Brooke and brother of Charles' wife Margaret; he served from 1871–73 and would have been familiar with the life and business of Kuching.[93] However, the Ken Wat 'brothers' in fact are on record before 1882 – they appear in the *Straits Calendar and Directory* of 1870 as *Keng What and Co*, Gambier and Pepper Merchants.[94] Ken Wat appeared in several subsequent editions of the Directory, owning a sago factory in Kuching. In the Directory of 1877, Ken Wat & Co is listed as Sarawak's Farmers for Opium, Spirit, Pawnbrokers and Gambling. The Sarawak Gazette refers to improvements at Kuching near the buildings of 'Ken What & Co' in 1871 so they appear well established there as well by that date.[95] There is a familiar pattern about this evidence. Ken Wat, or Law Kian Huat as he is known by his descendants, was a Teochiu. Like the gutta percha merchants of Singapore,

91 Combanaire, p. 151.

92 H. De Windt, *On the Equator*, London, Cassell, Petter, Galpin & Co, 1882, p. 39.

93 N.S Talib, *Administrators and Their Service. The Sarawak Administrative Service under the Brooke Rajahs and British Colonial Rule*, Selangor Darul Ehsan, Oxford University Press, 1999, p. 52.

94 *Singapore Calendar and Directory (Including Sarawak and Labuan)*, Singapore, 1870, p. 6.

95 SG 18 June 1871.

in Sarawak the connection was formed between the Teochiu and the gutta percha-gambier-pepper trade and opium farming. Also, as in the case of the Low family in Singapore, the gutta percha merchant in Sarawak was a wealthy and well-regarded figure – possibly the leading *towkay* of Kuching and a friend of Charles Brooke.[96] See Plate 11 for an example of a Kutching gutta percha *Towkay*.

In the 1880s, Chinese trade expanded into the newly acquired Baram river area, formerly part of Brunei. The following report from the Sarawak Gazette about trade with local groups is expressed in terms that suggest that one group of traders was well enough known not to require further identification:

> Ken Wat's men who sold them superior blachu for $2 the kayu, they have always been accustomed to pay $3 for blachu of an inferior description ... A number of Brunei Malays ... were angry ... saying they would be ruined if our Chinese traded at those prices and were particularly angry when Ken Wat's men paid without beating down the sum demanded for a pikul of gutta, namely $32.[97]

Blachu was calico or coarse cloth – St John defined it as 'grey shirting'.[98] Cloth was a major trade item in Singapore and Sarawak indeed throughout the region as noted in Chapter 4. The extensive nature of the 'Ken Wat' gutta percha trade network – spreading across the full extent of Sarawak territory and moving quickly to establish the network in the new cession as soon as possible – emphasizes the nature of the Chinese trade. The strategy was multi-pronged – move quickly into the new areas, offer the gutta percha collectors the products they wanted and at good prices, and in turn pay a good price for the gutta. This would beat the competition and establish trade relationships for the future that would displace existing trade relationships. Chee writes that low margin/high volume strategies, intertwined with complicated networks of credit relationship, continue to be a feature of Chinese 'economic rationality' in rural Southeast Asia.[99] The technique was a valuable method to lay the ground for the bonds of social capital. The competitive pricing that 'Ken

96 C.A. Lockard, *Chinese Society and Politics in Sarawak: Historical Essays*, Jalan Salim, Sarawak Chinese Association, 2009, p. 53; C.A. Lockard, *From Kampong to City: A Social History of Kuching, Malaysia 1820–1970*, Athens, Ohio, Ohio University Monographs in International Studies, Southeast Asia Series, no. 75, 1987, p. 69.

97 SG 1 May 1883, p. 50.

98 St. John, *Life in the Forests of the Far East*, Two Volumes, London, Smith Elder and Co., 1862, Vol. II, p. 29.

99 Chee, p. 182.

PLATE 11 *Tea time with a Chinese gutta percha trader (Towkay), Sarawak.*
Note: The staff and ornate surroundings suggest this was a successful gutta percha dealer, or *Towkay*; visited by an engineer from the German cable company, Norddeutsche Seekabelwerke (NSW), in the early twentieth century.
Source: Archive NSW GmbH, Germany, with permission.

Wat's men' practiced was also a form of predatory pricing, designed to force out the existing Brunei Malay traders and to make it difficult for other Chinese traders to form trading relationships with collectors. That is, trade practices build the linkages of social capital, but can work to exclude others as well.

Chew has surveyed the overall structure, role and contribution of Chinese traders in Sarawak, listing the Foochow, Hakka (Kheh), Hokkien, Teochiu and Cantonese as the predominant groups. Of those groups, the most dominant merchant groups and those most involved in the forest produce trade were the Teochiu and the Hokkien.[100] Lockard notes the role of Teochiu and Hokkien in the forest produce trade and their position in Sarawak life at that time.[101] They were influential in advisory groups like the Chamber of Commerce: one of the first matters that group considered was the export duties on gutta percha.[102]

100 Chew, p. 38.

101 Lockard, *Chinese Society and Politics in Sarawak: Historical Essays*, p. 51.

102 BCL archives MS27 293 A: 17 December 1897; 7 January 1898.

While Hokkien traders were also involved with gutta percha trading, the weight of evidence suggests a more significant involvement by Teochiu traders. Given the dominant role the Singapore Teochiu had in the gutta percha trade, this supports the likely importance of that group in the trade in Sarawak.

Shipper/Carrying Traders

> The trade grows, most of its carriage still depends upon native crafts which sail direct to and from the river stations or ports on the coast[103]
>
> CONSULAR REPORT ON THE TRADE OF SARAWAK FOR 1899

These are the last linkage in the trader network. The role of the *Shipper/Carrying Traders* falls into two main categories; firstly, as the means by which trade functioned, that is, the transport network that underpinned the trade of Sarawak and, secondly, their role as traders – those who traded in gutta percha as well as carrying it. Shippers and the carrying trade played a significant part in the trade of Sarawak because geography and resources shaped the way produce was both imported and exported. In 1864 the British Consular report noted that there were sixteen rivers that formed the means of communication with the interior; two rivers were navigable by large ships, the rest by 'native craft' that could travel great distances.[104] The report also noted that there was little industry, with home trade depending upon foreign transactions, making Kuching the entrepot where goods were collected.[105] Trade statistics in the Sarawak Gazette showed that considerable amounts of gutta percha arrived at Kuching via 'coastal shipping' from elsewhere in Sarawak, which illustrates the significance of the shipping network to the economy and to the trade.[106]

Aside from the various larger steamers and schooners of European nationality, some of which were chartered by Chinese traders, there were many so-called 'native craft'. In the 1865 commerce statistics, for example, of 164 vessels entered, 130 were 'native craft'. The year 1865 appears to have been a significant

103 Consular Report on the Trade of Sarawak for 1899, BSP 1900[Cd.1] [Cd.352], p. 306.

104 Consular Report on the Trade of Sarawak for 1863, BSP 1865[3478], p. 27.

105 Consular Report on the Trade of Sarawak for 1863, BSP 1865 [3478], p. 30.

106 See Treasurer's Annual Reports to the Rajah, Sarawak Gazette, various issues.

one for the expansion of the trade activity of Sarawak – the Consular Report for Sarawak noted that one of the changes noticed during the year was in the 'carrying trade ...that the Chinese merchants ... who are partners with some of the principal Chinese houses in Singapore, have lately commenced chartering British vessels for the purpose of conveying their goods from Sarawak to Singapore'.[107] Decades later, the Trade Report in 1901 noted that the trade of Sarawak still depended for most of its carriage upon small craft which sailed to and from the river stations or ports on the coast.[108] Shippers and small craft had been the engine room of Sarawak's economy for decades – like the 'bandongs', vessels that carried so much of the extra gutta percha in 1900.[109] Small regional shipping was crucial to the expansion of trade generally and to gutta percha trade; conversely, the lack of it was a serious inhibitor to trade. In years when the monsoon was severe and many 'native craft' were wrecked, the effect on trade was commented upon in trade reports.[110]

Riverine shipping was the lifeblood of Sarawak trade since it was the only means of ensuring exports moved down rivers to coastal ports and outwards to Kuching and points beyond. The Sarawak Gazette printed reports from Residents over decades about gutta percha awaiting shipment being stored by the Chinese traders and how pleased the traders were when the vessels arrived.[111] Shipping delays in turn meant lack of goods arriving for re-supply to the bazaar and river traders – and delays in being able to forward produce stored in the bazaars affected traders' cash flow.[112] By 1870 three steamers carried freight and passengers between Kuching and Singapore. In 1875, the Sarawak Steamship Company was formed; later there was a fortnightly service to Singapore but services were the subject of complaint. The Chinese merchants of Sarawak often ran or hired smaller shipping to manage their needs.

Like the four other categories of actors, shipper/carrying traders required professional skills and knowledge, awareness of seasonal and market demands, and a capacity to deal with other actors. Most of these shippers profited not just from carrying freight, but by acting as traders in their own right, especially those travelling to more remote points where people had little access to markets and where there was only limited competition. Larger vessels were registered at Sarawak's main ports such as Kuching or in later years at Sibu.

107 Consular Report on the Trade of Sarawak for 1865, BSP 1866 [C.3669], p. 271.

108 SG 1 March 1901, pp. 42–43.

109 SG 2 January 1900, p. 14.

110 For example, SG 1 June 1874.

111 SG 12 May 1887, p. 90; 1 July 1892, p. 131; 1 October 1892, p. 187; 1 April 1897, p. 74; 1 May 1894, p. 74; 2 August 1897, p. 148; 1 March 1901, p. 62.

112 SG 3 March 1903, p. 57.

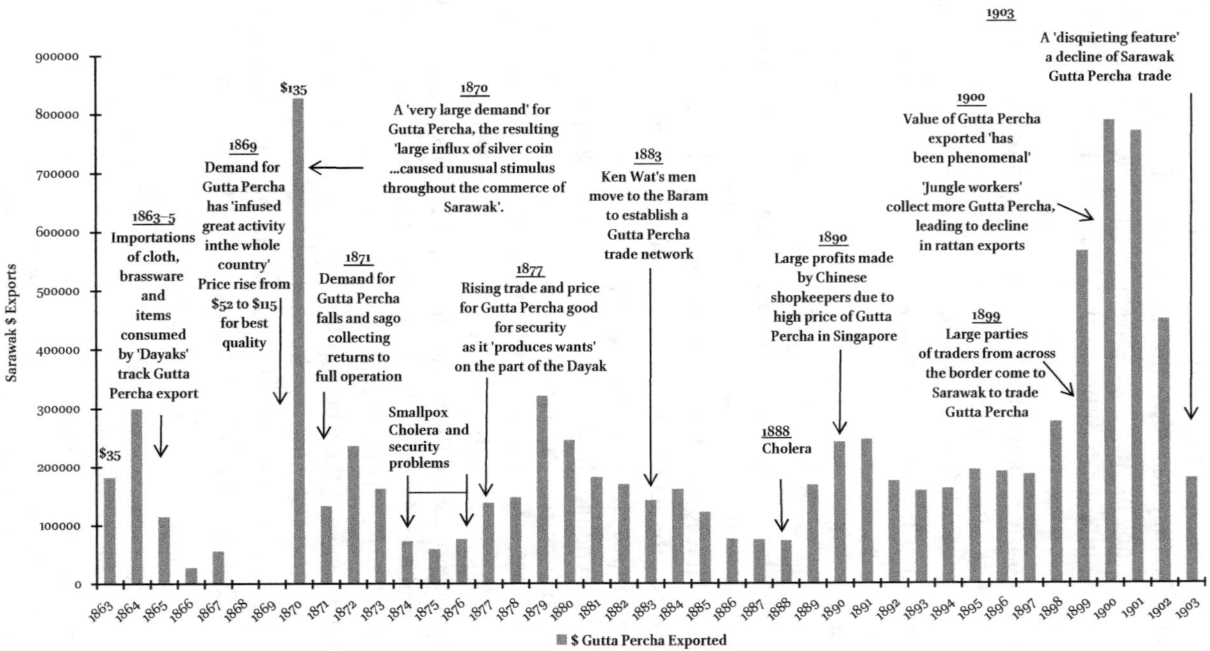

FIGURE 10 *Sarawak and the Gutta Percha trade – impetus and responses.*

Note: Examples of local prices of gutta percha (Sarawak $) and snapshot of the evidence from primary sources.

Source: GP figures Sarawak Gazette (SG), various issues 1870 – 1904, and Consular Report in British Session Paper (BSP) (1871 [C.429], p. 809. Figures for 1868–69 are not available. See Appendix 2, Table 1 for details. for statistical details. For other comments: 1863–65: BSP 1866 [3669], p. 271; 1863 price of GP: BSP 1865[3478], p. 25; 1869: BSP 1871[C.343], p. 8; 1870: BSP 1871 [C.429], p. 805; 1871: SG 30 March 1871; 1874–76: SG 2 March 1874 and 20 May 1876; 1877: SG 16 Jan; 1883 : SG 1 May 1883, p. 50; 1888 SG 1 May 1889, p.63; 1890: SG 1 March 1890, p. 38; 1899: SG 1 June 1899, pp. 121–122 and SG 2 April 1900, pp. 61,64; 1900: SG 2 January 1901, p. 1; 1903: SG 3 May 1904, p. 82. Ken Wat was a major Chinese gutta percha trader.

The commodity chains of Sarawak trade depended upon the continued success of the *Shipper/Carrying Trader* and the risks they took. It is possible that there was quite a deal of activity associated with river transport and coastal shipping that has been overlooked in terms of assessing its contribution to the economic and social development of Sarawak. Kaur has recognized the significance of shipping given the lack of other transport infrastructure in Sarawak and Chew has provided an analysis in relation to his examination of Chinese boat trade in Sarawak.[113] Wadley has noted the importance of Borneo's rivers and river networks as a means of transport and communications, especially between the coast and the interior and that further study of these networks 'would be of immense value'.[114]

The interaction between the demand from the external economy and the local Sarawak trade network outlined to this point can be seen by matching the documentary evidence with the trade data as illustrated at Figure 10. In a sense, this is a graphical snapshot of the connections between the global and local linkages in the world-system.

The extent and effectiveness of the local trading network described above can be seen when it is compared with the difficulties faced by the European-owned attempts begun in the 1890s to establish a factory to produce gutta percha from leaf material. In 1895, a new trader arrived in the Sarawak gutta percha market – and it was not a Chinese connection. The Hourant factory was set up in 1895, some product having been exported the previous year for testing in Paris.[115] It was hoped that this would be a profitable, but sustainable way of collecting gutta percha (see background discussion in Chapter 3). The location of the factory in Sarawak reflected features of the gutta percha commodity chain. Firstly, the role of the political regime in that Brooke's attitude to the promotion of trade meant that he granted the company a monopoly.[116] Secondly, proximity to Singapore and the market and agency houses there that were linked to the external market, made the Sarawak gutta percha trade commercially attractive. Finally, the fact that Sarawak was well placed to connect with the gutta percha resources of the whole of Borneo, an area recognized as being one of the largest sources of gutta percha, and one that was not subject to restrictive official controls on collection of forest produce such as that seen in the Malay peninsula. Clearly, the geographic proximity of the shipping lines

113 Kaur, *Modern Asian Studies*; Chew, *Chinese Pioneers on the Sarawak Frontier*.

114 Wadley, *Histories of the Borneo Environment*, p. 12.

115 SG 1 April 1895, p. 57 and see Chapter 2 above.

116 P.L. Sherman, *Report of the Special Agent of the Forestry Bureau Sent to Investigate Gutta-Percha and Rubber in the Straits Settlements, Java and Sumatra*, Manila, Forestry Bureau, 1901, p. 345.

from Kuching rather than from those from the eastern side of Borneo was an additional advantage: this reflects the significance of geography in the shaping of commodity chains and trade networks.

Eventually, however, the company became insolvent, and Brooke was said to have sworn he would never again allow a monopoly of this kind.[117] One possible reason for Brooke's decision was reports very early on that collectors were in fact felling young trees to obtain leaves, prompting Brooke as early as October 1895 to issue an Order banning the practice – with the leader of any group doing so incurring a potential fine of up to $100 or imprisonment.[118] The problem of the leaf gutta trade in terms of return on labor and its lack of sustainability was discussed in Chapter 3, but another weakness of the Hourant enterprise was the fact that it sought to operate in competition with the very considerable entrepreneurial skills of the Chinese traders and trade network that effectively ran from the inland river ports through to Singapore itself. Hourant's factory was not linked into the same personal relationships with indigenous collectors – and did not have the extensive connections of barter and credit to negotiate with collectors. The operation lacked the social capital that was at the core of the trade. That is, the strengths of the commodity chain – factors that enabled the shift from dissonance to consonance among actors – were not present in the Hourant endeavor. Between 1894 and 1902, the entire period during which attempts were made in Sarawak to trade in leaf gutta, just $5,935 of product was exported.[119] While it was unlikely the factory was ever intended to oust the local Sarawak trade, the factory's exports compared with $3,582,612 of gutta percha sold through indigenous-Chinese trade network in the same period. In 1903, the Gazette reported that Mr. G. Hourant had been appointed the Sarawak Inspector of Police.[120] See Appendix 2, Table 3 for details.

Local indigenous knowledge and Chinese trade practices, and the establishment of relationships based on social capital, were behavioral mechanisms that facilitated the operation of trade. There were other, practical, mechanisms which enabled the trade to operate – and for traders to work together.

There were three key functional mechanisms that facilitated trade in Sarawak. The first reflected the responses and unique skills of local indigenous collectors; this was the use of a complex network of overland paths that traversed a physically challenging environment. The second feature was the rise of bazaars and trade venues along rivers and at key port towns. The third was the development of credit networks and facilities for the barter of goods from

117 Sherman, *Report of the Special Agent of the Forestry Bureau*, p. 345.

118 Order No. VII 1895 in SG 1 October 1895, p. 178.

119 See Appendix 2.3 for details.

120 SG 3 March 1903, p. 46.

the external market. These latter two mechanisms were developed largely by Chinese traders. That is, each group provided essential structures in the local trading architecture according to their resources: this included local or trade knowledge, capital, and social network resources. These three mechanisms are examined next.

Overland Trade Paths

The river transport network described previously was just part of the gutta percha trade route. Today's Sarawak has a system of 55 rivers comprising around 3,300 km of navigable waterways.[121] In the Brooke period, the cost and difficulties of road construction severely limited the expansion of roads or even forest paths. By 1938 the nation still had only 1094 km of roads, less than ten percent of which were sealed, all-weather surfaces.[122] Healey concludes that communication in Borneo is often located on high ground, with the orientation to rivers as a means of communication not an ecological imperative; many rivers may not always be navigable.[123] While the rivers of Sarawak are famous, the significance of overland networks to trade and social interaction in Sarawak has received little scholarly attention, yet many of the rivers of Sarawak extend within close proximity of each other and were linked by a network of *overland* tracks used or built by collectors who carried produce to the rivers for onward transport to the bazaars and outports.[124] In the nineteenth century, most reliance for overland connections (even those requiring weeks of trekking), involved the use of either existing forest paths, or else local travellers moved along paths cleared as they went. Indeed, records show that such on-the-spot clearing occurred well into the twentieth century and was an accepted, indeed necessary, means of moving across Borneo. A 1942 report described paths as varying between 'from 3 or 4 feet wide, to an apology for a track that tried one's temper when the weather was bad ... one sometimes uses tracks made by pigs and very often cuts one's own way ... one unloaded coolie [person not carrying a load] can with judicious use of his *parang* keep

121 T. Thiessen, *Borneo: Sabah, Sarawak, Brunei (Bradt Travel Guide)*, Chalfont St. Peters, Bradt Travel Guides, 2016, p. 3.

122 Ooi, p. 206.

123 C. Healey, 'Tribes and states in "pre-colonial" Borneo: Structural contradiction and the generation of piracy', *Social Analysis*, No. 18, 1985, p. 11.

124 See Combanaire's description of opening of forest paths at Combanaire, pp. 187–190.

the party in fair progress except in actual swamp'.[125] Elsewhere, paths were constructed of trunks of trees or bamboo laid lengthwise, with the rounded side up. The difficulty of walking on them became legendary among European writers who had visited the country (for example, even a 'tall stout man' walking through such tracks could disappear into the long grass and walkers could emerge after an hour's walk 'wet to the waist'[126]). The ad hoc nature of such paths, the informality, even the invisibility, of these networks of tracks makes it impossible to estimate how lengthy the networks might have been, but the historic record gives strong evidence of just how important and widespread overland trading was, especially for valuable forest produce like gutta percha.

Charles Hose noted that the Baram and Rejang river networks in Sarawak and the Batang Kayan river (in neighbouring Dutch territory) all have their sources within a comparatively short distance of each other.[127] A map which Hose provided illustrated several 'jungle paths' that tracked between various river tributaries, including a lengthy track from the Limbang river to Claudetown on the Baram river. The tracks were used to provide more direct routes to better trading points or to avoid navigational difficulties in rivers – or avoid security problems – or local tax collectors. Overall his map includes dozens of paths yet does not cover the whole of Sarawak. It does give an indication of just how many paths there may have been across Sarawak, or indeed the island generally.[128] Haddon wrote of the trade route in the north of Sarawak that was used by gutta percha collectors: from the north, they travelled down the Limbang and Madalam rivers then, walking overland, they transferred down the Malinau, Tutau and Baram rivers to the major bazaar at Claudetown where shipping and traders were ready to trade and ship the gutta percha out to the rest of the world.[129] This would seem to relate to the lengthy 'jungle path' marked on Hose's map. Another track, marked on Baring-Gould and Bamfylde's map, linked the Baram and Iwan rivers (in Dutch Apo Kayan territory); and in the south of Sarawak, a path linked the fort and bazaar at Lubok Antu to Dutch territory just a couple of miles away, one of the easiest cross-border access points.[130]

---

125 Leach, D.L., 'The Survey of Sarawak', *The Geographical Journal*, Vol.100, No.3, Sep. 1942, p. 100.

126 See Roth, H.L., *The Natives of Sarawak and British North Borneo*, 2 Vols, London, Truslove & Hanson, Vol II, p. 26.

127 Hose, *The Geographical Journal*, 1893, p. 197.

128 Hose, *The Geographical Journal*, 1893, p. 288.

129 Haddon, p. 297.

130 R.M. Pringle, *Rajahs and Rebels: The Ibans of Sarawak under Brooke Rule, 1841–1941*, Ithaca, Cornell University Press, 1970, p. 210.

This meant that overland border crossings were required to make river trade effective – the one needed the other:

> The overland trade from the head of this river with s'kayan and Kapuas, Netherlands Indies, consisted chiefly of treasure and brassware, chiefly for the purpose of trading for jars and gutta-percha across the frontier.[131]

Overland trade was not simply a Sarawak affair and collectors were willing to ignore national boundaries whenever necessary – many of the people living on the other side of the border were from a similar heritage. The territory between the Dutch Kapuas river and the Batang Lupar river in Sarawak was known to the Iban as the 'Emperan' or 'flat land'. Iban ancestors originated from that region before many of them moved across to (pre-Brooke era) Sarawak. While Iban movement from Sarawak back into the Kapuas was usually peaceful, both warfare and trade were common occurrences according to Wadley.[132] There were the peoples of Kayan and Kenyah origin with similar language and traditions, trading produce with the Iban for fabric and brass gongs, or travelling across to Sarawak bazaars to deal with Malay or Chinese traders, or have those traders travel across to them.[133] The better exchange rates and trade in goods such as salt, firearms and ammunition offered by Sarawak traders drew trade away from Dutch interests.[134]

There was, then, increased impetus to open new forest paths as trade became more attractive. The Kayan and Kenyah living in the region of the Upper Rejang and the Baram rivers of Sarawak and in nearby Dutch territories were well placed to take advantage of the gutta percha resources. They also had relationships with the Punan groups who resided there and who had skills in collecting/trading gutta percha and in guiding other collectors to the source of gutta percha. As a result, on one occasion a large party of Kayan traders who travelled from the 'watershed' in Dutch territory to the up-river bazaars in 1899 numbered in the hundreds.[135] Another expedition arrived several years

131 SG 1 August 1892, p. 153.

132 R.L. Wadley, 'The history of displacement and forced settlement in West Kalimantan, Indonesia', in *Conservation and Mobile Indigenous Peoples: Displacement, Forced Settlement, and Sustainable Developments*, eds D. Chatty and M. Colchester, New York, Berghahn Books, 2002, pp. 315–317.

133 Eghenter, *Bijdragen tot de Taal-, Land- en Volkenkunde*, p. 762.

134 See R.L. Wadley, 'Boundaries, territory, and resource access in West Kalimantan, Indonesia, 1800–2000' in R.L. Wadley, ed., *Histories of the Borneo Environment*, pp. 137–158, Leiden, KITLV Press, 2005, pp. 141–143.

135 SG 1 June 1899, pp. 121, 122.

later, bringing thousands of dollars of good quality gutta; the Kayan said that the Rejang markets were preferable to those in their own country, and that they would continue going to the market, despite the great distances 'as long as they were on good terms with the *hivans* (Ibans) ….'[136] The Kayan, like the Iban, had a history of migration within Borneo going back generations; some movements resulted in the displacement of other populations, but often they involved working with other groups of hunter-gatherers.[137] The engagement with the overland trade environment was well within their capability and cultural perspective.

The Bintulu river could be linked along its inland sections to the important Rejang river by an overland journey taking a day and a half, a route used by traders during the N E monsoon when the two rivers would normally be isolated from each other. This track was also used by Tinga river traders.[138] Low wrote in 1848 that forest paths between the villages of all the Sakarran and Saribas (rivers) tribes, enabled constant communication.[139] An overland trade route used by the Kayan effectively enabled them to cross Borneo from west to east.[140] Lumholtz also writes about the Kenyahs walking extensively in central Borneo; groups trekked overland and then, when they could use rivers or streams, made 'new prahus and then continue the journey'.[141] Brunei Muruts ignored the border and travelled to the Sarawak bazaars when the Trusan became part of Sarawak.[142] In 1887 the Sarawak Gazette reported that a group of Muruts arrived from Dutch territory after an overland journey of ten days; they were so impressed with the trade that they would induce others to follow their example. Another group came even though the return journey would require a month of walking.[143] Even in 1949 Muruts from the Trusan and Dutch territory visited the bazaars at Lawas and a few years later Murut collectors from Dutch

136 SG 16 February 1906, pp. 59–60.

137 See T.H.G. Mering Ngo, 'A new perspective on property rights: Examples from the Kayan of Kalimantan' in *Borneo in Transition. People, Forests, Conservation, and Development*, eds C. Padoch and N.L. Peluso, Kuala Lumpur, Oxford University Press, 1996, pp. 137–149.

138 SG 28 February 1871.

139 H. Low, *Sarawak. Notes During a Residence in That Country with H.H. The Rajah Brooke*, London, Richard Bentley, 1848, Reprinted 1988, Singapore, Oxford University Press, p. 169.

140 R. Burns, 'The Kayans of the Northwest of Borneo', *Journal of the Indian Archipelago and East Asia*, 3, 1849, p. 141.

141 C. Lumholtz, *Through Central Borneo An Account of Two Years' Travel in the Land of the Head-Hunters Between the Years 1913 and 1917*, 2 Vols, New York, Charles Scribner's Sons, 1920, Vol. II, p. 426.

142 SG 1 June 1894, p. 88.

143 SG 3 January 1887, p. 11.

territory still took a 16-day round trip over the border to trade at Lawas, which was easier than the 25-day round trip to the notionally closer trading point in Netherlands East Indies over what was a more difficult terrain, 'especially if carrying a load'.[144]

Essentially, forest people moved around their environment on foot or by water, constructing boats, shelter or paths as required.[145] It is easy to assume that river transport was simpler and so that must be the only system that people used, but this is an outsider's perspective. A harsh environment might exclude outsiders but not experienced locals. In any case, river networks were just not sufficiently close to the location of forest produce. The labor and skill required to engage in the collection of forest produce as a year on year activity requires endurance if trade were to be carried on. An example is given by Sutlive in his biography of Tun Jugah, a paramount leader of the Iban, who in 1918 went on a *bejalai* expedition for three years when he was fifteen years

PLATE 12
Overland trekking conditions.
Source: Four Kayan in their Camp c. 1896. Photograph by Charles Hose, V0037407EL.
Wellcome Library, London. With Permission.

144 SG 7 May 1949, p. 121; SG 31 May 1952, p. 112.
145 See Combanaire, 1910, p. 188.

old. During part of the *bejalai*, the group was collecting *bulu* resin, collecting at least one pikul (133 pound) *each*, they carried the resin on their backs every day for sixteen days, setting out at day break and walking till the middle or late afternoon. They would then stop and build a lean-to for the night. They did this for 16 days ('sixteen days, sixteen lean-tos') until they came to a river where they then constructed a boat so they could travel to Kapit.[146] Such expeditions replicated thousands of similar gutta percha treks in decades past. See Plate 12.

Collectors on gutta percha expeditions were at one with the forest, collecting as they went and living off the forest, constructing boats or paths as required, seeking advice from other collectors such as the Punan and, if necessary, ignoring the territorial or resource 'boundaries' of others. Constructing a path through the forest where there was no river system to travel or to facilitate trade for some other reason, particularly in the face of difficult security situations, did not present a problem if the value of the trade was worth the effort. When a large quantity of gutta percha was collected by Trusan Muruts they were concerned about being 'pillaged' if they travelled by the usual route – instead they cut a new path through the forest, even though this delayed them considerably. The extra time and effort involved had the added reward of opening up new sources of gutta percha and 'kayu-garu' (used for incense). It was expected that several others would follow as a result.[147] The actions represented an important example not simply of the value of the gutta percha trade, but also of how forest economies could rapidly expand commodity chains if the trade was sufficiently valuable. Other paths and tracks were constructed under the prompting of Brooke administrators.[148] Batu Bala, a Kelabit historian, writes of the need to combine multiple overland and downriver voyages to complete a single trading journey. When relatively large amounts of produce were to be moved, this process might include carrying repeat loads of produce: splitting loads into manageable sizes and carrying the first part of a load to the first staging point, then returning to collect the second load for delivery to the staging point, going back to collect the third bundle and so on. Then, once all the loads were positioned, the whole process was repeated again and again until all the multiple loads had been moved to a final point (the Kelabits called this method *suud*). This, just for the overland part of the journey, which might

146 V.H. Sutlive, *Tun Jugah of Sarawak: Colonialism and Iban Response*, Kuala Lumpur, Sarawak Literary Society, 1992, p. 55.

147 SG 2 July 1894, p. 107.

148 Such as SG 1 June 1901, p. 124.

be followed by repeated small journeys by river. The whole process often took place over sometimes difficult and dangerous terrain.[149]

While the rivers of Sarawak and the actors who traded on them were a vital feature of trade – in terms of gaining essential access to the valued forest produce demanded by the global market – the river networks were in one sense highways to nowhere. Ultimately, it was the labor and knowledge of the forest collector that enabled the trade to operate. This overcame the physical gap in the commodity chain to bring the valued forest resources overland to the trade and transport points on the river. The trade was sufficiently lucrative for traders to be motivated to develop systems and synergies to overcome such physical inhibitors. The geography and unsettled conditions in Sarawak were clearly a severe challenge to collectors, yet they were not insurmountable ones.

The economic and social role of the overland trekking routes may be considerably more significant than has been recognized, largely because they are difficult to conceptualize today and have never been fully mapped. Even the brief survey above suggests a network that was by no means unsophisticated. The paths mapped out by Hose and the narrative record of gutta percha traders recorded in the Sarawak Gazette and others set out here suggests a need to develop a deeper understanding of this feature of the forest economy. The lengthy, planned and well-resourced gutta percha expeditions were a break with Dunn's concept of the otherwise 'inescapable consequence' of living in the subsistence zone of Sarawak. At the same time, the opening of traditional societies to the demands of the external economy brought with it what must be seen as unstoppable consequences which saw those same societies swept into global trade networks and relationships. Geography shaped patterns of trade; the importance of overland routes was heightened by the interaction with the external economy. This forced people into contact with one another – making trade more profitable but increasing the risk of hostilities as will be seen in Chapter 8. These patterns also emphasize the fact that the forest trade of Borneo must be seen as more than just the river based trade as it is frequently presented. The significance of the overland trade paths on the one hand, and conversely the role of the river traders and shippers, has been a somewhat forgotten area of academic focus, but one benefit of a commodity chain style of analysis is that it draws attention to the role of these traders, and thereby draws a more accurate understanding of the way they engaged in trade both with each other and with the external economy.

---

149 See S. Batu Bala, *Kelabits Story. The Great Transition* (ebook), Singapore, Partridge Publishing, 2014, p. 32 and Ch. 2 'Business Routes' for a description, including of the various overland/river routes used by the Kelabits.

Trade Venues: Bazaars and Outports

Having collected gutta percha and traversed land or riverine routes, collectors traded their gutta percha either with Chinese boat traders, or brought it to bazaars where commerce and social interaction occurred.

Collectors and other traders needed to identify practices and opportunities to negotiate with each other for trade. The success of each of the groups of traders described above was shaped by features of the geographic, economic and political environment in Sarawak. In turn, the success of each was also dependent upon, and contributed to, the rise of bazaars throughout Sarawak: their scope and spread typified the expansion of Brooke rule. Initially, Malays occupied an important position in the bazaars; Boyle wrote in 1865 that the bazaar at Kanowit on the Rejang was run largely by Malays intermixed with a 'few Chinamen',[150] but that dominance did not last. The Sarawak bazaars were very much the product of Chinese entrepreneurship. This point has been touched on in discussion above; it was noted since the early days of the Sarawak Gazette, that the Chinese population in the outports was a sign of trade and energy, and that that the Chinese built all the substantial buildings at bazaars and outstations.[151] Associated with these accounts were very encouraging reports of large quantities of gutta coming in to new stations. Bazaars were a means of attracting collector/traders with produce and a means of enticing the gutta percha collector to commit himself to future trade by the offer of credit in exchange: the goods on display at the bazaar offered a wider choice of products than could be displayed by river boat traders.

> At the first part of the month a great deal of gutta was sold, the sellers almost entirely clearing out of the bazaar of what few things the Chinamen had in stock, at present there is literally nothing to be bought; such things as rice, tobacco, sugar &c have been out for some time...[152]

An important development that was crucial in improving the conditions of trade was the establishment of forts – they provided facilities from which law enforcement could occur. The forts and associated bazaars were an important means by which James, and in particular Charles Brooke, could establish controls over the movement of people, especially the Iban. These trading points reflected both Brooke political controls and attempts at economic development.

150 Boyle, 1865, p. 98.

151 SG 15 December 1871; SG 1 December 1874; SG 10 October 1876.

152 SG 1 April 1890, p. 51.

The forts and administrative infrastructure provided by the Brooke administration, aside from providing (generally speaking) more peaceful conditions for the improvement of trade, also provided some practical support for its operation. By the turn of the century, Residents and Assistant Residents had a role providing cash to local Chinese traders who presented letters of credit taken out by their Kuching connections with the Sarawak Treasurer. By this means the official administration functioned as a bank to the local traders, in addition to the unofficial network of Chinese traders who used their own connections, although shortage of cash was always a difficulty.[153]

In the bazaar, traders interacted and connected with the outside world. In discussing the increased demand for imported articles, the Consular Report for 1865 noted that the trade 'by means of the depots of commerce ... [is] gradually forcing its way up these rivers among the tribes of the interior'.[154] As the larger outports were established with regular visits by shippers, they attracted more bazaar facilities and more visits by regional shippers, and they became 'attractor ports', like small entrepot ports. In 1875, the Auditor's Report to the Rajah encapsulates a snapshot of a trade network of Sarawak at the time:

> Bintulu ... Going down to the new bazaar which is being erected, I was struck with the apparent briskness of the trade; Kayans, Punans, and people from wild and distant tribes could be seen in the Chinamen's shops, as well as several Brunei traders; in fact, there is an air of general prosperity about the place ...[155]
>
> [BINTULU WAS A MAJOR COASTAL STATION ESTABLISHED IN 1868].

Elsewhere, the Batang Lupar, Saribas and the Skrang rivers were in the heart of Iban territory. The area possessed gutta percha and other forest produce resources and Teochiu Chinese traders were located there.[156] Simanggang, for example, was a Teochiu bazaar and so too were the smaller shop-houses up river with credit links to it. Teochius travelled up the river and established shops in Lubok Antu and further up river.[157] As noted above, there was a forest path from there to settlements in Dutch territory, so the area attracted gutta percha traders from that territory as well. Elsewhere, the Rejang river was navigable

153 SG 2 May 1903, pp. 105–106.

154 Consular Report on the Trade of Sarawak for 1865, BSP 1866 [C. 3669], p. 271.

155 SG 16 June 1875.

156 Chew, p. 67.

157 Chew, p. 67.

for hundreds of miles inland and this made it ideal for trade, and Kanowit was the most important bazaar of the Rejang. There were also major trading centres a Sibu, Song and Kapit. At Sibu by 1870 'already the active Chinamen are penetrating into the interior in search of gutta'[158] – no doubt encouraged by 'boom' prices achieved for gutta percha in that year for the lengthy production of submarine cables being undertaken at that time. Further north, the Baram river and its many tributaries twisted back and forth, and with forest paths (like those mapped by Hose) this provided important connections through Sarawak and with Brunei and Netherlands Indies from where gutta percha collector/traders were keen to travel because of the perceived attractions of the Sarawak bazaars and the goods on offer there. Claudetown (see Plate 13) – later

PLATE 13 *Bazaar at Claudetown (Marudi).*
Note: Furness' annotation for this photograph advised that the bazaar sold 'everything from glass beads to sewing machines, silk scarfs to calico, from arrack to white-seal champagne, from Chinese confections to patent medicines'. Trade took place in either Mexican dollars or in forest produce such as gutta percha, rattans, edible birds' nests and so on. The photograph was taken during the rainy season when shipping could not travel up river as far as Claudetown; as a result large piles of rattan awaiting transport can be seen in front of the stores.
Source: W.H. Furness, *The Home-Life of Borneo Head-Hunters: Its Festivals and Folk-Lore*, Philadelphia, J.B. Lippincott Company, 1902, p. 22.

158 SG 24 January 1871, p. 38.

called Marudi – was the principal town of the district, where Chinese merchants had erected 'magnificent shops' and warehouses and the trade of the river and overland regions was attracted to it and in turn attracted steam shipping.[159] One Resident said that the Chinese bazaars tended to stock the same goods, which included biscuits.[160] De Windt described the Kuching bazaar in 1882 and mentions that the most favoured item of food for indigenous customers was *Huntley and Palmer's* biscuits that were imported to Kuching in great quantities – BCL was the Sarawak importer of these biscuits as its advertisements in the Sarawak Gazette reminded readers even decades later.[161]

Interaction among the very different populations of Sarawak was a significant, indeed essential means of establishing, building and maintaining the trade network – and outports and bazaars provided venues to facilitate this interaction. That is, they provided an important venue for social capital, building trust and coping with the risks that otherwise existed in Sarawak. These needs were particularly significant given the sparse and hostile populations of Sarawak and the cultural differences that existed among them. Bazaars were in part a practical manifestation of growing demand for forest produce and the goods that forest traders sought. Tagliacozzo proposes that historically in northern Borneo the growing importance of Chinese demand for forest goods necessitated more formal arrangements in the transportation of ecological products from the forest canopy and the mountains of the interior; they were a means to allow for mutual exchange.[162] By the time of Brooke administrations, improved security arrangements, however ad hoc they may have been, and the expansion of the trading infrastructure generally, provided a mutually re-enforcing environment for social relationships to be formed. Ranjit Singh notes that one benefit of holding regular markets (and this could be said of bazaars) was that they provided an important occasion for 'formalised peaceful exchange'.[163] Charles Hose, for example, often arranged meetings to bring hostile groups together at Claudetown, invoking traditional practices such as feasts, exchanging oaths and opportunities for trade; that is, combining

---

159 Hose, 1893, p. 194.

160 A.B. Ward, *Rajah's Servant*, Cornell University Southeast Asia Program Data Paper No.61, Ithaca, Cornell University Press, 1966, p. 29.

161 De Windt, p. 38; see SG 1 December 1908, p. 306.

162 E. Tagliacozzo, 'Onto the coasts and into the forests; Ramifications of the China trade on the ecological history of northwest Borneo, 900–1900 CE', in *Histories of the Borneo Environment*, ed. R.L. Wadley, Leiden, KITLV Press, 2005, p. 36.

163 1984, pp. 398–399.

traditional rites and commercial exchange to promote better relationships, and improved conditions for trust, since all those factors were interdependent. In April 1899, Hose arranged a meeting with a group of around 6,000 Madang (Kenyah) border peoples and Baram Kayans. He recorded that much 'borak' was drunk; even a 'rice and paddy competition' was organized with the aim of encouraging local cultivators. Hose wrote that in recognition of the need to end old blood feuds, he again arranged a race of 'war canoes' as well as other land and water races. He considered it necessary to replace old forms of fighting and head-hunting with 'some other form of violent competition'.[164] After the formalities, a 'great deal of business' was done in gutta percha and rattan that was sold to Chinese traders.[165] The Claudetown peacemaking regattas were the antecedents of the famous races held at the Baram to this day. The opportunity for different groups to meet on neutral ground with an intermediary and with opportunities for the exchange of formalised mechanisms of goodwill (feasts, oaths, killing of pigs, trading) presented an important means to develop trust and social capital. In effect, they were the venues at which to stage occasions to facilitate the development of good relations upon which trade flourishes; such patterns of trust and the development of entrepreneurial relationships in the marketplace of subsistence economies have been studied elsewhere.[166]

Bazaars and the outports of Sarawak, and the entrepot of Kuching where much of the country's produce was centralized before subsequent export, were then focal points for both trade and social interaction. One way in which the capitalist world-system should be considered, according to Stanley, is not by analysing structures at the state level, but rather through its organization around and through cities[167] (again, the message is to look local to understand the global). The role of bazaars and outports in Sarawak, given its relatively more rudimentary level of economic development during this era, was less dependent upon the 'city' as such, but rather bazaars and outports (Kuching, etc) that filled the role of cities by linking with the external economy – and, as Stanley writes of cities, it was at this level where citizens utilize their 'transnational'

164 Hose and McDougall, Vol. II, p. 293; see pp. 293–300 for Hose's detailed description of the activities.

165 SG 1 May 1899, p. 85.

166 M. Viswanathan, S. Sridharan and R. Ritchie, 'Understanding consumption and entrepreneurship in subsistence marketplaces', *Journal of Business Research*, 63, 2010, pp. 570–581.

167 B.E. Stanley, ' "Going Global": Wannabe world cities in the Middle East' in *Emerging Issues in the 21st Century World-system: New Theoretical Directions for the 21st Century World-System*, ed. W.A. Dunaway, Westport, Praeger, 2003, p. 168.

or 'extra-local networks' for exchange and empowerment'. In Sarawak, in the more distant hinterland, the presence of Chinese traders effectively brought the city (and its extra-local network) *to* the hinterland. Chinese boat traders had a role in facilitating peaceful exchange: their role was to enable exchange to occur in a way that accommodated the trade practices and cultural beliefs of collectors. For many in the collecting community, access to bazaars was not easy and the more remote communities or groups like the Punan might be visited by Chinese boat traders who provided similar services, on a more modest scale, to those seen at bazaars.

Credit and Barter

As with many societies of the periphery, the trade of Sarawak depended as much for its success on importing and distributing products to collectors on credit as it did on actually collecting and shipping produce – such as gutta percha. While rice and cloth were important for subsistence trade and as a medium of exchange as noted earlier, the Singapore Market Reports identified items important for the Borneo trade; such as 'Brass Wire ... is in demand by Borneo traders'.[168] Ceramic Chinese jars, beads and salt were also important items for barter to the indigenous market.[169] All these items were usually provided on credit in return for future supplies of produce. The credit network linking Chinese *towkays* with more remote Chinese bazaar and Chinese boat traders was at least as significant a network as the credit provided by Chinese traders to produce collectors. That is, there was a credit network linking a hierarchy of Chinese traders, and another credit network linking Chinese traders with the produce collectors. Credit underpinned the relationship between actors and the articulation of trade of Sarawak. It was supplied most commonly not by cash but by the advance of goods by barter. Trade was not simply a question of physically managing the commodity chain – that is, bringing produce to bazaars, loading and unloading ships and so on as outlined above. It also required relationships among actors, and the balance between trust and obligation.

It is difficult to overestimate the impact of credit in the development of the Sarawak economy. Whatever the shortcomings – and there were many – in the economy and the polity as a whole, the Chinese credit system had two continuing advantages to credit providers. One advantage was the obligation and

168 *Straits Times and Singapore Journal of Commerce*, (hereafter *STSJC*), Singapore, 13 June 1849, p. 422.

169 See STSJC 28 June 1848, p. 287.

relationship established between creditor and debtor. The other advantage was that the goods that were supplied on credit were offered at a profit mark-up. The risks of offering credit were many – debtors, be they other traders or collectors, were sometimes poor credit risks – but overall the credit system was based on the idea of credit providers establishing long term relationships. Chew notes that success in trading in hinterland Sarawak depended upon not simply how much could be bought and sold, but the extent to which traders were willing to take the risk of extending credit.[170] One Chinese trader died in 1889 being owed $2900 and was himself in debt to other *towkays* and 'Dayaks' for a total of $3300. By way of comparison, European cadets entering Brooke service at around that same time, in 1880, were paid $960 per year.[171] In 1903, Chinese traders of the resource-rich Rejang were in debt for up to $30,000 to collectors and the debts were registered at the Government's offices and 'guarantees given'.[172] Traders of the Baram were in debt to collectors for large sums, which the collectors lost when the Chinese traders went bankrupt, which occurred more than once.[173]

Traders took a risk – gutta percha collectors could trade their produce to whomever they wished, even if they already were in debt to other traders to whom they should have provided the produce to repay their debts.[174] As noted earlier, the Chinese river trader was often advanced credit on the assumption that he could obtain produce at a certain price from collectors – but collectors could hold back gutta percha if the market price did not suit them (for example, Muruts did this)[175] or simply ignore the market altogether if they chose to. The trader might be left without any produce – and both trader and collector were vulnerable to changes in market price: this occurred when the gutta percha market fell so seriously in 1903 and some found themselves with large amounts of produce unsaleable even at low prices.[176] Chinese river traders existed by credit they obtained from other Chinese bazaar traders, who in turn existed on credit from Kuching traders (*towkays*) linked to Singapore Chinese traders. One Resident, who reported to the Rajah in 1893 about how the Kuching consignees used unfair weights, said that the local traders, even if

---

170 Chew, p. 91.

171 S. Runciman, *The White Rajahs: A History of Sarawak from 1841–1946*. Cambridge, Cambridge University Press, 1960, p. 162.

172 SG 2 May 1903, p. 85.

173 SG 1 July 1904, p. 144.

174 See SG 2 April 1900, p. 73.

175 SG 1 December 1887, p. 201.

176 SG 4 June 1904, p. 128.

their complaints were justified, were unlikely to take action because the traders were all on borrowed capital and would be afraid to say anything for fear of the consequences from their creditor.[177]

To mitigate risks, interest rates were high. The Consular Report of 1863 said that, as money was scarce and security was questionable, the rate of interest in Sarawak was 'exorbitant' and as high as 25 and even 35 percent per annum, compared, for example, with 12 percent at Singapore in 1865.[178] In 1893, Order No IX by the Rajah (Charles Brooke) reiterated the position of the Order of 2 May 1873 – that interest in credit contracts should be no higher than 12 percent per annum, and that credit contracts should be 'fair and just to both parties'.[179] However, enforcing the law in the informality of the forest economy was not always feasible. If the illiteracy of collectors and fraudulent means such as false weights or adulterated gutta percha were utilized, then the system as a whole was open to abuse – all actors took risks. The system operated on trust – or it operated on risk, depending upon one's perspective. John Dill Ross, the gutta percha trader discussed in Chapter 3, described his relationship with Chinese merchants. It was a system that thrived on avoiding the use of 'complicated forms' even if there was some 'wastage' and that 'Chinese and native traders' responded badly to being treated by strict and 'untrusting standards'.[180] What Ross emphasized was the importance not just of the credit system itself, but the relationship that lay behind it. It was a matter of establishing and maintaining a system of networks involving flexibility and 'give and take' that enabled the trade of the region to function and to resolve other difficulties in the trading environment. A credit arrangement not only requires a relationship, it can build and promote a relationship between participants.

The credit system operated in part to obviate the wider regional problems associated with shortages and inconsistencies in the supply of currency,[181] but more importantly it was part of a wider solution to the needs of the Sarawak environment where resources of all sorts were undeveloped. Such 'debt-credit arrangements have typified extractive economies in thinly inhabited, resource-rich

177 SG 1 August 1893, p. 135.

178 Consular Report on the Trade of Sarawak for 1863, BSP 1865 [3478], p. 2; ST 5 September 1865, p. 3.

179 SG 1 July 1893, p. 109.

180 J.D. Ross, *Sixty Years: Life and Adventure in the Far East*, 2 Vols, London, Hutchinson & Co, 1911, Vol. II, pp. 53,59.

181 See J.H. Drabble, *An Economic History of Malaysia, c. 1800–1990 The Transition to Modern Economic Growth*, Basingstoke, Macmillan, 2000, p. 79 and W.G. Huff, 'Bookeeping barter: Money, credit, and Singapore's international rice trade, 1870–1939', *Explorations in Economic History*, Vol. 26, 1989, pp. 161–189.

"peripheries" ' according to McSweeney, who notes the role of such systems as crucial for linking international markets for forest goods, trading posts and 'indigenous extractors'.[182] That is, in problematic environments like Sarawak, moving from dissonance to consonance involved various mechanisms like credit, bazaars and other trading infrastructure – at its heart, however, was the reality that the economy was a balance between trust and risk – in effect, using social capital to achieve that balance, or taking the risk regardless.

The gutta percha trade in Sarawak reveals the inter-dependencies between the administrative and policy framework of Brooke rule, the relative strengths of other actors in the trade, and the geographic features of Sarawak. These parameters all shaped the formation and operation of the trade, just as external factors such as the demand from cable manufacturers in turn had an impact. Miller and Greenhill defined a commodity chain as 'a series of linked nodes through which a product is transformed and transferred along a chain from primary extraction to final consumption'.[183] As proposed earlier, the different traders in the network performed roles akin to nodes. In effect, the gutta percha trade of Sarawak can be seen as a process of increased engagement with the world-system. If we are 'agnostic' about the flow of profit in the commodity chain, it can be seen how the gutta percha trade accords with the argument put forward by Marichal et al. about the inter-connection that embeds the global into the local and vice versa.[184] Gutta percha, although not answering all the economic needs of Sarawak, had features that suited the economic conditions and policy framework of the Brooke administrations. In turn, the existing features of the forest environment, collectors' and traders' practices and beliefs, and the commodity itself presented conditions which enabled the 'nodes' of the chain to be established almost immediately – and in a manner that overcame the many barriers that the physical, economic, politico-administrative, and social settings would otherwise have put in its way.

Accessing and working the Sarawak forest for produce required expert knowledge and this required skills and adaptations which outsiders were

182 K. McSweeney, 'The dugout canoe trade in Central America's Mosquitia', *Annals of the Association of American Geographers*, Vol. 94, Issue 3, (September) 2004, p. 645.

183 R. Miller and R. Greenhill, 'The fertilizer commodity chains: Guano and nitrate, 1840–1930' in *From Silver to Cocaine. Latin American Commodity Chains and the Building of the World Economy, 1500–2000*, eds S. Topik, C. Marichal and Z. Frank, Durham, Duke University Press, 2006, p. 230.

184 C. Marichal, S. Topik and Z. Frank, 'Conclusion: Commodity chains and globalization in historical perspective' in *From Silver to Cocaine. Latin American Commodity Chains and the Building of the World Economy, 1500–2000*, eds S. Topik, C. Marichal and Z. Frank, Durham, Duke University Press, 2006, p. 360.

not capable of or indeed interested in taking up. As a result, Europeans and even Chinese were largely excluded from the collection of gutta percha. Local knowledge was at the core of the trade. Skills that included an ecological as well as commercial sense of forest resources were embedded in a cultural and spiritual outlook that envisaged the forest environment and the collector as a continuum between nature and spirit – ceremony and omen were part of the collectors' world. The condition whereby collectors were at one with the forest and the spiritual world they envisaged should not, however, be seen as limiting or contradictory to the idea that these actors could be effective or astute traders.

Others in the Sarawak trade had their own skills that complemented those of the collector. Chinese traders may have lacked the intimate ecological skills of the forest collector, but they had commercial and trading skills – and links with the external economy. Chinese traders possessed a capacity to negotiate with the collector and parts of the Sarawak trade environment, and with the commercial realities of the local or regional market. Certain traders such as the *towkays* also understood the external trade network. The demands of regional trade shaped the Chinese traders' world just as the forest environment shaped the collectors' world. Each group developed skills and mechanisms to bridge the gap with other actors. For those groups to interact either with each other or with the collecting environment, practices evolved to facilitate the operation of trade – these solved some of the considerable physical, economic and security challenges facing those who worked to provide a commodity of interest to the 'civilized world'. These mechanisms facilitated very different actors in moving from 'dissonance to consonance' in their relationships. The development of bazaars and credit arrangements provided for the articulation of trade and the development of relationships. The arrangements provided what Curtin described as special institutional arrangements for the mutual security of both sides.[185]

Motivations for trade were focused by the very lucrative prices that gutta percha sometimes commanded. The trade contributed to the pace of change in local communities. To succeed, all actors required the ability to negotiate and be prepared to take the risks associated with the use of credit and interaction with others. That is, they balanced both the challenge and opportunity of engagement in the trade environment at the local level. There were several ways in which collectors and traders worked together to manage risks and solve the differences that separated them – groups contributed in different ways according to their skills and resources. The forest paths were in fact migratory trade

185 P.D. Curtin, *Cross-Cultural Trade in World History*, Cambridge, Cambridge University Press, 1984, p. 1.

paths that were an essential part of the trading infrastructure that solved the practical problem of bringing produce to market. Chinese traders did not have the resources and the skill to move beyond the river systems where they were stationed. That said, on those rivers the Chinese traders provided other essential features including bazaars where produce was collected and sorted, where trade-goods were stored and set out for barter and sale, and where network relationships were formed. Rivers and the boat traders that moved on them could move only so far in the hostile Borneo environment.

Dunn described land in subsistence zones as an area that members of a community regard as exploitable and familiar territory, not necessarily owned but being subject to simultaneous exploitation by different peoples.[186] In Sarawak, the movement of collectors, especially Iban, into subsistence zones beyond those with which they had ancestral connections was not especially in contravention of traditional trade patterns or beliefs. But the demands of the external economy for produce, together with the arrival of Chinese trader networks, provided new opportunities for trade and consequently new challenges to old ideas of shared exploitation of resources. Nonetheless, knowledge of the forest ecology, trade demands and acknowledging and obtaining benefit from omens were all a feature of the collectors' world.

In the new trade environment, the relationships among actors – including between collecting groups and between collectors and intermediary traders – arose from the responses each developed to the trade. Power adaptations and challenges at the periphery were then influenced by demand from the external economy but very much altered by local factors and shaped by local actors. How naïve collectors were in matters of trade is a matter of conjecture. Chinese traders dominated links to the external trade environment: as a result, collectors were cut off from sources of information about the external economy. Most had less experience with the commercial world of the external economy due to lack of access, and less awareness of current prices for either the produce they were trading or the goods they sought to buy – this would make them relatively easy targets. That is, at the local level, collectors were managing a balancing act between the old ways and the needs and opportunities of the new. Conversely, Chinese boat and smaller bazaar traders were to some extent at the mercy of collectors who could ignore credit obligations or develop exchange relationships with other traders, and hold back produce based on the knowledge they had of local, if not regional, market prices for produce.

The impact of these changes is considered in the next two chapters.

186 Dunn, p. 41.

CHAPTER 7

Impact of the Gutta Percha Trade – Opportunity

> The dyaks have for many months past been actively engaged in searching for gutta and rattans … Many of these Dyaks had never come downriver so far in their lives before and, as they obtained good prices for their gutta, they went away much pleased with their visit.[1]
>
> SARAWAK GAZETTE, 1 September. 1890

Who benefited and who suffered when local economies were drawn into the world-system in the colonial era? What evidence might be used to find answers to questions about whether globalization was a good or a negative feature of the nineteenth century world? Such questions are addressed in this and the following chapter. One way to determine the effect is to identify and measure the things people traded, reflecting both the extent and possible causes of change and the motivations to trade. For example, this reveals that collectors traded gutta percha for subsistence items like rice or cloth, but also sought opportunities to trade for prestige goods that enhanced social standing and the sense of spiritual wellbeing. In this chapter, evidence from the historical record is matched with trade data to reveal connections between traders and the external economy. While much is made of the way peripheral economies and populations have been affected by the force of global trade and the spread of empire, surprisingly little use has been made of the statistical trade evidence to identify patterns of change, a previously missed opportunity noted by Cleary.[2] That opportunity is taken here. Statistical and narrative evidence also gives some voice to otherwise unheard communities, provided it is used with caution when interpreting that evidence.[3]

1 Sarawak Gazette (hereafter SG), 1 September 1890, p. 118.

2 M.C. Cleary, 'Indigenous trade and European economic intervention in North-West Borneo c. 1860–1930', *Modern Asian Studies*, Vol.30, no.2, 1996, pp. 301–324; M.C. Cleary, 'From hornbills to oil? Patterns of indigenous and European trade in colonial Borneo', *Journal of Historical Geography*, Vol. 23, No. 1, 1997, pp. 29–45.

3 For historiographical reasons, this material is presented as annual trade figures which can be compared with specific evidence from the documentary record – as a means of establishing,

 | DOI 10.1163/9789004357280_009

Interaction with the External Market

Collectors alternated between collecting gutta percha and other forest produce according to changes in the market place – this can be seen in the documentary record which provides examples that report how collectors quickly adapted to changing market prices or collecting conditions – as Dunn suggested, they were well aware of market demands.[4] One example of this can be seen when comparing the connection between the gutta percha and rattan trades. Rattan was one of the most valuable of the forest produce trades after gutta percha. Rattans (sometimes known as *rotan*) were types of cane collected from palms growing in Sarawak and most of Southeast Asia (they are also found in other equatorial regions of the world). Rattan was useful because of its flexibility, strength and lightness and featured in items such as basketry, furniture, and structural support. The international trade in rattan expanded in the nineteenth century, although it had a long history in the Southeast Asian region before that time.[5] Rattans from Borneo were, according to Hugh Low, 'esteemed finer' than those produced anywhere else in the world.[6] As an important forest resource, rattan was a likely choice for collectors to turn to if the external market or local conditions changed. Collectors switched to working rattans instead of gutta at Lundu, for example, when the price of gutta fell relative to that of rattan, but elsewhere at the same time others were collecting gutta to pay for padi when their crop had failed:[7] that is, each group was responding to local market, collection and crop conditions and balancing their choices accordingly. In 1887, the Sarawak Gazette noted that Muruts were holding back

and at times measuring, connections in a process of globalization. This process was not always a parallel and steady process of integration and trade relationships over the four decades of data shown here, that would necessarily result in strong theoretical, statistical correlations. Therefore, correlation graphs are not used. Rather, certain significant or spasmodic events are connected that link various types of evidence, both documentary and trade, from various sources. The intention is to point to inter-connections rather than to look for long term trends in the form of strong statistical correlations.

4 F.L. Dunn, *Rain-forest Collectors and Traders: A Study of Resource Utlilization in Modern and Ancient Malaya,* Monographs of the Malaysian Branch, Kuala Lumpur, Royal Asiatic Society, No. 5, 1982.

5 Summarised from J. Dransfield and N. Manokaran, (eds), *PROSEA: Plant Resources of South-East Asia 6, Rattans,* Bogor Indonesia, PROSEA, Yayasan Obor, 1994, 'Introduction', which includes an historical overview of the global rattan trade; see also F.W. Foxworthy, 'Minor forest products of the Malay Peninsula', *Malayan Forest Records* No. 2, 1922, pp. 154–155.

6 H. Low, *Sarawak. Notes During a Residence in That Country with H.H. The Rajah Brooke,* London, Richard Bentley, 1848, Reprinted 1988, Singapore, Oxford University Press, p. 42.

7 SG 1 June 1874.

gutta percha waiting for the price to improve and instead began bringing rattans to the bazaars in the Trusan area.[8] Other comments reflecting collectors choosing between rattans or gutta percha as a better option are also found.[9] The rattan trade of the Baram in 1885 was such as to draw specific comment in the Consular Trade Report, and Charles Hose describes how rafts of up to 250,000 canes were brought down the Baram.[10] In 1900, with the dramatic rise in demand for and price of gutta percha as a result of the Pacific cables being produced, 'the majority of jungle workers' of the Rejang gave preference to gutta rather than rattan collecting, causing a consequent decline in exports of the latter in that region; and in 1902 falling gutta percha prices and the 'great distances' that had to be travelled to collect it led some collectors to choose rattan collecting.[11]

The graphs presented earlier, at Figure 3 and Figure 4, that compared the production of submarine cables with exports of gutta percha from Singapore and Sarawak respectively showed that in the 1880s there was a steady production of cables, with a peak in 1884 – exports of gutta percha from Singapore were consistently good in 1881–84, fell in 1885 (when cable production fell), then began to improve for several years. Yet these good export figures for Singapore were not repeated in Sarawak gutta percha export figures. Why would Sarawak trade have failed to take advantage of the good conditions arising from cable production at this time? As seen Figure 11 below, comparing gutta percha and rattan exports, gutta percha exports from Sarawak were low in this period, but there were increases in rattan exports from 1883 to 1888. Had Sarawak collectors turned their back on gutta percha, despite external demand for cable production? Falls in gutta percha prices and better access to rattan may have been the reason. The graph shows how collectors were affected by or took advantage of changes in local trade conditions or how they responded to changes in the global commodity market. Also of note also are the extremes in gutta percha exports when compared with rattan.

The market price for gutta percha used for telegraphic purposes varied widely and actual market prices are not available in regular annual form; but for purposes of providing a context for trends in demand and market price, it is worth considering that the notional average import prices paid per hundredweight

---

8 SG 1 December 1887, p. 201.

9 For example, SG 1 June 1882, p. 40 – rattan preferred over gutta.

10 Consular Report on the Trade of Sarawak for 1885, 1887 [C.4915] [C.4923], p. 194; C. Hose, 'A journey up the Baram river to Mount Dulit and the highlands of Borneo', *The Geographical Journal*, Vol.1, No.3 (March), 1893, p. 195.

11 SG 1 March 1901, p. 61; 3 March 1903, p. 54.

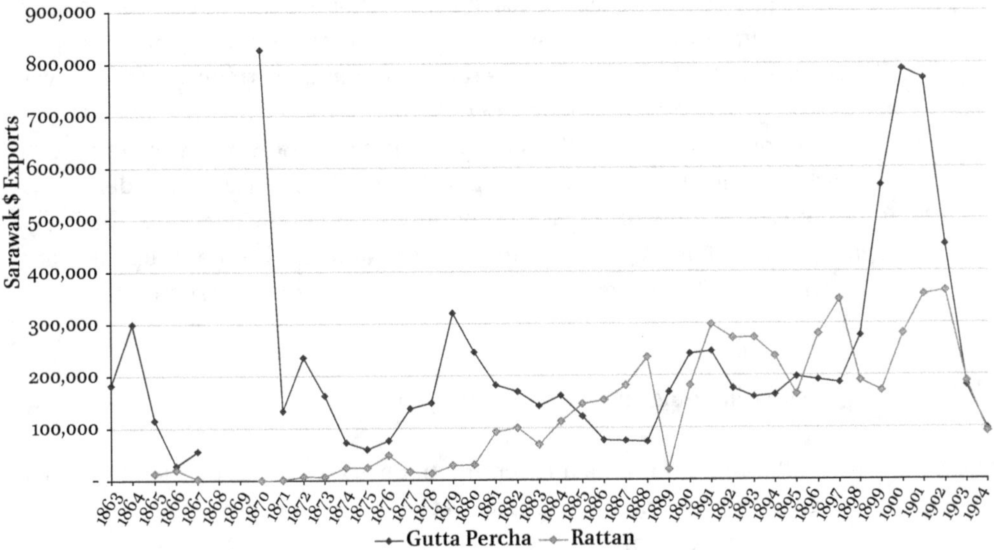

FIGURE 11 *Sarawak gutta percha related to rattan exports.*
Source: Treasurer's Reports reprinted in Sarawak Gazettes (SG), various issues 1870–1904; and Consular Reports in British Sessional Papers (BSP): 1865 [3478], p. 28; 1866 [3582], p. 805; 1866 [3669], p. 266 for the years 1863–65 respectively. Figures for 1868–69 are not available. See Appendix 2, Table 3 for details.

of gutta percha imported into Britain during the early 1880s was around £6 to £7 – yet in 1889 the price almost doubled from its low level, to achieve £12 per hundredweight. It is noticeable that, once the market price for gutta percha in London rose significantly in 1889, so too did gutta percha collection in Sarawak – and rattan collecting dropped significantly. The Murut collectors mentioned above who held back their produce in 1887 may well have done the wise thing if they managed to hold onto their gutta percha until 1889. Prices remained at similarly high levels for the next three years.[12] Figures at Appendix 2 (Figure 23) show that gutta percha exports for Sarawak, in both Sarawak Dollars and pikuls, reflect these trends, so the changes were not simply a result of exchange rate variations. Prices for gutta percha fell again in the early 1890s and rose significantly at the end of the decade because of cable production, and the alternation between collecting rattan and gutta percha re-appears.

The gutta percha – rattan trade figures reflect both local trade practices and the links between the global and the local economies. The question that arises

12 Statistical Abstract for the United Kingdom 1869–83, BSP 1884 [C.4117], pp. 42–43,106–107; Statistical Abstract for the United Kingdom 1885–99, 1900 [Cd.306], pp. 60–61; 70–71; 150–51.

from these responses is whether this reflected an opportunity or a weakness for the local communities? The answer was one seemingly answered by collectors, who chose whichever of the two was the more profitable trade at the time, yet in the process became increasingly tied to the external economy. Such interdependence could present wealth and opportunity, or the loss of it when the external market changed. This paradox – opportunity yet increased dependence – is at the core of the study of trade at the periphery and in a wider sense, the study of global history. Such a paradox can be seen in other trade data when these are matched against those of gutta percha, as seen in the next two sections.

The Changing Nature of Sarawak Imports

> The Resident of the Rejang reports trade active in this district, there is a great scarcity of rice, and this leads to extra work in gutta[13]
>
> SARAWAK GAZETTE, 15 July 1871

Working gutta percha to barter for or buy rice was especially important when the padi harvest had failed or if padi was expensive to buy, and this was a regular point of comment in the Sarawak Gazette for decades. At times the Gazette was critical of collectors for this practice, saying that the gutta percha trade prompted some to live in semi starvation for months of the year because they would 'not plant in the proper manner'.[14] Yet the paper also recognized the reverse of the trade, seeing that a response to shortfalls in cultivation was to find relief in collecting gutta percha.[15] Such was the paradoxical result of the prosperity achieved by the trade: it was a matter tackled in the Gazette in 1874. The cover page of the July 1 issue noted that the scarcity of rice in Sarawak could be measured directly from the trade returns as having arisen when the high price achieved for gutta percha caused the farming of rice to be neglected in favour of the 'wealth so easily obtained in collecting [gutta percha] not only for rice but also for luxuries'. The consequent falling off in the gutta percha trade had meant that there was a resulting 'large increase in the production of rice ... we

13 SG 15 July 1871.

14 SG 16 April 1877.

15 Such as SG 1 June 1874, SG 2 January 1892, p. 15.

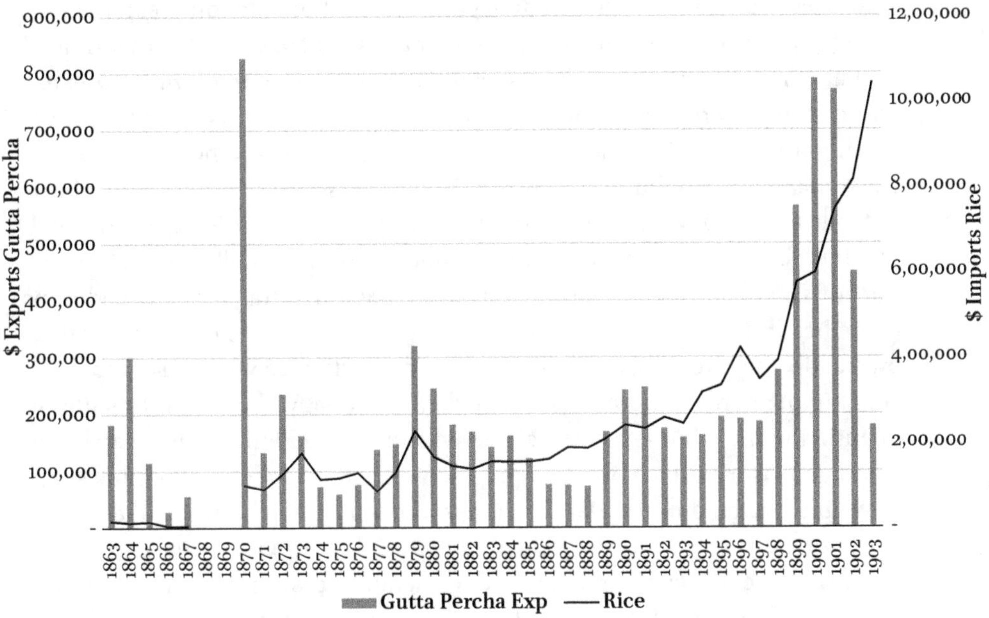

FIGURE 12 *Sarawak gutta percha exports related to importations of rice* (*Sarawak dollars*). *Source*: Treasurer's Reports reprinted in Sarawak Gazettes, various issues 1870–1904; Consular Reports in British Sessional Papers (BSP): 1865 [3478], p. 28; 1866 [3582], p. 805; 1866 [3669], p. 266 for the years 1863–65 respectively. Figures for 1868–69 are not available. See Appendix 2, Table 4 for statistical details.

consider it a sign of prosperity far healthier than that attained in 1871 and 1872 by the rush after gutta'.[16]

It is clear from narrative evidence that trading gutta percha for rice was an important way for gutta percha collecting communities to buy food and trade figures link the two trades, as is illustrated in Figure 12.

The figures for 1872 and 1873 back up the Gazette's comments about high levels of rice importations in those years, and the connection the newspaper made between the fall in gutta percha exports and imports of rice in 1874 is also evident. The variations in the figures for the two commodities continue tracking each other through the 1870's with the increases in 1877 to 1879 and falls in 1880 and 1881 most noticeable. A similar pattern is repeated between 1897 and 1901, when a remarkable increase in both gutta percha exports and importations of rice can be seen. The overall expansion of Sarawak trade at the end of the century may have been due in small part to population growth but, as noted earlier, population growth is regarded by most sources as being very modest. If the commentary from the Sarawak Gazette and the trade figures are compared,

16 SG 1 July 1874.

it suggests that gutta percha trading gave the local population greater choice and opportunity for meeting their need for the key item of food at that time.

It also laid the ground work for greater dependency on trade with the external economy. Were these choices rational? In economic terms, it is not feasible to compare the yield per hours work for rice agriculture on the one hand, and collecting gutta percha on the other. This is for several reasons. Firstly, it is difficult to quantify labor allocations across these sectors. Swidden rice cultivation, for example, involved women and children as well as men – which the gutta percha trade did not. The risk involved in gutta percha expeditions (from headhunters or others looking to fend off gutta collectors entering their own collecting areas) also meant that the two activities were based upon incomparable parameters in terms of risk. Additionally, there is a difficulty in quantifying the labor costs associated with the collection of gutta percha including measuring, even in an average sense, factors including the time spent travelling overland or upriver to sources of supply (which might take days or weeks or months).[17] Also, the yield obtained, and the time to taken to cut down a tree to collect the gutta, might vary considerably according to the season, size of the tree and local forest conditions. Von Gaffron reported that a collector working quickly could cut two trees per day but it cannot be known if this was common 'average' practice or not and Bright notes that the 'exudation' process generally occurred over several days'.[18] Obviously, the longer time that was allowed for this process the more gutta percha might be collected and collectors could in theory have spent this time searching for and cutting other trees in the locale. Or they may have engaged in other produce collecting activities, social life or even headhunting. Finally, the latex was boiled, cleansed and coagulants (and sometimes adulterants to improve appearance and market value) were added.[19] This process might be repeated to improve the product obtained.[20] Again, the number of times this was done, or the labor that each group decided to put into this phase of collection, would influence how much labor went into

17 See for example C. Hose and W. McDougall, *The Pagan Tribes of Borneo*, 2 Vols, London, Macmillan and Co, 1912, Vol. I, pp. 150–151, also L.M. Potter, 'A forest product out of control. Gutta percha in Indonesia and the wider Malay World 1845–1915', in *Paper Landscapes: Explorations in the Environmental History of Indonesia* eds. P. Boomgaard, F. Combijn and D. Henley, Leiden, KITLV Press, 1997, p. 286.

18 See W.G. von Gaffron 'Nota Omtrent den getah-pertja-boom', *Nutuurkundig Tijdschrift voor Nederlandsch-Indië*, 16, 1858, p. 226 and Potter pp. 286–287; C. Bright, *Submarine Telegraphs Their History, Construction and Working*, London, Crosby Lockwood and Son, 1898, p. 257 n.†.

19 C. Bright, p. 257.

20 T. Seeligmann, L. Torrilhon and H. Falconnet, *Indiarubber and Gutta Percha. A Complete Practical Treatise*, 2nd English edn, London, Scott, Greenwood & Son, 1910, p. 330.

the process. Conversely, the swidden process itself is based on different phases and occupied different groups of the community at different times; women and children provided much of the effort required, for example planting, weeding etc., while men cleared and burned the land (women do not appear in the written record as having been part of the gutta collecting expeditions). Gutta percha collecting expeditions often occurred during the 'down time' that occurs as part of the cycle of swidden agriculture; in that sense, there was no opportunity cost in collecting expeditions since the needs of subsistence agriculture were not otherwise being interrupted.

Moreover, to the extent that gutta collecting expeditions were an opportunity to obtain prestige items (*pusaka*) discussed below, or as an opportunity to prove valour by going on risky expeditions, gutta percha collecting cannot be measured against a eurocentric concept of cost-benefit to indigenous communities earned by gutta percha trading. This particularly applied to the Iban, for whom warfare and trade existed 'in a social context' with reward going to those who gained a 'mastery of the external world' such as might be demonstrated by successful trading with the external economy.[21] In this regard, the global-local connections promoted by the demand for gutta percha highlight the extent to which this trade should be seen as more complex and nuanced than merely a series of linked commercial transactions. We would be comparing different returns on labor: food, financial returns for rice cultivation on the one hand, compared with food, financial, prestige goods and social standing as a return for gutta percha collecting expeditions which did not accrue from swidden agriculture.

The gutta percha – rice trade statistics do, however, raise questions. Why were the rice importations so comparatively low in the 1860's (especially 1864) and in 1870, given that gutta percha exports were so high in those years and presumably people could have afforded to buy more rice? At that time, gutta percha was the key export for Sarawak, and major commodity most indigenous communities had to trade, so why did they not buy rice at the levels seen in later years? Conversely, while the figures towards the end of the century may in part reflect price rises and exchange rate variations, the increases in rice importations from 1898 onwards, appear too high simply to reflect the fact that gutta percha collectors could now afford to buy more rice – could people have eaten all that much more rice compared with levels of the past? Yet, clearly, both sets of statistics increase noticeably around that time. Indigenous produce collectors made up the largest proportion of the population and were not directly profiting from the other areas of improvement in Sarawak trade at the

21 See Sather, C., ' "All Threads are White" Iban egalitarianism reconsidered' in *Origins, Ancestry and Alliance: Explorations in Austronesian Ethnography*, eds C. Sather and J.J. Fox, Canberra, ANU E Press, 2006, p. 105.

end of the century (pepper and gold), so the market for rice would not seem to have been a result of their buying rice.

In fact, the connection between these two trades may be associated not just with the greater wealth of gutta percha collectors and their communities, but in part because of the trading practices of Chinese traders. In his analysis of the international rice trade, Huff concludes that the region's economic development was promoted by the Chinese barter system based on rice as the medium of exchange – the 'pecunia of the east', which had the advantage of avoiding disputes about 'differing coins of varying reliability', and of course for Chinese traders to 'profit from both sides of the exchange' as Huff puts it.[22] In this case they could obtain a beneficial value in the exchange of both rice and gutta percha. The significance of this in relation to Singapore's role in regional economic development was discussed in Chapter 4. Latham and Neal have noted too that the region's rice trade was dominated by the Teochiu, and that rice was used as a medium of exchange for Singapore traders to pay for hinterland produce, including gutta percha, even as late as the 1930s.[23] The history considered earlier showed that gutta percha was also a Teochiu trade in Sarawak. One advantage of rice as a medium of exchange relates to a point made by Jevons in his classic *Money and the Mechanism of Exchange*, about the benefits of barter goods, which is that such goods can subdivided; that is, broken down into amounts small enough to be traded in exchange for (in this case) forest produce such as gutta percha.[24] This was helpful for trade both within the various layers of the Chinese trade networks – where goods of necessity had to be broken down into more readily transportable and more affordable configurations – and especially in subsistence communities where the amount that might be traded varied according to the local trading conditions. To test the proposition that the demand for Sarawak gutta percha influenced its rice trade, it is important firstly to identify in what years the gutta percha exported from Sarawak may have been paid for mainly with currency – or treasure as it was known in the trade statistics. The relationship between gutta percha and imports of treasure is set out at Figure 13.

In the 1860's, the connection between gutta percha exports and the import of treasure appear to link closely. This suggests that in those years it was mainly treasure, rather than rice, that was the means of exchange used to buy Sarawak gutta percha. The enormous level of treasure imported in 1870 (not equaled

22 W.G Huff, 'Bookeeping barter: Money, credit, and Singapore's international rice trade, 1870–1939', *Explorations in Economic History*, Vol. 26, 1989, p. 175,185 and p. 176.

23 A.J.H. Latham and N. Neal, 'The international market in rice and wheat, 1868–1914', *The Economic History Review*, New Series, Vol. 36, No.2 (May), 1983, p. 274 and p. 182.

24 W.S. Jevons, *Money and the Mechanism of Exchange*, New York, D. Appleton & Company, 1919, p. 6.

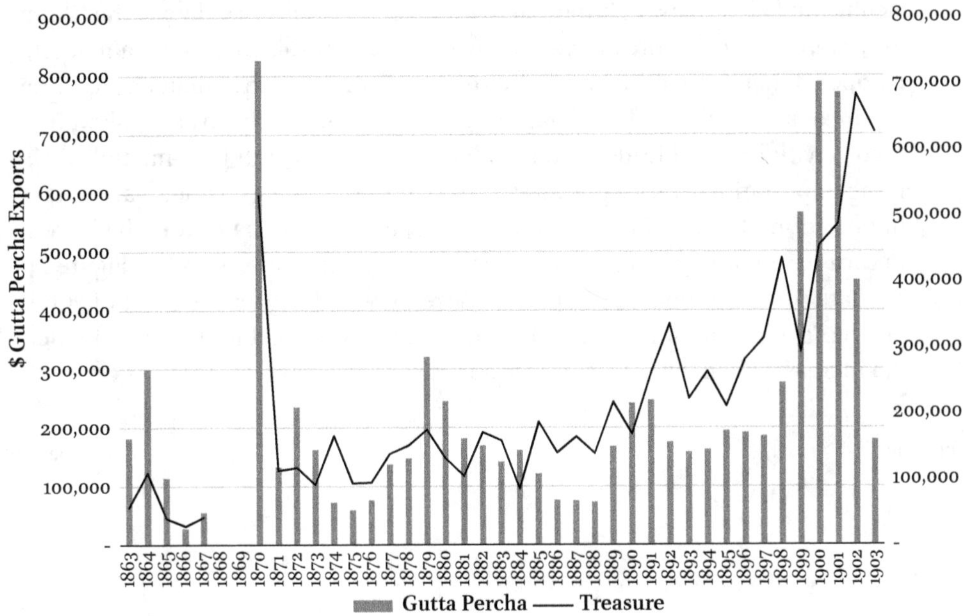

FIGURE 13 *Sarawak gutta percha exports related to importations of treasure.*
Source: Treasurer's Reports reprinted in Sarawak Gazettes, various issues 1870–1904; Consular Reports from British Sessional Papers (BSP): 1865 [3478], p. 28; 1866 [3582], p. 805; 1866 [3669], p. 266 for the years 1863–65 respectively. Figures for 1868–69 are not available. See Appendix 2, Table 4 for statistical details.

or surpassed until 1902) occurred around the time that gutta percha was purchased mainly by BCL for the West Indian cable and other ventures it had with Silvertown. It was in describing the trade of 1870 that Vice Consul St John had said there was a large influx of 'silver coin consequent upon the gutta percha export'.[25] As a European operation, the company had access to sources of capital to obtain produce; that is, it could obtain treasure in the considerable amounts required to pay for gutta percha exports. This suggests that, when there was more direct European engagement in the trade (BCL), there was a greater tendency for treasure to be used to pay for the produce. After 1869 BCL was no longer the monopoly trader in Sarawak gutta percha and, apart from the BCL exports in 1870, it was Chinese traders from Singapore who were now becoming more active in the market.

It is from those years onwards that change begins to appear in the statistics: variations in the importations of treasure no longer seem to link as closely to the levels of gutta percha exported. How was the increase in exports of gutta percha in 1872 paid for when importations of treasure hardly changed? Similarly, how was the increase in gutta percha exports in 1899 paid for given the very

25 Consular Report on the Trade of Sarawak for 1870, BSP 1871[C.429] p. 806.

significant fall in treasure imported in that year? If Chinese traders, rather than BCL, were now the more active presence in the Sarawak gutta percha market, it appears that these traders must have relied more on barter than on treasure as a means of exchange,[26] and bartering rice for gutta percha in Sarawak would accord with regional trade patterns identified by Huff, and Latham and Neal. If the importations of rice grew as a means of 'paying' for gutta percha, can the importation of rice to Sarawak and the rise in gutta percha exports be linked in turn to seemingly low importations of treasure in these years? At Figure 14 the data compares importations of both treasure and rice with exports of gutta percha. The results suggest that in some years the importation of rice in fact did increase when levels of treasure decreased.

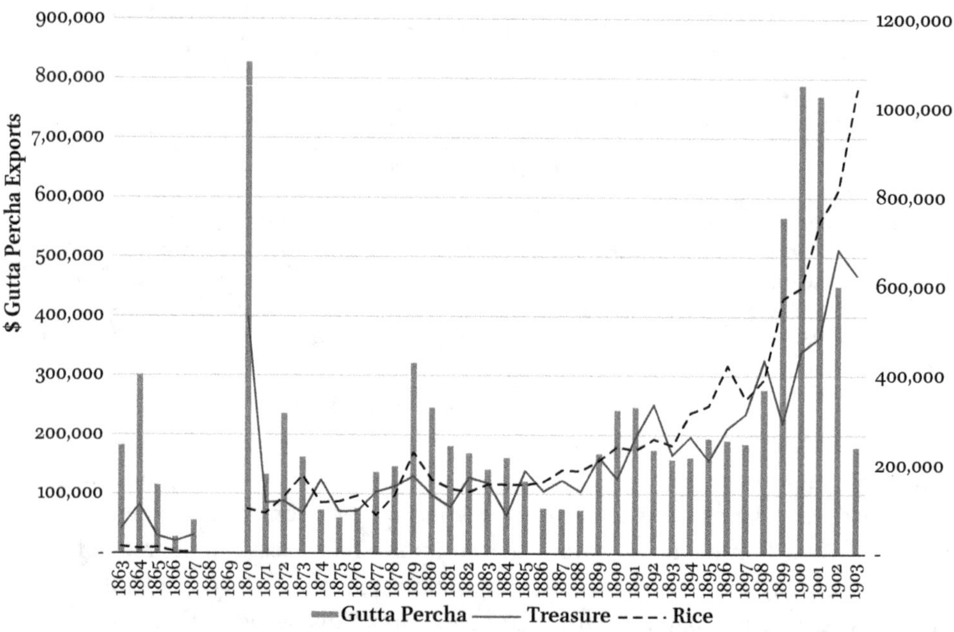

FIGURE 14 *Sarawak gutta percha exports related to importations of rice and treasure: entwined commodity trades?*

Note: The data presented suggests a connection between the demand for gutta percha and the Sarawak rice trade. The connection reflected the nature of Chinese trade networks, their use of rice as a medium of exchange, and the domination of Teochiu traders in both the rice and gutta percha trades.

Source: Treasurer's Reports reprinted in Sarawak Gazettes, various issues 1870–1904; Consular Reports from British Sessional Papers (BSP): 1865 [3478], p. 28; 1866 [3582], p. 805; 1866 [3669], p. 266 for the years 1863–65 respectively. Figures for 1868–69 are not available. See Appendix 2, Table 4 for statistical details.

26 Consular Report on the Trade of Sarawak for 1869, BSP 1871 [C.343], p. 8.

Rice imports seem comparatively low in the 1860s, when it was not utilized to pay for gutta percha because the Chinese were not active in the gutta percha trade but are noticeably higher in later years, once Chinese traders do become more active in the trade. Thus, from the 1870's onwards, the levels of rice imported lift noticeably. Of note are the levels in 1872, 1873 and 1879 when rice increased, rather than treasure, to pay for greater gutta percha exports. The other years in which this relationship is most noticeable are in 1890 and 1899 when gutta percha exports rose, and yet there was a significant fall in treasure importations seemingly made up for by a dramatic rise in rice importations. Also in 1900 and 1901, the rise in treasure imported seems insufficient to have paid for the great increase in gutta percha exported but the increase in rice made up the shortfall. At this time too, the rise in pepper exports added pressure to the need for exchange.

In part, the increase in rice importation at the end of the century could relate to growth in the Sarawak economy. An increase in population growth could have been a contributing factor but such growth, as noted earlier, is generally considered modest and of its own would not seem to explain the increase. Another cause for the increase in the levels of rice imported may be due to the absence of collectors from their padi growing activities as they exploited the opportunities to collect gutta percha, thereby requiring, or enabling, those remaining at home to buy more rice to make up the shortfall (as the Sarawak Gazette had been pointing out for decades). Greater wealth in the country also gave people an increased ability to buy rice. This could be considered a great gain for impoverished communities, or conversely it could be seen as an increased dependency upon the external economy. In parallel with that cause, there is an additional reason to propose that the gutta percha trade contributed to a demand for rice as a medium of exchange. The 1902–03 increases in treasure and rice importations were likely also fueled by the significant increase in Sarawak's pepper exportations which placed added pressure on the demand for treasure or some other means of exchange: recalling too that pepper, the region's rice trade and the gutta percha trade were all dominated by Teochiu traders, it seems plausible this was yet an additional reason for rice becoming so important in the local economy. In effect, there was an inter-relationship between these commodity trades as forest collectors and traders linked with the wider regional economy through barter trade networks.

If the linkages are valid, they illustrate the interconnectedness between global trades, as actors within a commodity chain develop systems and relationships that provide the 'logic' that enables the trade to function: in this case through the operation of Teochiu trade connections and the demand for rice from gutta collectors. In Sarawak, the inter-relationship between these various trades emphasizes the dependencies between the way in which actors traded, society as a whole, and the impact of the global economy. Again, this suggests that studying local features of global trade enables a fuller understanding of the world-system – the one influencing the other and not all features of peripheral trade are shaped by demand from the core. In fact, the trade patterns of the Chinese 'world-system', using rice as a means of exchange and relying heavily upon clan networks, appear very much to be shaping local and regional trade. The connections between the gutta percha and rice trades shown above do not necessarily indicate that the rice trade in Sarawak was simply a manifestation of the Chinese trade networks and somehow divorced from indigenous economies. European metropolitan centres generated the demand for gutta percha that provided local communities more opportunities to buy rice. For rice to be used successfully as a medium of exchange, it had to be sought and consumed by collecting communities. The entry of Chinese traders into the Sarawak gutta percha trade enabled the distribution of rice into the forest economy and into forest communities. In comparison, BCL relied more on treasure; the Company was neither interested in, nor capable of, running a supply chain that could ensure the movement of rice into indigenous communities where it was a valued and transformative aspect of the local economy and society. There, as Ooi has noted, the trade in 'jungle produce' was in any case a more lucrative and attractive alternative to subsistence rice production.[27]

Another set of trade statistics that suggests an interconnection between the external and local economies are those that compare the importations of cloth (or 'shirting' as mentioned previously) and the exports of gutta percha. Cloth had been produced traditionally by most groups in Sarawak, either woven, or processed from tree bark,[28] but commercially manufactured cloth was a very common item of trade, as seen earlier when 'Ken-Wat's men' sold *blachu* cheaply to gutta percha collectors as a way of building up good trade relationships. It was traded for gutta percha in Singapore in the early years of the trade, reflecting the ease of use for exchange in the regional markets. Like rice, cheaper cloth was so commonly bartered it was itself a medium of exchange with indigenous traders and European travellers in Sarawak often carried it for trade or to pay for services.[29] Charles Hose wrote that in the Baram district, before trade became more monetized, the more common exchange for forest produce

27 K.G. Ooi, 'For want of rice: Sarawak's attempts at rice self-sufficiency during the period of Brooke Rule 1841–1941', *Journal of South East Asian Studies*, Vol. 29, No.1, 1998, p. 20.

28 Hose, C., 'The natives of Borneo', *The Journal of the Anthropological Institute of Great Britain and Ireland*, Vol. 23, 1894, p. 165; SG 1 September 1894, p. 146.

29 Discussed by D. Chew, *Chinese Pioneers on the Sarawak Frontier*, Singapore, Oxford University Press, 1990, pp. 108–109; F.W. Burbidge, *The Gardens of the Sun or A Naturalists*

was 'cotton goods' and Chew lists it as an item frequently carried by Chinese traders for exchange with forest produce traders.[30] There was, then, in a sense nothing unusual about cloth being an important part of Sarawak's imports, it was part of an historical and regional pattern of trade, but an interlinkage between the export of gutta percha and the importation of cloth, like the rice trade, reflects the nature of indigenous trade patterns and the role gutta percha played in linking the local economy with that of the near region and beyond. In linking the trade statistics for gutta percha and cloth, the response of collectors to the profits they made from the demands of global telegraphy can also be seen. Note the linked and dramatic rise and fall in the years between 1864 and 1866, and 1870–1871 and the continuing rises from the mid-1880s up to 1901, coinciding with the peak years of cable production as shown in Figure 15 below.

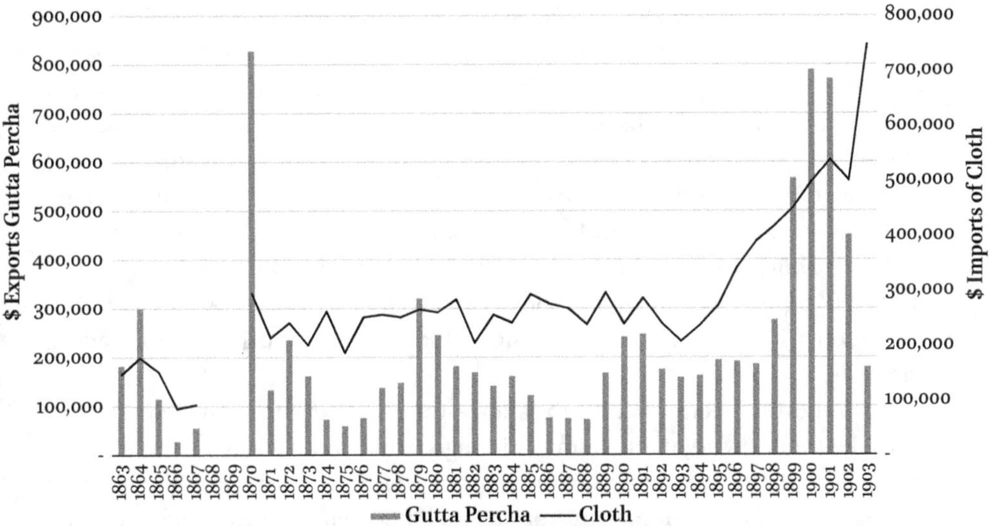

FIGURE 15 *Sarawak gutta percha exports related to importations of cloth* (*Sarawak dollars*). *Source*: Treasurer's Reports reprinted in Sarawak Gazettes, various issues 1870–1904; Consular Reports in British Sessional Papers (BSP): 1865 [3478], p. 28; 1866 [3582], p. 805; 1866 [3669], p. 266 for the years 1863–65 respectively. Figures for 1868–69 are not available. See Appendix 2, Table 4 for statistical details.

Journal on the Mountains and in the Forests and Swamps of Borneo and the Sulu Archipelago, London, John Murray, 1880, p. 330.

30 C. Hose, 'The natives of Borneo', *The Journal of the Anthropological Institute of Great Britain and Ireland*, 1894, Vol. 23, p. 168; see also D.D. Daly, 'Explorations in British North Borneo, 1883–87', *Proceedings of the Royal Geographical Society and Monthly Record of Geography*, New Monthly Series, Vol.10, No.1, (January) 1888, pp. 7, 15–16 and D. Chew, *Chinese Pioneers on the Sarawak Frontier*, Singapore, pp. 83,84,88.

The importations of cloth compared with exports of gutta percha show trends similar to that of rice. In part, the increases in both cloth and rice importations at the end of the century reflected improvements in the overall terms of trade for pepper, gambier and gold, but the parallels with the gutta percha trade seen from the graphs in this chapter emphasize the role of the trade and how it connected with indigenous communities in satisfying local needs. As with rice importations, cloth was largely directed at the indigenous population which was not in large measure participating in the pepper, gambier or gold trades. Chiang pointed to the significance of cotton piece goods in regional trade, but notes that, from being the single most important commodity of Straits Settlement trade in 1870, there was a noticeable decline by 1900 to only 5 percent of either imports or exports.[31] Yet this fall is not in evidence, in fact it is reversed, in Sarawak as shown in Figure 15. Cloth was a significant item of trade valued by indigenous groups both for its practical uses and because it provided a durable and fungible commodity within their own exchange networks. In that regard, the importation of cloth (like rice) into the Sarawak economy may have been a manifestation both of collectors meeting their subsistence needs and Chinese traders and local traders using it as a medium of exchange to pay for gutta percha. Given the economic significance, the role of rice and cloth as media of exchange in regional and Sarawak trade remains a topic worthy of further study.

A snapshot of three other trade connections that may highlight the impact of the gutta percha trade on the society and economy of Sarawak is set out below. In turn, it highlights the economic and social impetus driving local trading interests.

An important item in the subsistence trade was ironware which was imported either as readymade items or in ingots that could easily be transported to the hinterland and then transformed into axes, agricultural implements and weaponry. It was an important item of exchange with the external economy over many centuries.[32] Again, the statistics link gutta percha trading in helping collectors to buy ironware. Figure 16 illustrates the connection – as with rice and cloth , the general improvement in Sarawak exports towards the end of the century may have contributed to increased imports, but the generally linked trend across the period parallels the movements in gutta percha exports.

Of any of the products that indicated a shift from local traditional, subsistence patterns to those introduced by the European trade networks, it may be sugar that was the most representative. It was increasingly traded and

31 H.D. Chiang, *A History of Straits Settlements Foreign Trade 1879–1915*, Singapore National Museum, 1978, p. 118.

32 See D. Freeman, *Report on the Iban*, University of London, London, The Athlone Press, 1970, p. 175 and C. Padoch, *Migration and Its Alternatives Among the Iban of Sarawak*, PhD Thesis, Columbia University, 1978, p. 193.

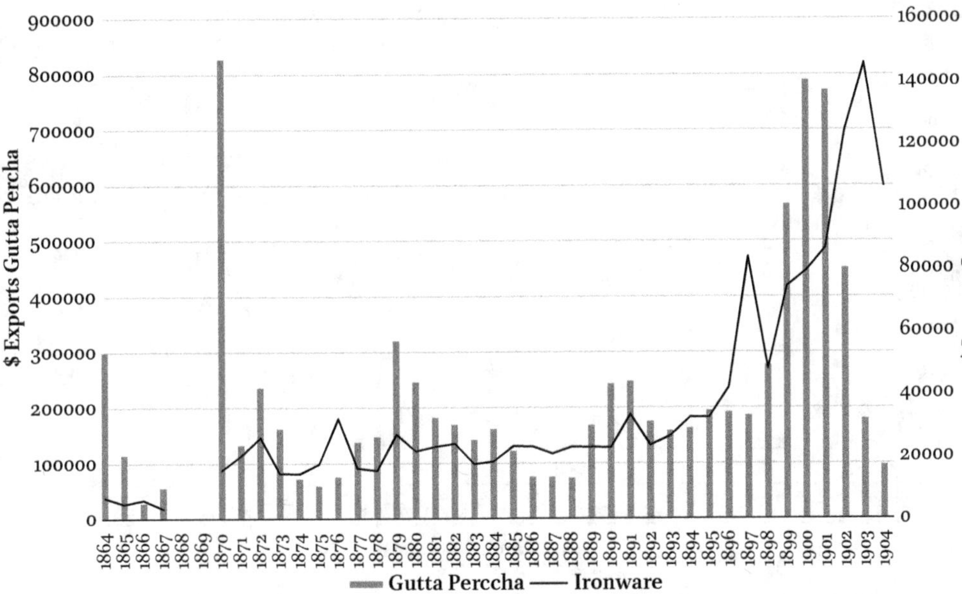

FIGURE 16 *Exports of gutta percha related to importations of ironware (Sarawak dollars).* *Source*: Treasurer's Reports reprinted in Sarawak Gazettes, various issues 1870–1904; Consular Reports in British Sessional Papers (BSP): 1865 [3478], p. 28; 1866 [3582], p. 805; 1866 [3669], p. 266 for the years 1863–65 respectively. Figures for 1868–69 are not available. See Appendix 2, Table 5 for statistical details.

Chew lists it as an item frequently carried by Chinese traders for exchange with forest produce traders.[33] The gutta percha trade made it far more accessible to many, as the apparent paralleling trade figures in Figure 17 show.

Another example where trade figures may shed light on the role of the gutta percha trade was its possible links with opium importations. As Chinese traders enabled the distribution of gutta percha from the hinterland to the external market, the question arises as to what else might Chinese traders have done with the profits of the gutta percha trade? One possibility may have been to buy opium. The links between gutta percha and opium trading in Singapore and in Sarawak have been noted earlier, so profits from gutta percha trading could have been directed to investing in the opium trade. The tax upon it was an important part of the Brooke revenue base. Ken Wat [Law Kian Huat], Sarawak's leading gutta percha towkay, along with another Teochiu merchant, also owned the government opium concessions for several years.[34]

33 Chew, 1990, for example pp. 83, 84, 94, 96.

34 C.A. Lockard, *Chinese Society and Politics in Sarawak: Historical Essays*, Jalan Salim, Sarawak Chinese Association, 2009, p. 53.

FIGURE 17 *Exports of gutta percha related to importations of sugar (Sarawak dollars).*
Source: Treasurer's Reports reprinted in Sarawak Gazettes, various issues 1870–1904; Consular Reports in British Sessional Papers (BSP): 1865 [3478], p. 28; 1866 [3582], p. 805; 1866 [3669], p. 266 for the years 1863–65 respectively. Figures for 1868–69 are not available. See Appendix 2, Table 5 for statistical details.

Opium was an important factor in the local economy but its use was largely restricted to the Chinese community: the Sarawak Gazette, for example, reported that increases in Chinese immigration were usually accompanied by increases in opium imports.[35] Chew discusses the credit debt that Chinese boat traders – important actors in the gutta percha trade – incurred because of their addiction to opium supplied by their *towkay*.[36] In the years when these boat traders benefited – on however limited a margin – from a better gutta percha trade (or the reverse), some connection might be expected in the opium trade figures since the peripatetic traders had direct contact with the gutta percha trade. Wealthier merchants may have used gutta percha profits either for their

35 SG 24 June 1878, p. 42; see also H. Low, *Sarawak. Notes During a Residence in That Country with H.H. The Rajah Brooke*, London, Richard Bentley, 1848, Reprinted 1988, Singapore, Oxford University Press p. 134; O. Beccari, *Wanderings in the Great Forests of Borneo. Travels and Researches of a Naturalist in Sarawak*, London, Archibald Constable & Co, 1904, Reprinted Singapore, Oxford University Press 1986, pp. 19, 68; N. Tarling, *Britain, the Brookes and Brunei*, Kuala Lumpur, Oxford University Press, 1971, p. 322; Chin, p. 29; Chew, p. 77 and A. Kaur, 'The babbling Brookes: Economic change in Sarawak 1841–1941', *Modern Asian Studies*, Vol. 29, No.1, 1995, p. 106.

36 Chew, p. 85.

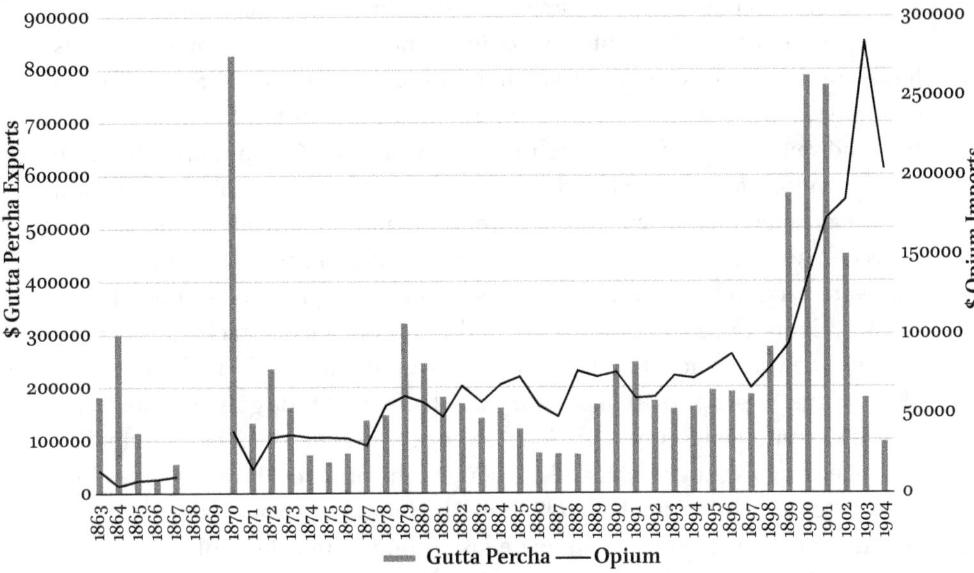

FIGURE 18 *Sarawak gutta percha exports related to importations of opium*
Source: Treasurer's Reports reprinted in Sarawak Gazettes, various issues 1870–1905; and Consular Reports in British Sessional Papers (BSP): 1865 [3478], p. 28; 1866 [3582], p. 805; 1866 [3669], p. 266 for the years 1863–65 respectively. Figures for 1868–69 are not available. See Appendix 2, Table 5 for statistical details.

own opium as some were also users,[37] or a certain *towkay* may have invested profits from the gutta percha trade into opium for onwards trade, either within Sarawak or regionally to Singapore. Figure 18 suggests that the gutta percha and opium trades of Sarawak may well have been connected. Certainly, a sufficient linkage depicted in the Figure suggests that the possibility of such a relationship is another fertile area for further research.

Prestige Goods or *Pusaka*

> One article of importation appears for the first time this year, viz., jars. These are of a peculiar kind, manufactured in China, to resemble the jars which, together with gongs and brass guns, form the old treasure of the Dyaks...[38]
>
> Consular Report on the Trade of Sarawak for 1866

37 Beccari, p. 154.

38 Consular Report on the Trade of Sarawak for 1866, BSP 1867–8 [3953-I-VIII], pp. 437–438.

The indigenous population of Sarawak at this time is best described as living a near subsistence existence, but collectors did not trade produce only for subsistence goods – the *pusaka* trade was an important feature of the Sarawak economy. *Pusaka* were prestige items that had significance as heirlooms or because they had spiritual significance such as if they were inhabited by *antu*. They were also an important display of social standing, wealth or power in the community. *Pusaka* could include expensive Chinese ceramic jars, beads, ornate cloth or brass items such as gongs, boxes or canons. These items were also investments and were used in times of hostilities as deposits, as guarantees of good faith, to repay debts, or settle disputes to avoid violent revenge attacks. Kreps notes that *pusaka* was critical to maintaining kinship ties and links to the authority of ancestors.[39] The pursuit of *pusaka* was a hallmark of the gutta percha trade.

The Commercial Reports on Sarawak for 1864 noted that the principal exports were sago, gutta percha and timber and the principal imports were cloth, brassware (*pusaka*) and tobacco.[40] From the following year, ceramic heirloom jars like those referred to above were also an important item of trade for exchange. Almost three decades later, little had changed:

> The articles in which the export duties shew [sic] the largest increases are timber and gutta percha ... The import duty which shews the greatest increase is that on brass-ware, iron and steel. The increase in the duty on jars – which is greater than during the previous years, as well as that on brassware &c – may be attributed to the greater profits which have fallen to the interior tribes who purchase these articles, owing to the energy displayed by them in the collection of jungle produce, consequent on the better prices obtained for such produce...[41]

Connections between the gutta percha trade and that of *pusaka* items such as gongs and jars were often referred to:

> The falling off shown by gutta ... is reflected in the import duties by the smaller amount of duty received on gongs and jars as these are purchased mostly by the interior tribes who are the gutta hunters.[42]

---

39 C. Kreps, 'Non-Western models of museums and curation in cross-cultural perspective' in *A Companion to Museum Studies*, ed. S. MacDonald, Malden, Blackwell Publishing, 2006, p. 467.

40 Consular Report on the Trade of Sarawak for 1864, BSP 1866 [3582], p. 803.

41 SG 1 June 1892, p. 97.

42 SG 1 June 1893, p. 82.

Furness wrote that, after hostilities, peace and goodwill was gained not by words but by the practical exchange of items such as beads, jars, brass gongs or similar indemnities. He also cites one such example settled by negotiation with the payment of *tawaks* (bronze gongs). A large gong could be worth sixty Mexican Dollars.[43] Harrisson concludes that *pusaka* were held from one generation to the next and fostered social cohesion and cultural ties and the pursuit of them encouraged contact and trade between groups, while Eghenter finds that amongst the Kenyah and Kayan, prestige items could be procured through regular commercial expeditions.[44] The attraction of gutta percha for collectors was in part commercial gain but also these social and cultural factors. Some Chinese traders specialised in luxury items, and Tagliacozzo suggests that markets and the expansion of commerce and the importation of prestige items had the effect of mutating local systems of culture and altering status relationships.[45] In this sense, a key result of the expansion of global commodity trade – markedly the gutta percha trade – into the forest economy of Sarawak was improved access to prestige items that in turn provided mechanisms for dealing with hostilities and demonstrating social standing. That is, the trade in *pusaka* not only had a material and spiritual significance, it had the potential to alter the nature of social relationships within and between communities. In this regard, the expansion of global telegraphic connections had an unforeseen and far reaching impact because, as will be seen below, the gutta percha trade made a significant contribution to the *pusaka* trade in Sarawak.

In 1873, gutta collectors along the Sadong river obtained gongs, guns, brassware, elaborate costumes and coins for ornamentation in exchange for their produce.[46] The trade figures repeatedly reflected this type of pattern. The trade returns for the first half of 1891 for the Trusan area showed a significant increase in exports very largely based on gutta percha: in turn, imports rose, especially articles for 'Murut consumption' – brassware almost doubled, and heirloom jars increased considerably.[47] In 1899 the Resident of the

---

43 W.H. Furness, *The Home-Life of Borneo Head-Hunters: Its Festivals and Folk-Lore*, Philadelphia, J.B. Lippincott Company, 1902, p. 118.

44 B. Harrisson, *Pusaka. Heirloom Jars of Borneo*, Oxford, Oxford University Press, 1986, p. 1 and C. Eghenter, 'Histories of conservation or exploitation? Case studies from the interior of Indonesian Borneo' in *Histories of the Borneo Environment,* ed. R.L. Wadley, Leiden, KITLV Press, 2001, p. 754.

45 Chew, p. 94; E. Tagliacozzo, 'Onto the coasts and into the forests; Ramifications of the China trade on the ecological history of northwest Borneo, 900–1900 CE', in *Histories of the Borneo Environment*, ed. R.L. Wadley, Leiden, KITLV Press, 2005, p. 37.

46 SG 17 July 1873.

47 SG 1 September 1891, p. 141.

Bintulu concluded that he expected a considerable improvement in gutta percha exports due to the large number of Iban settling in his region, noting that already there was a four-fold rise in the importation of brassware. Freeman has suggested that of the various prestige items held by Iban, brassware was more likely to be exchanged than jars.[48] Trade data for the gutta percha trade supports a close link between brassware and gutta percha trading, as seen in Figure 19. The occasional rises (and falls) such as in 1891 and 1895–96 were also a result of variations in the rattan trade, which appears to have been affected by changes in demand for gutta percha (see above in Figure 11). Figures for 1904 are included to reflect the impact of falls in gutta percha trade in 1903 which severely reduced the importation of brassware the following year. In effect, the chart is illustrating how the demands and commerce of global telecommunications affected the importation of prestige items into Sarawak.

With regard to Chinese heirloom jars, according to Sandin, Iban chiefs or warriors of good families would achieve recognition by proving bravery,

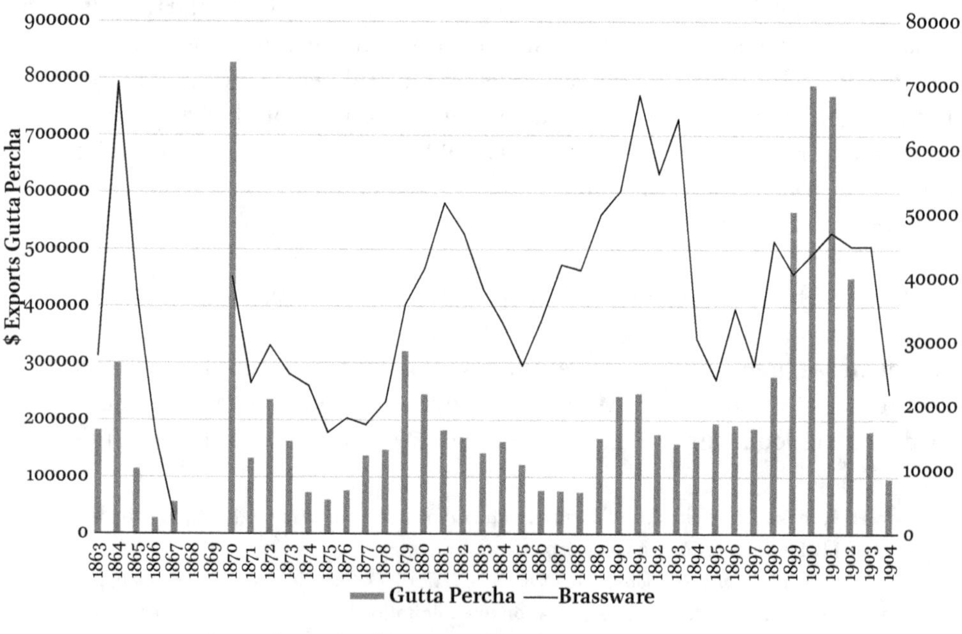

FIGURE 19 *Sarawak gutta percha exports related to importations of brassware.*
Source: Treasurer's Reports reprinted in Sarawak Gazettes, various issues 1870–1905; and Consular Reports in British Sessional Papers (BSP): 1865 [3478], p. 28; 1866 [3582], p. 805; 1866 [3669], p. 266 for the years 1863–65 respectively. Figures for 1868–69 are not available. See Appendix 2, Table 5 for statistical details.

48 Freeman, *Report on the Iban*, p. 267.

obtaining a head (headhunting) or owning valuable jars.[49] Sandin, himself an Iban, wrote of his ancestors' quest for jars, headhunting and the significance of *bejalai*. Many of the overseas *bejalai* expeditions he connects with the rise of the gutta percha trade and the opportunities it provided to buy jars. This, he notes, was because Iban chiefs needed jars to be 'recognised as influential or powerful', with certain jars seen as the equivalent of having killed one enemy or holding the equivalent bargaining power of two captives. A warrior who had taken a head or held captives or jars was recognized as *raja berani* or 'rich and brave'.[50] Standing in the community grew from ownership of heirloom jars inhabited by *antu*.[51] Furness writes of one of his photographs (Plate 14) showing

PLATE 14 *Furness' photograph of Iban bargaining over jars.*
Source: W.H. Furness, *The Home-Life of Borneo Head-Hunters: Its Festivals and Folk-Lore*, Philadelphia, J.B. Lippincott Company, 1902, p. 125.

49 B. Sandin, 'Sources of Iban traditional history', *Sarawak Museum Journal*, Kuching, No. 67, Special Monograph no. 7, 1994, p. 235.

50 Sandin, p. 235.

51 See J.H. Walker, *Power and Prowess. The Origins of Brook Kingship in Sarawak*, Hawai'i, Allen & Unwin, University of Hawai'i Press, 2002, p. 20; St. John, S., *Life in the Forests of the Far East*, Two Volumes, London, Smith Elder and Co., 1862, Vol. 1, p. 300.

two Iban bargaining over jars, which were being offered for fifty Mexican Dollars and, in a later expedition, he showed a copy of the photograph to some Kenyah, and this led to very considerable discussion, in particular a debate as to whether the sum offered was foolishly high – some thought it was, others thought it a 'mean sum' but the image of the jars elicited 'unstinted admiration'.[52]

Various types of *pusaka* were kept as security deposits by officials of the Brooke administrations as surety for 'good behaviour' or for Iban who were to embark on *bejalai* expeditions, or by others who had been involved in previous hostilities.[53] Groups of Iban, for example, who were suspected of being likely to cause problems in the future sometimes had their jars, brassware and even financial deposits held as security.[54] Pringle writes that such measures were a favourite device that was used to control Iban migration under Charles Brooke.[55] For Charles Brooke, imposing 'fines' of jars or requiring Iban *penghulus* to deposit jars as a pledge of future good behaviour became an important practice of his political repertoire (that had both symbolic and financial credibility).[56] In July 1893, for example, the Sarawak Gazette (a Government mouthpiece) wrote about His Highness the Rajah's 'necessary actions' against the Iban leader Munan, who was 'interfering in the affairs of rivers in which he held no authority'. Munan was held in custody in Kuching and fined ten jars, which were to be held for ten years as a pledge of good faith and then returned provided no disturbances took place. In so doing, Munan's alleged 'overbearing pride has been humbled before his fellows' according to the paper.[57] Jars, brassware and heirloom items continue to be valued today – although money and other modern items such as chain-saws and motorcycles are more important.[58] Kedit adds that many *bejalai* expeditions were designed specifically to obtain old Chinese jars and, even by the beginning of the twentieth century, a jar was sufficiently valuable for one Iban to sell it to pay for his son's education.[59]

Most indigenous groups across Borneo revered such jars, and they held a valued place of spiritual significance in most communities. Some were used

52 W.H. Furness, *The Home-Life of Borneo Head-Hunters: Its Festivals and Folk-Lore*, Philadelphia, J.B. Lippincott Company, 1902, p. 125.

53 For example, SG 2 April 1900, p. 75; SG 1 June 1882, p. 40.

54 SG 1 July 1898, p. 145.

55 R.M. Pringle, *Rajahs and Rebels: The Ibans of Sarawak under Brooke Rule, 1841–1941*, Ithaca, Cornell University Press, 1970, p. 181.

56 For example, SG 1 Sep 1898, p. 182.

57 SG 1 July 1893, p. 108.

58 P.M. Kedit, *Iban Bejalai*, Kuala Lumpur, Sarawak Literary Society, 1993, pp. 56, 75, 86.

59 Kedit, p. 142.

to bury the dead (ossuaries) as well as to demonstrate social standing.[60] Such benefits must have provided a very considerable impetus for people to pursue the gutta percha trade – an opportunity that had rarely been available on such a scale in years gone by. Jars varied by size and type, and across Borneo different groups valued different styles of jars. One result of this was that certain traders could buy jars in one region, say at Sintang in the Kapuas area of the Netherlands Indies (see Map 5) and sell or exchange them for profit for different types of jars elsewhere in the region.[61] The Iban crossed from Sarawak to Dutch territory to attend an annual jar fair at Sintang, where a written Manual had been printed to assist traders.[62] The publication listed and sketched various styles of jars, and likely dated from 1839. It was written in Malay text, a copy of which remains extant today.[63] Later reports of jars being bought over the border appear in the Sarawak Gazette.[64] Other groups also keenly sought gutta percha to channel profits in this way. Nyiri's detailed study of the importance of dragon jars to the Sarawak Kelabits notes, for example, the increase in access to jars because of better trade with the external economy and resulting improved transport in the late nineteenth and early twentieth century.[65]

Sandin's collection of Iban oral histories set out numerous examples of Iban trade expeditions throughout the region, when groups went on expeditions to collect and trade forest produce such as gutta percha to obtain jars and trade jars with the many populations that also valued jars. These included purchases in the over-border areas of the Netherlands Indies, but also in Brunei, British North Borneo, Sumatra and the Mindanao region of the Philippines – suggesting

60 See a detailed study of the importance and history of jars in the Kelabit highlands of Sarawak in B. Nyiri, *Chasing Dragons through Time and Space: Martabani dragon jars in the Kelabit highlands, Sarawak, East Malaysia*, Ph. D thesis, University of Leicester, School of Archaeology and Ancient History, 2016. An early study by Grabowsky of jars in southeast Borneo includes several sketches and market prices obtained in the mid-1880s: F.S. Grabowsky, Ueber die ‚djawet's' oder heiligen Töpfe der Oloh ngadju (Dajaken) von Süd-Ost-Borneo, *Zeitschrift für Ethnologie*, 17, 1885, pp. 121–128.

61 B. Harrisson, *Pusaka. Heirloom Jars of Borneo*, Oxford, Oxford University Press, 1986, Ch. 4; also, Sandin, Part Four.

62 Harrisson, p. 18.

63 Described by C. Kater, 'Iets over de bij de Dajaks in de Wester-afdeeling van Borneo zoo gezochte tempajans of Tadjau's, door den Assistent-Resident, *Tijdschrift voor Indische Taal-, Land-, en Volkenkunde*, TBG, 1867, 16, pp. 438–449.
See H. Chambert-Loir Henri, M-F Dupoizat. 'Les jarres de Bornéo d'après un texte malais de 1839', *Archipel*, Volume 66, 2003. pp. 113–160, for a study and presentation of the 1839 text.

64 For example, SG 1 August 1892, p. 153.

65 See especially Nyiri, 2016, Ch. 7.

not only the extensive geographic and economic reach of Iban jar and gutta percha trading, but the geographic spread of the jar trade itself.[66] Given the significance of jars to most indigenous groups of Sarawak, and the economic benefits of the gutta percha trade, the trade data should depict the practical manifestation of the way in which groups embraced the opportunities that the trade offered. Figure 20 below appears to bear that out.

Unfortunately, import statistics for jars were not collected before 1866, but the first dramatic rise in the importation of jars can be seen in 1867 – gutta percha exports in that year more than doubled and the value of jars imported almost tripled. The next rise seen in the importation of jars occurred in 1871 – following the peak in gutta percha exports in 1870, likely a reflection of gutta collectors using the profits of their trade to buy imported jars. Similarly, a rise in gutta percha exports in 1879 may be reflected in the increased importation

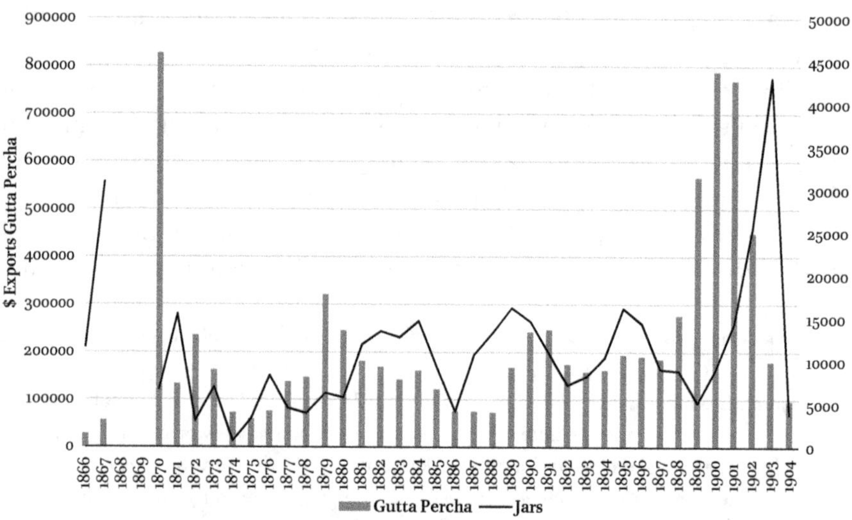

FIGURE 20 *Sarawak gutta percha exports related to importations of heirloom* (Pusaka) *jars*
Note: As with importations of brassware at Figure 19, the connection between the gutta percha exports and importations of jars can be seen here. Statistics for jars were not collected before 1866, hence the gap in the data.
Source: Treasurer's Reports reprinted in Sarawak Gazettes, various issues 1870–1905; and Consular Reports in British Sessional Papers (BSP): 1865 [3478], p. 28; 1866 [3582], p. 805; 1866 [3669], p. 266 for the years 1863–65 respectively. Figures for 1868–69 are not available. See Appendix 2, Table 5 for statistical details.

66 Sandin, Part Four.

of jars in 1881. The other rises and falls in later years may be due to the collection of other forest produce (rises in 1887 and 1888, for example, parallel increased rattan exports in those years, see Figure 11 above).

In 1889, it was reported that thousands of Iban had left Sarawak and gone to Sumatra to collect gutta to buy up old jars and that they had in fact been around the whole of Borneo for such purposes.[67] Sarawak gutta percha traders working offshore, as well as those remaining at home, were exploiting the trade for the opportunity to obtain jars and the prestige that went with them. The following year, it was reported that the Murut in the Trusun area of northern Sarawak traded large amounts of gutta percha for jars:

> During the month, a considerable number of Muruts have been down to trade, and between them between 35 and 40 piculs of gutta percha had been brought to the bazaar by them ... There is an increasing demand for the coloured jars "kaboks" and doubtless a large number will be imported this year; the duty on these during the month amounted to $123 whereas in January last year the total was $33...[68]

Clearly, the very dramatic rise in the importation of jars in 1900 – 1902 was a direct response to the dramatic rise in the exports of gutta percha. While other exports from Sarawak increased noticeably in those years, that is in pepper and gold, neither of these trades involved indigenous groups who sought *pusaka*, so the significant rises do not appear to be linked to these other exports. Some of the other years of increases in the importation of jars may have been a result of increased levels of Iban expeditions to other countries in the region specifically aimed at bringing back jars during the period of great demand for gutta. Rises in 1901 and falls in 1903 were reported in the importation of Jars (and 'dyak property') associated with the arrival or departure of large numbers of Iban gutta collectors.[69] The noticeable increase in the 1903 importations and falls in 1904 seen in the graph could be in part explained by gutta percha collectors returning from expeditions to the region after the gutta percha boom up to 1902. Of significance, though, is the fact that in 1903 the government decided to take advantage of what was likely seen as an opportunity to raise revenue from activity in the jar trade – by increasing the import duties on jars to $5.[70] The decision was taken after a period of a few years in which healthy exports

67 SG 1 July 1889, p. 95.

68 SG 1 April 1890, p. 51.

69 SG 1 February 1901, p. 34 and SG 3 March 1903, p. 54.

70 SG 3 March 1903, p. 52; SG 1 April 1903, p. 63.

of gutta percha and a strong jar and *pusaka* trade suggested that the government was missing out on an important revenue opportunity. The increase in duty led to a 'tremendous rush' to import jars ahead of the price rise, and a consequent significant fall the following year as locals were reluctant to pay the new higher price.[71] Also contributing to a reluctance to buy imported jars in 1904 was the continued lackluster gutta percha trade, and the fact that a local Chinese company had decided to begin producing cheaper, so-called heirloom jars which of course were not subject to import duties.[72]

Scott, in his *Moral Economy of the Peasant*, proposed that a fear of food shortage and living close to the margin gave rise to a 'subsistence ethic', making the peasant risk-averse.[73] Yet Freeman's detailed study found, for example, that Iban might occasionally trade the valued *pusaka* of gongs or jars; but most remained highly reluctant to part with such goods even in times of serious food shortage, preferring to seek other sources of sustenance such as by growing sago, or obtaining wild palms from the Punan, and such findings would likely hold true for all groups.[74] Given such difficulties, why then would gutta percha collectors trade it for anything other than subsistence items and why was the *pusaka* trade so closely connected with gutta percha exports? A 'subsistence ethic' might suggest traders would be keen to trade gutta percha for rice, salt, ironware and so on, rather than prestige items. The quest for *pusaka*, as it relates to prestige and social standing and to religious anchoring of communities, is a much broader philosophical and ontological question than might be clarified here in a single publication about the changing face of local economies in a globalizing nineteenth century world. Nonetheless, a few considerations are suggested here. Kopytoff makes a point in his discussion of *subsistence* and *prestige* trade, where he concluded that the 'problem of value and value equivalence has always been a philosophical conundrum in economics.[75] It involved the mysterious process by which things that are patently unlike are somehow comparable to and exchangeable…'. In a contemporary analysis, Banerjee and Duflo found that the extremely poor (living on less than $1 per day) were often underfed, yet continued to spend significant amounts

71 See SG 3 March 1904, p. 49.

72 SG 2 January 1903, p. 11.

73 J.C. Scott, *The Moral Economy of the Peasant, Rebellion and Subsistence in Southeast Asia*, New Haven, Yale University Press, 1976, p. 4.

74 D. Freeman, *Report on the Iban*, University of London, London, The Athlone Press, 1970, p. 267.

75 I. Kopytoff, 'The cultural biography of things: Commoditization as process' in *The Social Life of Things: Commodities in Cultural Perspectives*, ed. A. Appadurai, Cambridge, Cambridge University Press, 1986, p. 71.

of money on alcohol, tobacco or on festivals (weddings, funerals, or religious festivals).[76] In other communities surveyed, the preference was to spend on consumer goods such as televisions or radios. *The Economist* concluded in an analysis of a World Bank study of extreme poverty that, 'the poor do make choices ... Even those in absolute poverty care about their relative standing...'.[77] All these findings emphasise that, even in the most difficult circumstances, people make choices and judgements. The choices in turn motivate the need to trade. Whether it was rice or cloth, or *pusaka,* these choices were reflected in the gutta percha trade of Sarawak. Also, *pusaka* were an investment with a recognized value that could be 'traded' to settle feuds, pay reparations or to negotiate political settlements with the Brooke administration; and, of course, the more valued *pusaka* were often owned by elites who could in any case afford them. Heirloom jars might be inhabited by *antu,* giving them spiritual significance as well, and endowing increased social standing on their owners.

Another factor that played a part in the quest for *pusaka* was that *pusaka* had significance for power relationships in local social networks. As will be seen in a later section, the near anarchy that broke out at times in the gutta percha resource disputes across Sarawak (see 'Gutta Percha Wars' below) suggests that the trade was transformative both in material terms and in terms of power relations between groups. Old enmities and resource disputes were revisited in disputes over gutta percha resources, for positioning within the new Sarawak society, and for winning influence with the Brooke administration. These were effectively power disputes in the commodity chain. In a trade sense, groups who were best able to win a more powerful and influential position in the gutta percha trade and its associated 'wars' stood to gain spiritually, economically, politically and in social standing. Risk taking, power seeking or power positioning were then interlinked motivations in the trade.

A further factor that explains why the trade was not used only to satisfy subsistence needs may be that, in the sense of risk management, an advantage for Sarawak communities was the ability to collect forest produce for food or commerce as an adjunct to cultivation; that is, they did not have to depend entirely upon cultivation to survive.[78] This may have altered concepts of 'safety

76 A. Banerjee and E. Duflo, 'The economic lives of the poor', *Journal of Economic Perspectives*, Vol. 21 (1), (Winter), 2007, p. 147.

77 *The Economist,* 'Economic Focus, Another day, Another $1.08. Even those in Absolute Poverty have Choices', April. 28, 2007, p. 98.

78 See H. Knapen, *Forest of Fortune? The Environmental History of Southeast Borneo, 1600–1880*, Leiden, KITLV, 2001, pp. 182–183 on options in Borneo to avoid starvation in times of famine.

first' raised by Scott. More importantly, however, is another factor that may explain the importance of the gutta percha – *pusaka* trades. This relates to a point made by Knapen, who thought that in Borneo it was in fact the very risk associated with obtaining 'scarce' goods from the external economy – the very hazards involved in the trade – that was part of the attraction.[79] Sandin, in introducing his commentary on Iban oral history, noted that influential Iban chiefs all possessed jars and that the quest for these in foreign lands (*bejalai*), often by parties collecting and trading gutta percha, caused the loss of thousands of lives.[80] Certainly, Knapen's theory seems consistent, for example, with the Iban idea of *bejalai* (achieving wealth and proving valour, perhaps on headhunting or dangerous expeditions as gutta percha collecting efforts might well be) and it is clear from the narrative of the gutta percha 'wars' that are considered in the next chapter, that indigenous traders (and not just the Iban) were by no means afraid of embracing the trade despite the risks of which they must have been aware. Theoretical approaches to the impact of risk in subsistence economies have tended to deal with the topic as a feature of rational choice theory or the theory of the optimising peasant. That is, in subsistence economies, actors make choices that aim to maximise their limited resources or to *manage* risk.[81] In Sarawak, indigenous trade networks – both risk *taking* and even risk *seeking,* as well as risk *managing,* appear to have been very much part of life. In Sarawak, trade relationships and the success of the forest produce trades, were shaped by the threat of risk and the way in which actors either managed or avoided it or sought it – that is they actively *took* risks. Whatever the motivating factors, risk taking and negotiation were a key part of social interaction. For the Chinese trade network, a capacity for risk taking and negotiating and knowledge of local conditions were also a key element of success, and even survival. Chinese traders, in particular the boat traders, took on risks when trading in the more remote areas. And all traders, indigenous and Chinese alike, were aware of the fluctuations of the gutta percha market and all knew that they could be caught with unsaleable produce or lose out in credit relationships. That was the risk of trading in such a difficult

79 Knapen, p. 376.

80 Sandin, p. 235.

81 M. Lipton, 'The theory of the optimising peasant', in *Development Studies Revisited: Twenty Five Years of the Journal of Development Studies*, eds. C. Cooper and E.V. K Fitzgerald, Frank Cass and Co Ltd, London, 1989, pp. 411–435; S.L. Popkin, *The Rational Peasant: The Political Economy of Rural Society in Vietnam*, Berkeley, University of California Press and F. Ellis, *Peasant Economics Farm Households and Agrarian Development*, 2nd edn, Cambridge, Cambridge University Press, 2000.

environment. One balanced trust and risk against the likely gains: commercial, subsistence, prestige or spiritual.

One way of assessing whether it was submarine telegraphy and the gutta percha trade, rather than other trades or unknown factors that were changing the face of Sarawak society and its economy is to compare some of the trade figures with the production of submarine cables. Such developments should not otherwise be interrelated except for the fact of the exports of gutta percha. In effect, this process pulls together the two extremities of the gutta percha commodity chain. Did demands of submarine telegraphy have an impact on the indigenous economy and cultures of Sarawak? By comparing world submarine cable production (rather than gutta percha exports) with selected Sarawak importations, the linkages between metropolitan and local factors, of globalization in operation could be revealed. A periodic snapshot (Figures 21 and 22) provides evidence to link the external (pull) and local (push) factors in this trade. Or, put another way, it measures the impact of the broad arc of the commodity chain from the metropolitan to the periphery. The graphs below are presented in biennial format for the sake of clarity, but the overall patterns in each of the graphs at the very least suggest a strong likelihood of a causal connection. The two trades chosen are those which appear most frequently in the historic record as relating to gutta percha collectors – rice and brassware *pusaka*:

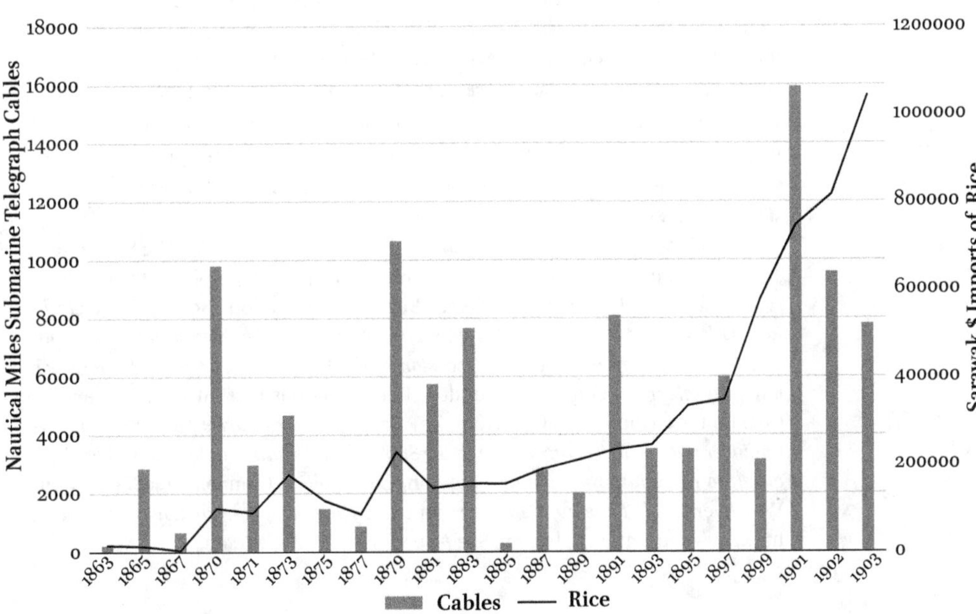

FIGURE 21 *Global production of submarine cables related to Sarawak importations of rice 1863–1903 (Biennial).*

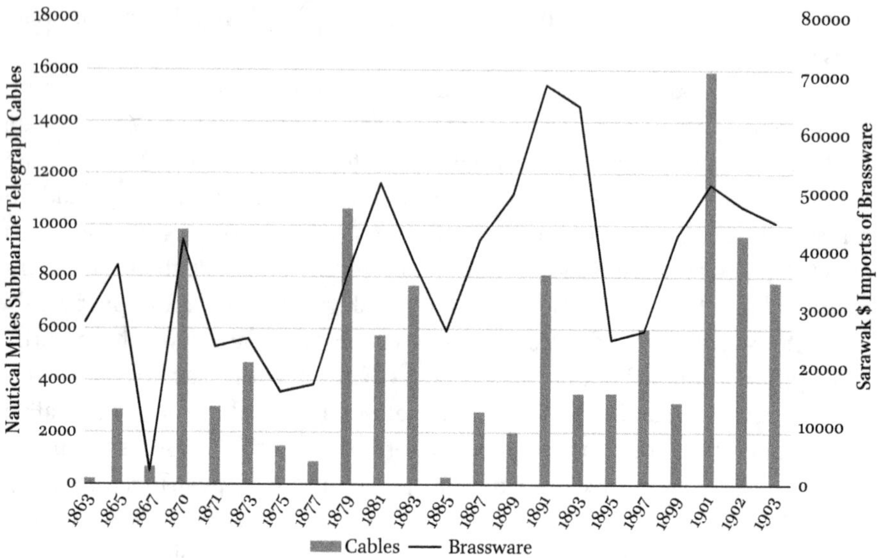

FIGURE 22 *Global production of submarine cables related to Sarawak importations of brassware* Pusaka *1863–1903 (Biennial).*

Note: For example, Sarawak peoples were enabled to purchase brassware because of demand for gutta percha associated with the Atlantic cable of 1865; the lengthy cables of 1870, including West Indian Cables of the same year, associated with the Borneo Company's engagements with submarine telegraphy and exploitation of Sarawak gutta percha resources; and, of course, the Pacific Cables being manufactured at the turn of the century. Note also, the falls in purchasing apparently mirrored the falls in the manufacture of cable in other years.

Source: For Sarawak Statistics: Treasurer's Reports reprinted in Sarawak Gazettes, various issues 1870–1905; and Consular Reports in British Sessional Papers (BSP): 1865 [3478], p. 28; 1866 [3582], p. 805; 1866 [3669], p. 266 for the years 1863–65 respectively. Figures for 1869 are not available. See Appendix 2, Tables 1, 4 and 5 for statistical details.

For N.M. Miles Cable (Est'd), both Figures: *Pall Mall Gazette*, 'The West Indian Telegraph Expedition', 1870, July: n.p.; United States Hydrographic Office, *Submarine Cables*, Washington, Dept. of Navy, Bureau of Navigation, 1892, pp. 41–59; *India Rubber World*, New York, 1 February 1902, p. 136; Telcon, *The Telegraph Construction and Maintenance Co. Ltd on the Occasion of the Visit of the Delegates to the International Telegraph Conference*, London, Telegraph Construction and Maintenance Co. Ltd, 1903, Appended; C. Bright, *The Life Story of Sir Charles Tilston Bright Civil Engineer, with Which is Incorporated the Story of the Atlantic Cable and the First Telegraph to India and the Colonies*, London, Constable & Company, 1908, p. 313 and W. Glover, *Cable Timeline 1845–1900* and *Cable Timeline 1901–1950*; 2017/Continuing: http://www.atlantic-cable.com. See Appendix 2 Table 1 for data.

The noticeable rises and falls in the gutta percha exports shown in earlier graphs are unlikely to have arisen because of changes in Sarawak population or migration, which as noted earlier in large measure did not occur, and would certainly have occasioned commentary in the local press and Consular reporting had it done so. The connections in the documentary record and the trade data also reveal the close relationship between submarine telegraphy and the gutta percha exports of Sarawak: the evidence links the activities of Telcon/Gutta Percha Co, the Borneo Company and Sarawak's gutta percha exports, as clearly documented in corporate records (see Chapter 4).

This comparative study matching the documentary record with trade data reveals that the gutta percha trade operated in a way that reflected the skills and concerns of the local actors. Within the forest environment, collectors used the profits from gutta percha to meet shortfalls in their ability to produce or purchase rice. The new trade made buying rice an attractive proposition compared with the difficulties and seasonal failures that might come with local rice production. The connections made here between gutta percha exports and the importations of cloth are not intended to suggest that cloth imports were something new in themselves; as noted, they were a part of the trade landscape of the whole region and had been so for some time. Rather, the linking of the trade figures with the export of gutta percha reveals that the gutta percha trade was embedded in local trade patterns and facilitated their growth because of the profitability of gutta percha for local collecting communities. The graphs connecting the gutta percha, rice, cloth and treasure trades revealed interlinked commodity chains in Sarawak because of the resources, the collecting environment, skills and trading patterns of both the collectors on the one hand, and the use of various forms of exchange by Chinese traders on the other. The extent and impact of the gutta percha trade was also revealed in the figures for the import of *pusaka* items since these connect the external economy with the local social, cultural and the spiritual milieux: additionally, it reveals how the trade underpinned shifts in social relationships within and between communities as they sought to protect and enhance social standing by obtaining *pusaka* through trading gutta percha. Collectors could turn to rattan rather than gutta percha in certain years when market demands made those choices more attractive. The strategies of different actors reflected a linkage with the global economy on the one hand, and the capacity to respond to local market restrictions and opportunities on the other.

Close relationships between certain trades are not required to prove that global linkages exist. Rather, periodically strong interrelationships can be inferred from comparing documentary and quantitative evidence to show that several commercial, political and social factors in the environment shaped the commodity chain. This is why commodity chain analysis can provide a methodology that may encompass both qualitative and quantitative evidence – so that world events and human motivations are not overlooked because they cannot be forced into theoretical or average patterns of behaviour. The statistical material shows how global and local factors interact. For a few of those years, these trades may have been influenced by other exogenous factors – increases at the end of the century connected with the overall growth of the Sarawak economy through the rise of the pepper trade for example. Yet the *pusaka* imports were unlikely to have been strongly influenced by improvements in pepper exports as that trade was not associated with indigenous communities. Population growth was limited and likely only a minor factor.

When societies are drawn into a period of change and economic development, there are challenges and opportunities and adjustments to be made. In some cases, these changes related to the mechanisms of trade itself, some of which have been described earlier. Yet the scale of change in nineteenth century Sarawak, occasioned by the gutta percha trade, had other features. These other changes are considered next.

CHAPTER 8

Impact of the Gutta Percha Trade – Change and Challenge

> As a rule, traders bringing down gutta get so pillaged on the road that barely half the amount originally started with reaches the station ... from all accounts things are as usual in a very unsettled state in Ulu Trusan and numbers of complaints have been made ...[1]
>
> SARAWAK GAZETTE, 1 March 1892

∴

Continuing the analysis of who benefited and who suffered when local societies were drawn into the global economy, this chapter considers two quite different developments promoted by the rise of the gutta percha trade. One is at the commercial level in the mechanisms of trade. The trade encouraged the move to monetization – a hallmark of communities embracing the external economy. On the other hand, community relations were changed. Resource hostilities arose – a somewhat neglected aspect of regional history. The gutta percha 'wars' were disruptive to communities and even facilitated the cession of parts of Brunei territory to Sarawak. The findings set out in the chapter emphasize a violent aspect of this trade, and trade at the cross roads of local-external relationships. While monetization and warfare are at two extremes of trading interaction, the disparity between the two underlines the process of upheaval and the importance of local factors in the response to the process of change.

Monetization

> At the end of the month ... a party of nearly two hundred Dyaks [departed] on a trading venture, and in search of work, jungle produce &c left ... taking with them upwards of $10,000[2]
>
> SARAWAK GAZETTE, 1 Jan. 1890

1 Sarawak Gazette (hereafter SG), 1 March 1892, p. 56.

2 SG 1 January 1890, p. 12.

 | DOI 10.1163/9789004357280_010

Like many Southeast Asian societies, Sarawak communities relied mainly upon barter trade prior to interaction with the nineteenth-century global economy, and barter continued to be a key means of trade throughout the century.[3] In parallel with the role of rice and cloth and other items in the barter trade described above, as Sarawak became more closely connected with the external economy, monetization came to occupy a greater role in the local trade network. This process had a chequered history, in part reflecting different cultural preferences in the forest economy, and also because of problems of shortages of currency supply. In some areas, wealth was still calculated in brass guns rather than currency.[4] The Consular Report for 1863 noted that Dayaks near Kuching preferred payment in money to that in beads, gongs and other articles – while further away there was a shortage of currency and preference for beads, brassware and traditional forms of exchange.[5] In 1875 the Resident in Upper Sarawak reported that very few groups paid their taxes in rice, as had traditionally occurred, but eked out their rice with money payments, raised by neglecting their farms and undertaking employment or seeking jungle produce.[6]

The bazaar was a focal point for monetization, indeed bazaars were necessary for monetization to develop. For example, at Oya, the Chinese bazaar was so successful by 1878 that traders introduced 'ready money purchase' as agents for Kuching bazaar merchants.[7] Bazaar traders could handle larger amounts of produce when they had cash available. Cash was a useful addition to exchange of goods, particularly if there happened to be a shortage of goods to trade – as occurred during periods when trade in forest produce was brisk and bazaars sold out of the things that gutta percha collectors were interested in purchasing.

Chew believes the arrival of Chinese traders in Sarawak provided an impetus for monetization; prompted, he believes, by the 'relentless' extraction of forest product by the indigenous population for sale to Chinese traders, with an increasingly sophisticated use of money as a medium of exchange.[8] The Iban quest for jars, as suggested in Chapter 7, also contributed to the growing rise of a monetized economy in Sarawak, as Iban sought currency in order to purchase

3 See also W.G. Huff, 'Bookeeping barter: Money, credit, and Singapore's international rice trade, 1870–1939', *Explorations in Economic History*, Vol. 26, 1989, pp. 161–189; R.A. Cramb, *Land and Longhouse: Agrarian Transformation in the Uplands of Sarawak*, Copenhagen, Nias Press, 2007, and D. Chew, *Chinese Pioneers on the Sarawak Frontier*, Singapore, Oxford University Press, 1990.

4 SG 16 April 1875.

5 Consular Report on the Trade of Sarawak for 1863, BSP 1865 [3478], p. 30.

6 SG 17 August 1875.

7 SG 24 June 1878, p. 44.

8 Chew, p. 108.

jars in neighbouring countries. In 1877, collectors at one outport in just a few months sold $4,000 worth of gutta percha – a considerable sum – which constituted an incentive to people to form more parties to go 'in quest of that produce'.[9] To put that sum in context, the entire state revenue for Sarawak in that same year was just over $185,000.[10] One can imagine times when Charles Brooke must have wished he had set aside gutta percha, not antimony, as a state monopoly. The Sarawak Gazette wrote in 1885 that many Iban valued 'both jars and dollars', travelling in groups of forty or fifty and collecting produce to realize money to purchase jars.[11] In 1872, the Gazette informed readers that some 'Dyaks were amassing riches' and that money was taking the place of jars and gongs and that some who possessed 500 or 600 Dollars 'put it out to interest among the Chinese traders', but this was risky if the trader went bankrupt.[12]

Charles Hose wrote in 1894 that more than a decade of Brooke rule and the consequent growth of trade in the Baram district of Sarawak had meant that 'dollars and cents have found their way far into the interior, so that even the Punans [Penan] know the purchasing power of dollars'.[13] The major traders were using more cash, certainly by the end of the century. In 1893 one Assistant Resident reported that gutta percha, rattans and sago were traded from the area as Kuching firms kept their agents supplied with money to buy all the produce offered.[14] In 1900, the year in which demand led to very good prices for gutta percha, came reports from the Trusan that a great number of Muruts had brought down large quantities of gutta percha to the bazaar and that at one time Chinese traders ran out of cash and had considerable difficulty convincing the collectors not to go elsewhere to trade as they insisted that they would 'not take goods in exchange'.[15] Hundreds of Iban who returned from *bejalai* were reported to be carrying large sums of cash.[16] Cramb noted that the 'gutta trade linked the Iban to a monetized export market ... the revenue from gutta ... permitted considerable accumulation of wealth'.[17] This appears to be supported by the oral history Sandin retells of the great Iban gutta

9 SG 25 August 1877, p. 64.

10 SG 27 May 1878, p. 35.

11 SG 1 December 1885, p. 114; and see 1 December 1900, p. 226.

12 SG 1 April 1872 and Chew, p. 92; 1 July 1904, p. 144.

13 C. Hose, 'The natives of Borneo', *The Journal of the Anthropological Institute of Great Britain and Ireland*, Vol. 23, 1894, p. 168.

14 SG 1 September 1893, p. 150.

15 SG 2 April 1900, p. 75.

16 SG 1 June 1897, p. 111.

17 R.A Cramb, 'The commercialization of Iban agriculture', in *Development in Sarawak: Historical and Contemporary Perspectives,* eds R.A. Cramb and R.H.W. Reece, Melbourne, Centre of Southeast Asian Studies, Monash University, 1988, p. 109.

expeditions where the participants chose to return with silver coins rather than jars. In a sense all of the above suggests a pattern consistent with Jevons' point about monetization having a beneficial advantage over barter which requires the 'double coincidence of wants'.[18] The coins had financial as well as prestige value, and were readily convertible to the expensive jars that the Iban sought.[19] As gutta percha collectors were also traversing considerable distances to collect gutta and then obtain jars or brassware, the simple convenience of monetary exchange is clear.

By 1903, the Resident of the Rejang area, probably the most important trading region of Sarawak, summarised the changing nature of trade in his district. He noted the importance of 'phenomenal' exports of gutta, the levels of rattan and other forms of export, shipping and the role of Chinese traders and various forms of shipping – along with the great quantities of jars and brassware being bought into the district. Yet he also noted that, increasingly, business amongst the 'dyaks' was centered around the deposit of money with the Chinese, and that in the year 1902 alone more than $10,000 had been deposited.[20]

A point worth noting is that these were conceptually significant changes in that the local population was now assessing the exchange process in quantitative as well as qualitative terms – that is, assigning a numerical value to the medium of exchange. Charles Hose noted, for example, that the Kayan leader Taman Bulan learned from a Chinese trader how to use an abacus.[21] The historic record as reflected in this study shows that groups gradually adopted a monetized system of exchange, but at the same time maintained an adherence to a value system that revered *pusaka*. Local communities had adopted a new medium but not to the point of rejecting their traditional ones. The use of cash is a manifestation of the process of change and incorporation into the world economy, but the changes were also reflected in the use of forest produce to expand the avenues to obtain *pusaka* and subsistence items like rice, cloth and salt. In that regard, monetization may have been of more significance to intermediaries – Chinese traders – than to many indigenous traders, except perhaps in the case of the Iban. The Iban tendency to invest cash with traders or to establish shops in bazaars represented a potentially significant modification

---

18 W.S. Jevons, *Money and the Mechanism of Exchange*, New York, D. Appleton & Company, 1919, p. 3; the 'double coincidence' occurs when both seller and purchaser each have something to barter that the other desires and for the agreed quantity, p. 4.

19 See Sandin, p. 235 ff.

20 SG 3 March 1903, p. 52.

21 C. Hose and W. McDougall, *The Pagan Tribes of Borneo*, 2 Vols, London, Macmillan and Co, 1912, Vol. II, p. 211.

of the traditional ideas of exchange compared with other Sarawak indigenous groups. As Cramb concludes, transformations and the 'experience of engagement' with the external economy must be considered in this context of 'local histories, social relationships and cultural perceptions'.[22]

Overall, the variants in preferred choice of exchange by collectors, and other patterns of indigenous trade and engagement with the external economy described in this analysis, re-enforces this suggestion. These factors are more easily analysed when surveying the local conditions, a finding pointed to by Sverrisson and others.[23] Sandin gives an example of one successful Iban who had returned with his 'crew' from British North Borneo after a successful expedition buying jars and valuable brassware, but that others from his district did not follow this example since they were 'employed as fortmen' so they were paid money with which to buy jars.[24] Such opportunities were relatively limited and the freedom that gutta percha collecting and trading offered and the expeditions that it involved offered opportunities that would not have been available otherwise. Expeditions also increased social standing within the community in comparison with local employment even if it were available. As Cramb concludes, micro-level questions cannot be separated from macro-level questions of political economy.[25] By understanding better the nature of these responses, in turn we gain a clearer, more satisfying picture of the world-system itself.

Whatever the motivations, the attraction of the gutta percha trade led to social changes and hostilities. The responses indicate the nature of global commodity chains in that economies at the periphery were shaped in part by local trading, environmental and cultural motivations.

---

22 Cramb, *Land and Longhouse: Agrarian Transformation in the Uplands of Sarawak*, p. 35.

23 A. Sverrisson, 'Local and global commodity chains' in *Linking Local and Global Economies: The Ties That Bind,* eds C. Pietrobelli. and A. Sverrisson, London, Routledge, 2003, pp. 17–35; and see Cleary, *Modern Asian Studies* and *Journal of Historical Geography*; S.J. Stern, 'Capitalism and the world-system in the perspective of Latin America and the Caribbean', *American Historical Review*, Vol. 93, No.4, October, 1988, pp. 829–872; A.G. Frank, 'Immanuel and Me with-out Hyphen', *Journal of World-system Research*, VI, 2, Summer/Fall, 2000, pp. 216–231; J.D. Carlson, The 'Otter-Man' empires: the Pacific fur trade, Incorporation and the zone of ignorance', *Journal of World-system Research*, VIII, III, Fall, 2002, pp. 390–442; and R.W. Edwards, 'The North American Fur Trade World System', *Field Notes: A Journal of Collegiate Anthropology*, 1 (1), 2009, pp. 46–64 for other examples.

24 B. Sandin, 'Sources of Iban traditional history', *Sarawak Museum Journal*, Kuching, No. 67, Special Monograph no. 7, 1994. p. 241.

25 R.A. Cramb, *Land and Longhouse: Agrarian Transformation in the Uplands of Sarawak*, p. 31.

Changing Social Relations and the Gutta Percha 'Wars'

> Gutta hunters attacked a tribe of Muruts over the border beyond the headwaters of the Limbang, taking in all 32 heads and 12 captives
>
> SARAWAK GAZETTE, 1 July 1898, p. 148

⁘

While economic and commercial changes were occurring in Sarawak society – people were buying more rice, cloth, sugar and *pusaka,* Chinese traders were traversing the country, and the economy was becoming more monetized – at the same time the movement of peoples and the hunt for the valued gutta percha prompted significant social changes within Sarawak. These changes in Sarawak were also felt over the borders in other areas of Borneo.

Increased migration within Sarawak and across its borders associated with the gutta percha trade caused unprecedented interaction among indigenous groups. It also formed the basis for hostilities that cost lives and changed social relationships. The hostilities provided the spark for political arguments over state boundaries and for Brooke political interventions. Changing patterns of trade brought groups into contact with one another to a far greater degree than had ever occurred previously. The forces of the trade, the policies of the Brookes, and the presence of Chinese traders changed the nature of social interaction by drawing populations from the interior. Communities were becoming more 'globalized' both economically and socially.

As noted earlier, it was recorded that Iban, for example, travelled all around Borneo as well as to Sumatra, the Malay Peninsula and the Philippines in the hunt for gutta percha and the related goals of obtaining *pusaka* and the social standing arising from these expeditions.[26] If they were mapped, the location of these activities match the region marked out on Obach's map (depicted at Map 4) of those areas of interest to cable telegraphy. Different groups traded and moved throughout Sarawak and across its borders, with varying results. The role and impact of Iban gutta percha collectors is amongst the most influential of all those groups: in part because of Brooke policy, and because of their larger numbers, their greater migratory trading practices and their generally

26 SG 1 July 1889, p. 95, and S. Baring-Gould and C.A. Bampfylde, *A History of Sarawak under its Two White Rajahs, 1839–1908*, London, Henry Southeran, 1909, p. 376, and I.H. Burkill, 'Botanical Collectors, Collections and Collecting Places in the Malay Peninsula', *The Gardens Bulletin, Straits Settlements*, Vol. IV, Nos. 4 and 5, August 1927, pp. 153–154.

entrepreneurial and enthusiastic approaches to trade. In that regard, the gutta percha trade and the Iban had a peculiar relationship. The inter-relationship between Iban aggression and their trading activities was commented upon as early as 1872 in the Sarawak Gazette – it noted that the major trading stations had increased their revenues even though it was in those areas where the Iban were reported to be on 'headhunting expeditions: this may seem strange, but yet a fact which figures demonstrate is beyond dispute'.[27] The following year, the Gazette noted that, in the Iban regions, trade was very active 'notwithstanding the various brawls' and some were investing dollars with Chinese traders.[28] Despite the risk of hostilities, their energetic pursuit of forest produce meant that where the Iban went, better trade followed. The Rejang river, for example, was not Iban territory, but was famed for its old-growth forest suitable for swidden agriculture and for the wealth of its forest produce. The Resident for the Upper Rejang noted that the Iban migrations to the area in 1899 and 1900, would give immediate stimulus to trade and prosperity of the region.[29] That migration coincided with the dramatic increase in the gutta percha trade at the turn of the century and Freeman says that the Iban movement into the Rejang and beyond was 'the most momentous happening in the social and economic history of nineteenth century Sarawak'.[30]

Elsewhere, earnings from forest produce in the Baram[31] and the Trusan areas were crucial since they lacked the sago and mineral resources found elsewhere. After the cession of the Baram region to Sarawak in 1882, it took little time for Iban to become aware of the opportunities that existed there for collecting gutta percha and other forest produce: as early as January 1883, Iban collectors had arrived in the area from further south in Sarawak and this attracted the Chinese traders by May of that year. As a result, it was said, the local Kayan and Kenyah were keen to trade.[32] More Iban moved into the Baram to collect gutta – a group of 45 Iban had arrived in May 1884 seeking permission to collect gutta. The *Tuah* of the group reported that another 31 from his party were on their way and that another 200 intended following.[33] These activities suggest that Iban parties soon identified gutta trading opportunities and acted

27 SG 15 February 1872.

28 SG 1 April 1872.

29 SG 1 March 1901, p. 59.

30 D. Freeman, *Iban Agriculture: A Report on the Shifting Cultivation of Hill Rice by the Iban of Sarawak*, London, HMSO, 1955, p. 11.

31 See H.L. Godfrey, 'Trade and Transformation on the Borneo Frontier: towards a study of the Baram region of Sarawak in the late nineteenth century', *Borneo Research Bulletin*, 2014, Vol 45, pp. 87–135.

32 SG 1 May 1883, p. 50.

33 SG 1 July 1884, p. 75.

on them. By June 6, 1884 the Resident reported that a group of Iban had returned from the upper Baram with around three pikuls of gutta, being the 'first party that has succeeded in getting any gutta in the upper districts', and the diary records shipping regularly leaving with full cargoes of gutta after that.[34] The arrival of Iban and Chinese traders was reflected in the district's trade returns: in 1883, exports totaled $32,850, increasing to $114,357 in 1884, of which $55,402 was gutta percha. Not surprisingly, Chinese traders were pleased with the quantity of gutta percha and other produce being traded; their difficulty was in having sufficient cash to buy it all.[35] In 1885, writing about the Trusan area, the Sarawak Gazette felt that the Iban would be important to the forest produce trade, provided peace could be established among the local Muruts.[36] Clearly, the administration recognized the potential for conflict between groups (in this case between the Murut and the Iban), but saw hostilities more in terms of the effect on trade than any concern to recognize the impact on the Muruts who may have seen this as a transgression on their ability to exploit forest resources in their homeland.

Gomes, who spent seventeen years living with Iban, thought that the Saribas Iban were 'more anxious to improve themselves' than any other dyaks.[37] Pringle believes that Iban trading activities were more than 'mere commercial ventures', but a symptom of the wandering (*bejalai*) ethic, and a transmutation of the old habit of coastal raiding. By the 1890s, Iban from other rivers were involved in the search for jungle produce, he suggests, 'which so nearly resembled the old 'piratical' pursuits.[38] The impact of Iban gutta collectors and the associated import of *pusaka* when they moved to new parts of the country was remarked upon above.[39] Iban often traversed great distances into areas previously untouched by produce collectors[40] and by 1906 the forest produce of the Limbang was said to have been 'worked out' because the Iban had come up in hundreds and worked there for nearly ten years.[41] In some areas Iban were 'known as a race of 'traders', a reputation which they have never enjoyed in their own native land'.[42] It is a reputation also somewhat overshadowed by

34 SG 1 August 1884, pp. 85–86.

35 SG 1 August 1884, p. 87.

36 SG 1 December 1885, p. 113.

37 E.H. Gomes, *Seventeen Years Among the Sea Dyaks of Borneo*, London, Seeley & Co. Limited, 1911, p. 107.

38 R.M. Pringle, *Rajahs and Rebels: The Ibans of Sarawak under Brooke Rule, 1841–1941*, Ithaca, Cornell University Press, 1970. pp. 195, 199.

39 SG 1 February 1901, p. 34, SG 3 March 1903, p. 54.

40 SG 1 April 1903, p. 74.

41 SG 5 June 1906, p. 144.

42 Pringle, p. 196.

emphasis given by historians to the role played by the Chinese in that capacity. Yet Chinese traders knew that the presence of Iban meant better trade – in 1901, for example, Chinese traders began building shops at Balingian because they thought it was a 'very promising' location due to the 'influx of dyaks'.[43] As Iban became wealthier, they broadened their investments to include not just jars and brassware, but cash that was invested with Chinese traders,[44] and wealthier Iban also built shops. What these connections illustrate is a highly successful commercial network that existed between the Iban and Chinese, such as that reflected in the trade figures in the Baram quoted above, all underpinned by the gutta percha trade. All these changes were, of course, part of the increasing integration with the global economy.

In Pringle's *Rajahs and Rebels*, a book about the Iban and the Brookes, he surveys the many causes and effects of hostilities that arose during the Brooke era; he concludes that it was Brooke policy to encourage the Iban to seek 'jungle produce' and that motivations for doing so were political as well as economic.[45] It was in part a means of maintaining control over various groups or protecting interests in relations with neighbouring Brunei. Pringle concludes that Iban collecting of forest produce was essential in newly acquired territories where this was the most likely source of trade. He also notes that the Sarawak Government knew very well that Iban looking for gutta percha 'got into trouble', particularly, he suggests, when collecting provided an opportunity for headhunting.[46] As the most valuable of the forest produce trades, particularly in years of great demand by cable producers, this made gutta percha the more likely to generate hostilities. So, the advent of this valuable trade – and the impetus to trade provided by the spread of Chinese traders with their goods and credit – laid the groundwork for conflict to occur, and occur it did. Dove's call for an examination of the nexus between the quest for forest produce and the rise of hostilities in forest economies demands attention in the context of the gutta percha related hostilities.[47]

Indeed, some hostilities in Sarawak history may be viewed – ought to be viewed – in the context of the urge to trade valuable gutta percha resources. It is remarkable how the history of this trade and the narrative of hostilities in Sarawak parallel each other. The analysis set out below, however, provides further impetus for a re-assessment of how the instabilities in nineteenth century

43 SG 1 May 1901, p. 107.

44 SG 1 Sep 1898, p. 185.

45 Pringle, p. 269 ff.

46 Pringle, p. 268.

47 M.R. Dove, 'Transition from native forest rubbers to *Hevea brasiliensis* (Euphorbiaceae) among tribal smallholders in Borneo', *Economic Botany*, Vol. 48, 1994, p. 384.

Sarawak (and Borneo) should be viewed. These, it will be argued, were not simply a spasmodic series of aggressive acts or an excuse for Iban headhunting, but rather they were a response to an entirely new environment of change and opportunity. The hostilities were fueled by the gutta percha trade and by the changes that the trade helped to forge. These hostilities were not all initiated by Iban – on the contrary, many were the responses of other groups into whose ancestral lands the Iban had now moved, motivated by factors (political and economic) touched on above – often with the support or encouragement of Charles Brooke. The result, however, was just as violent regardless of who initiated hostilities. The history considered here is largely that relating to hostilities reported in Sarawak sources (with reference to recent scholarly surveys).

There was no 'declaration' of a gutta percha 'war' in Sarawak – but the history of the late nineteenth and early twentieth centuries was a history of bloodshed and uncertainty that was in some measure prompted or hastened by, interests that were connected to the gutta percha trade. Sather notes that what drew the Iban to the gutta trade was a desire to obtain traditional forms of heritable prestige wealth (*pusaka*) – as has been considered above.[48] Sandin's summary of the oral record of some of the famous Iban histories includes some of the better known Iban gutta percha expeditions resulting in hostilities in Sarawak and the wider region.[49] Sather, Pringle and Sandin each recognize some connection between Iban gutta percha collection and bloodshed – but the connection between violent gutta percha trade and its implications is only properly recognized when the trade itself is set down in a narrative against the wider history. The '*casus belli*' that arose between different groups were certainly well recognized in Sarawak at the time:

> This is rather a long-list of crimes of a sanguinary nature, but not more than might be expected when it is recollected that Dyaks by thousands are scattered, working gutta, &c., over an enormous area of territory, from the coast to the very interior, where they come into contact with many strange and timid tribes who invariably connect the name Dyak with head-hunting. When this is taken into consideration it is easily understood that fatal collision must occasionally occur and will continue to do so until such time as the bulk of the interior population comes more into touch with the civilized world.[50]

48 C. Sather, 'Introduction' to *Sources of Iban Traditional History*, B. Sandin, Kuching, *Sarawak Museum Journal*, No. 67, Special Monograph No. 7, 1994, p. 22.

49 Sandin, 1994.

50 SG 1 November 1894, p. 182.

Disagreements could easily escalate and it becomes clear that this peculiar trade altered the political history of Sarawak. The survey below is set out around the river systems rather than chronological order for ease of comprehension – it is by no means an all-inclusive survey of all the skirmishes and hostilities relating to the collection of gutta percha that were documented, for example, in the Sarawak Gazette. It is intended as an overview of some of the main hostilities to illustrate the issues and impacts that are of key concern.

Hostilities soon developed around the Rejang river, one of the great areas for gutta percha resources and for swidden agriculture. Both factors made it highly attractive to Iban who moved there from the Batang Lupar region, an area whose forest resources were reported in 1876 as being 'exhausted'.[51] By the end of that year came the following report from the Resident – it was to be typical of other reports in the period of gutta percha 'wars':

> Balleh Fort, 31st Dec 1876
>
> Sir- I have the honour to inform you that during the past month, the Tamans and Ukits have twice attacked parties of our Dyaks, who were working gutta in the Gat, killing Bugap of Meruan's house, and Luyoh one of Sarap's people. Three of the Dyaks were also wounded. The chiefs ask to be allowed to retaliate, as if these attacks are not put a stop to, there will be no security for anyone working jungle produce in the Balleh.
>
> G Gueritz[52]
>
> The Gat [Gaat] is a river on the upper reaches of the Rejang

The Ukit he referred to were traditional enemies of the Iban – resource hostilities such as these exacerbated old enmities; one problem made the other worse and gutta collecting expeditions increased the likelihood that old enemies would come into contact with each other. Disagreements between Ukit and Iban continued in this area for years to come. The Ukit were by no means passive in these activities and took the offensive and attacked Iban to defend their land or to seek retribution for perceived wrongs.[53] Hostilities were discussed in the Sarawak Gazette in March 1882, noting that the Ukits used poisonous

51 SG 20 May 1876.

52 SG 15 February 1877, p. 12.

53 See S. Thambiah, 'The emergence of the ethnic category Bhuket: Diversity and the collective hunter-gatherer identity in Borneo' in *Beyond the Green Myth: Borneo's*

blow-pipes to attack Iban gutta collectors who were then caught unawares, as occurred in 1882 when gutta and produce collectors were killed, making the Iban 'furious and determined to retaliate'.[54] Charles Brooke wrote that the Ukit were on friendly terms with the Kayan (who were also often in dispute with the Iban), and that the Ukit would often provide Iban 'heads' to the Kayan that they had obtained from parties searching for gutta percha.[55] In fact, the earliest gutta percha related dispute reported in the Sarawak Gazette concerned a disagreement between Kayan and Iban along the Rejang river, after a dispute over bargaining for gutta percha led to the death of the trader. Other traders in the group became involved and were about to 'set on the offensive' when the local leaders begged to be left to take control of the matter rather than have the issue escalate. In the end, the alleged murderer was sent to Sibu where he was tried and sentenced to death, with the approval of the Rajah.[56]

The Rejang area to which the Iban were so attracted had been a Kayan heartland. Sutlive says that Kayan resisted Iban migration into the upper Rejang in the late nineteenth and early twentieth centuries.[57] Certainly, the Batang Lupar region where the Iban had been based was becoming agriculturally less productive, but also it was less productive in the supply of valuable forest produce such as gutta percha when compared with the Rejang. The Kayan were very successful gutta percha collectors and were well aware of the resources they were seeking to protect. Other hostilities along the Rejang were carried out by Iban, although not just on their gutta percha expeditions. Eventually, after government expeditions against them, Charles Brooke expelled the Iban from the upper Rejang in 1881 and the bazaar at Balleh suffered – despite earlier optimism, it proved a disappointment in terms of trade for several years until the Iban eventually returned.[58]

In 1891, another attack occurred on a group of Iban working gutta percha near the Katibas river, an offshoot of the Rejang. Some were murdered in the dispute over gutta and an earlier head-taking incident, allegedly involving the assistance of some 'emperan Dyaks' from across the border in Dutch territory. A relative of one of the murdered Iban decided to form a large *bala* (war party)

Hunter-gatherers in the Twenty-First Century, eds. P.G. Sercombe and B. Sellato, Copenhagen, Nias Press 2007.

54 SG 1 March 1882, pp. 11, 15.

55 C. Brooke, *Ten Years in Sarawak*, 2 Vols, London, Tinsley Brothers, 1866, Vol. II, p. 250.

56 SG 15 July 1871.

57 V.H. Sutlive, *Tun Jugah of Sarawak: Colonialism and Iban Response*, Kuala Lumpur, Sarawak Literary Society, 1992, p. 9.

58 Chew, pp. 71–72.

to take revenge.[59] Swift official action was required to prevent matters escalating and Malay guards and a government vessel were sent up river; the 'murderers' were captured, tried and then executed. The Resident felt the argument resulted from disagreements over gutta trees and a subsequent theft of gutta and the attacks were a 'wanton act' that required exemplary punishment as retaliatory parties had been gathering in large numbers.[60]

Charles Brooke was ruthless in his determination to ensure the populace, especially Iban, was kept under control and often groups collecting produce such as gutta percha were blamed for problems that occurred. Again, because of gutta percha related hostilities, the government was prompted (or arguably found an excuse) to take punitive action against the 'up-river' enemy. On 1 June 1894, the editorial of the Sarawak Gazette wrote:

> In consequence of repeated attacks made by the ulu ai, or up-river Dyaks of Rejang on inland tribes … the Government has felt obliged to resort to strong measures. These took the form of an expedition against the Dyaks of the Chermin and Yong rivers, who … were guilty of murdering 31 Balui Dalam Punans … The murders are, as a rule, committed by the young men who are away in the jungles hunting for rattan, gutta and other jungle produce to trade with at the various markets on the Rejang river. These, in their desire for heads – an unaccountable fascination – care not whence or from whom they obtain such trophies.[61]

One of the difficulties associated with these sorts of hostilities was not merely the events themselves, or even the problem of reprisal attacks, but also the general uncertainties that spread through other communities. After the murders referred to above, groups in the Bintulu district, for example, said they were afraid to leave their homes, yet they were well up the coast from the area where the murders took place.[62] The general uncertainties and the dampening effect on the 'collecting' trade could be felt unless the Rajah was seen to be maintaining control. This prompted what became known as the 'Yong and Chermin Expeditions', involving a force of 5,000 men and 300 boats, when Brooke forces burnt the houses and destroyed goods and padi of those who had been involved, although no women or children were to be killed.[63] Mere

59 SG 1 July 1891, p. 107.
60 SG 1 July 1891, p. 107.
61 SG 1 June 1894, p. 79.
62 SG 1 June 1894, p. 85.
63 SG 1 June 1894, p. 83.

loss of life was not considered sufficient to induce fear in an Iban warrior in Charles Brooke's view. It was destroying villages, taking the valued collections of trophy heads (*tampun*) obtained from earlier headhunting events, jars and relics and other *pusaka* that made 'dyaks think of the consequences of their actions' and the loss of rice-stores and animals was a recognized way of reducing them to 'a condition of beggary'.[64]

As the borders of Sarawak expanded northwards, following the cession of territory from Brunei, so too did Iban settlements – and gutta percha related hostilities. Initially the territory acquired was close to Kuching and the Rejang, such as the Oya river. In 1872, the Resident reported that groups of Iban were continually arriving there from branches of the Rejang. At least one group consisted of 100 followers, with some groups arriving without official approval. While many of the settlements appear to have been agreeably accommodated, not all were.[65] In 1878, a group of Iban was on an expedition to the outskirts of the Oya when they allegedly killed three men. Some time later two boats of 'innocent Dyaks', who were in search of gutta, were attacked by friends and family of the murdered Oya people, who killed ten of the Iban.[66] Fearing further reprisals from the Iban, Brooke ordered 'all the Malay and Dyak Chiefs' to assemble and await his arrival'. He told the meeting that he deplored the murder of those who had been peaceably hunting gutta, but that any reprisals on their part leading to a war would only ruin trade, and close access to the country and the jungles and it is from 'those jungles they amass much money in seeking gutta'.[67]

Other hostilities related to gutta percha trading occurred further north, especially in the regions of the Baram and the Trusan rivers: the difficulties there were to have far more significant consequences. In these events, the occurrences took place when Sarawak people were trading in these areas when they were still part of Brunei; these gutta percha hostilities directly played into Charles Brooke's attempts to gain control over the areas – using the incidents as the pretense. The events involved not just Sarawak and Brunei but also Britain. The hostilities resulted in the cession of these regions to Sarawak and involved complex and tedious political negotiations among Charles Brooke, the Sultan of Brunei and the British Government. Hostilities arising from activities of gutta percha collectors were to prove the spark that helped bring changes in the borders of Sarawak that remain to this day.

64 Baring-Gould and Bampfylde, p. 170.

65 See Pringle, p. 265.

66 SG 26 February 1878, p. 11.

67 SG 26 March 1878, p. 21.

In 1873, six Sarawak traders from the Rejang were working gutta in the Baram area of Brunei when they were murdered by a group of Kayans in what was later described by a Sarawak official as a 'treacherous attack'.[68] Charles Brooke demanded that the Sultan of Brunei take action and that a fine be levied and paid to him. In response to Brooke's demand, the Sultan sent an agent to collect payment of the fine; the agent allegedly demanded an excessive thirty-fold increase in the fine and in fact levied it on a different, innocent group of Kayans. This act, and a generally unsettled condition amongst the Kayan population about treatment under the Sultanate, led to Kayan rebellions and the 'expulsion of Brunis [sic] from Baram' in 1874. The Sultan was said to be unable to protect his territory or draw revenue from it and 'he dare not even ask for it' according to the same Sarawak account.[69] Rather than attempting to grapple with the continuing problems that existed there, the Sultan decided to cede the Baram to Sarawak (or was cajoled into ceding) in return for an annual payment from Charles Brooke. Charles knew the area had great potential for trade but that it was an area where trade and stability had been restricted by the inability of the Sultan to exercise control over the local *pangeran*.[70] Charles had in fact been attempting to have the area ceded from Brunei for some years, and Sarawak gutta percha traders had previously experienced restrictions on their trade there.[71]

Years of negotiation and correspondence had not progressed the matter from the Brooke perspective, but the attack on the Sarawak gutta percha traders in 1873 had led to the domino-like series of actions that fell in Brooke's favour. Although it took a little time: British interests feared that a further expansion of Sarawak would weaken Brunei and, perhaps worse, draw Brunei trade into Sarawak and away from Labuan. Labuan was a British colony and an island off Brunei with few resources other than coal; it was dependent upon the sago, forest produce and barter trade with Brunei. As a result, the Foreign Office did its best to oppose any cession to Sarawak. Britain used an 1847 Treaty technicality to prevent the cession going ahead until 1882, after a period of what a Sarawak Resident described as 'somewhat ignoble spite, bred partly of ignorance and partly of jealousy'.[72] In the end, Brooke had his way; the troubles

68 SG 1 August 1873 and Baring-Gould and Bampfylde p. 334.

69 Baring-Gould and Bampfylde, p. 335.

70 Baring-Gould and Bampfylde, p. 334.

71 SG 1 August 1871; see Tarling 1971 on Charles' attempts over several years to find a way to gain a foothold in the Baram.

72 Baring-Gould and Bampfylde, p. 341.

sparked by the attacks on Sarawak gutta percha collectors in Brunei territory helped Brooke to win the cession of territory.

After it became part of Sarawak, the Baram district provided a clear example of the inter-relationship between Brooke policies and the gutta percha trade, and between the Iban and the Kayan and Kenyah groups of that area. The Kayan and Kenyah were traditional inhabitants of the Baram and were capable collectors and traders but, soon after the area was ceded to Sarawak, the Iban were given permission to trade there. By 1884, the movement of Iban to the Baram was a regular thing and they were going in large numbers.[73] Not surprisingly, hostilities over gutta percha arose. For example, a group of nine Iban in the area on a 'guttaing [sic] expedition' was attacked, three of whom were killed and their heads taken.[74] Iban collectors overwhelmed the Kayan and Kenyah in some areas of the Baram, leading the Resident to urge that they should be restricted to only the 'uninhabited streams' as they had already had the pick of the gutta producing districts.[75] Yet Charles Brooke had favoured the Iban and had always considered them the best produce collectors; although the Resident also noted in the same report that the Kayan and Kenyahs were quite conversant with the methods of working produce and needed to do so and to pay their tax.

In 1894, Charles Hose, the Resident responsible for the Baram, had written with optimism about the buoyancy of the trade along the Baram river – but then more hostilities broke out. In the view of Hose, it was not just this latest expedition that had caused hostilities, but rather it was *repeated* years of working gutta – this short extract from his report sets out the nature of the problems:

> RAJAH – I regret to have to report to Your Highness that the Kenniahs [Kenyahs] under Taman Bulan, Pengulu in the Pata river, have lately attacked and killed thirteen Dyaks who were working gutta ... I made Taman Bulan give up the heads and some gutta belonging to the Dyaks and have ordered him and the people engaged in these attacks to appear at the fort at Claudetown within six weeks' time.
>
> ...
>
> My own opinion Rajah is this, that for years past the Dyaks have worked gutta close to the Kenniah and Kayan houses; and that the Kenniahs have long wished to prevent the jungle produce being worked near their

73 SG 1 July 1884, p. 75ff.

74 SG 2 April 1883, p. 38; 1 May 1883, p. 41.

75 SG 1 June 1888, p. 79.

> doors ... I have called the Dyaks out of the river near Taman Bulan's people, and shall I hope, be able to prevent any further blood-shed but the Dyaks are very much enraged and have collected here from all parts ... about three hundred Dyaks are here at the Fort ...[76]

The murdered Iban were from several districts according to a later letter from Hose published in the Gazette, suggesting that the group was a large and well planned expedition.[77] Hose reported that the attack was made up of four different parties, involving 150 Kenyah; and in fact a second foray where another seven Iban were attacked and their heads taken also occurred in the near region.[78] Taman Bulan, the Penghulu referred to in the report as leader of this attack, was in fact the same Taman Bulan reported by Hose in another publication as having been progressive in the trade of the area and as having learnt to use an abacus, referred to in the previous section on monetization.[79] In that sense, he could be regarded as symbolic of the challenges and opportunities that the gutta percha trade brought to Sarawak. Haddon, who visited the Baram area in 1898, commenting on the hostilities says that the attacks by the Kenyahs followed from the fact that the Iban working of gutta had long been a grievance of the Kenyahs.[80] Despite these hostilities, Charles Brooke promoted the Iban presence in the Baram, in part because they were such good collectors of forest produce and there was so little other trade there, but also because it was a continuation of his practice of having them as a backup or what Pringle calls 'a trustworthy source of fighting power in the northeastern provinces'.[81] It was a risky policy, but one that Brooke had long used in Sarawak.

The history of diplomatic negotiation for the cession of the Baram to Sarawak was to be mirrored in the Trusan river district of Brunei in 1884. In that instance, the activities of gutta percha collectors were again a flashpoint in the negotiations that ultimately led to the cession of territory to Sarawak.

In 1884, several Iban were killed while on a trip hunting for gutta in the Trusan area.[82] The trip was one of the trading trips recorded in Iban oral tradition, retold by Sandin as being an expedition stopping at several places to

76 SG 1 December 1894, p. 207.

77 SG 1 December 1894, p. 212.

78 SG 1 December 1894, p. 212.

79 C. Hose and W. McDougall, *The Pagan Tribes of Borneo*, 2 Vols, London, Macmillan and Co, 1912, Vol. II, p. 211.

80 A.C Haddon, *Head-Hunters Black, White, and Brown*, London, Methuen & Co, 1901, p. 341.

81 Pringle, p. 270.

82 Baring-Gould and Bampfylde, p. 344 and SG 3 January 1885, p. 6 and see P.M. Kedit, *Iban Bejalai*, Kuala Lumpur, Sarawak Literary Society, 1993, p. 95.

collect gutta percha.[83] While the collectors of the Baram were mainly Kenyah and Kayan, along the Trusan it was Muruts who were the main collectors of gutta percha. At this time, the Trusan was still part of Brunei and because of this, and perhaps for fear of increasing hostilities, Brooke refused to give other Iban permission to take revenge for the killings. Brooke did not let the matter rest there; he demanded the enormous sum of $22,000 compensation from the Sultan for the Trusan murders. Eventually, after very considerable exchanges of correspondence and a repetition of arguments similar to those that had occurred over the Baram cession, the Trusan was ceded to Sarawak in January 1885.[84] Governor Treacher of North Borneo wrote that 'the unfortunate murder of Sarawak subjects in the Trusan … gave the Raja a hold on the Temenggong who has since become Sultan'.[85] As before, the wider political and economic arguments were complex, often personal, but the gutta percha hostilities provided Brooke with the excuse he needed to argue for cession of the Trusan. The hostilities were enough to add fuel to the already complex arguments among the Brooke administration, the Sultan of Brunei and various Brunei elite, the Foreign and Colonial Offices and the British North Borneo Company about the sharing of resources and the future 'carve up' of Brunei. The very future of Brunei and the fight for power and influence of the various personalities and governments was at stake. Also at stake was the future of Labuan. The gutta percha incident, however, emphasized the weaknesses of the Brunei sultanate and its difficulty in exercising control over the Trusan area.

At the very least the incident forced the need to resolve the question about cessions and borders. For Kedit, it was a result of Iban *bejalai* activities that the Trusan was ceded to Sarawak but the *bejalai* expedition required some impetus and the valuable gutta percha produce provided that impetus.[86] Sarawak history is often a history of small events writ large, small events that drifted out of hand or were used as the basis for later attacks, expeditions or negotiations. The weaknesses of the Brunei sultanate and its inability to control a vexed security situation laid the groundwork for a relatively small event to be used to open a wider argument. Both the 1873 Baram incident and the 1884 Trusan attack were hostilities related to the exploitation of gutta percha and trade, and were the spark to ignite the tinder of pre-existing political conditions. Brooke Sarawak put forward arguments about being the more stable and fairer system, better able to govern (and provide a better trade environment) in the interests

83 Sandin, 1994.

84 See Tarling, 1971 for a detailed discussion of this period.

85 14 June 1885, No. 81 CO 874/239 in Tarling, 1971, p. 330.

86 Kedit, p. 93.

of the inhabitants than was the Brunei Sultanate. As Pringle notes, the Trusan incident provided the Sarawak Government with a pretext to demand the cession of the district from Brunei.[87]

Following the cession of the Trusan to Sarawak, the Sarawak Gazette concluded that Sarawak collectors would now be able to collect and trade the valuable produce of the area, paving the way for 'development and civilization'.[88] Even under Sarawak rule, gutta percha continued to alter the dynamics of trade and stability in the Trusan. It was all very well for the Sultan and various Englishmen to divide territories; different groups of collectors were still interacting and clashes over resources and access still arose. Charles Brooke held a peace conference between the Murut and Iban,[89] but mutual suspicions continued, and the generally unsettled collecting environment was reported on in the Gazette in 1892.[90] In 1898, the Resident's report for the Trusan recorded, somewhat hopefully, that now that inland Muruts had been 'taken in hand', the road to the fort would be safer for some time to come and that all the gutta and other jungle produce would now come down to the bazaar.[91] Yet, years later, according to Sandin, in 1900 a group of Iban returned on another expedition to the Trusan, in a punitive expedition against Muruts to take revenge for the 1884 attack that is described here.[92] The 'war' sparked by the gutta expeditions of so many years earlier could not go unavenged.

Further Brunei territory was added to Sarawak when the Limbang district was ceded by the Sultan of Brunei in 1890 and as early as 1891 a group of Iban were active in the area. There were reports of disagreements between Iban and the local Tabun and Murut populations, though not all were blamed on the Iban.[93] The Tabun lived in the 'lower river' and the arrival of Iban collectors was a source of friction as both groups were competing for increasingly limited forest resources in that area. The same Report noted an increased export trade in gutta percha, and in the importations of brassware. The lower regions of the Limbang reportedly possessed only small amounts of rattans and gutta percha having been 'worked continually for some years past'; however, the subsequent years of increased gutta percha exports from the Limbang would suggest that the Limbang was a conduit for up-river and hinterland gutta percha

87 Pringle, p. 267.
88 SG 1 May 1886, p. 72.
89 SG 1 May 1886, p. 72 and Pringle, pp. 267–268.
90 SG 2 January 1892, p. 19.
91 SG 2 May 1898, p. 106.
92 Sandin, p. 243, see also Kedit, p. 95.
93 SG 1 April 1891, pp. 59, 60.

traders.[94] Further up-river were the Muruts and the gutta percha – and the potential for hostilities. In 1898 a group of Iban and Muruts attacked a group of Muruts living across the border, taking '32 heads and 12 captives'.[95] The Rajah issued orders for compensation to be paid and the offenders to turn themselves in, and the victims' heads and captives returned in order to prevent the likelihood of retaliation – and so that peace would be restored and future groups would be able to continue working produce in the area.[96] Eventually, after a visit from the Rajah, this occurred.[97] The Limbang, like the other districts of Sarawak, was subject to the same gutta percha hostilities that had been a reflection of damaged relationships between collecting groups, prompted by the attraction of the valuable gutta percha trade.

There were other violent attacks associated with the gutta percha trade; these occurred across the borders between Sarawak and the neighbouring Dutch territory of Netherlands Indies. By the mid-nineteenth century, the increased migrations and trade following the arrival of the Brooke and Dutch administrations, and the impact of global trade, changed the previous levels and patterns of human interaction across Borneo. The Brooke administration had to be mindful of its relations with the Dutch administration.[98] When collectors of forest produce crossed borders (from either side) administrations were anxious to retain control over their populace as far as was possible. However, only limited official control could be established given the isolation, limited administrative resources, and the determination of collectors to go where the trade was more attractive. Wadley concludes that, by the 1850s, Iban raiding was leading to reduced populations along the Dutch Kapuas area.[99] He suggests that it is possible that some movements of Iban into Dutch territories may have been to escape the suppressions of Brooke administration, particularly given Charles Brooke's practice of destroying valued assets (see above) making it difficult for families to remain once Brooke's forces had laid waste Iban settlements. On the Dutch side, by the 1850–60s, Sarawak Iban activities in the Kapuas region were so aggressive that the Dutch called the Iban the 'terror

94 SG 1 April 1891, p. 59.
95 SG 1 July 1898, p. 148.
96 SG 1 July 1898, p. 147.
97 SG 1 July 1898, p. 147.
98 See also SG 1 October 1895, p. 178.
99 R.L. Wadley, 'The history of displacement and forced settlement in West Kalimantan, Indonesia', in *Conservation and Mobile Indigenous Peoples: Displacement, Forced Settlement, and Sustainable Developments*, eds D. Chatty and M. Colchester, New York, Berghahn Books, 2002, p. 318.

of the Kapuas';[100] there were revenge attacks from the Dutch side through the 1870s and 1880s, leading to Dutch military patrols and resulting in destruction of long-houses, rice fields and the confiscation of 'heads' from Iban living in the Dutch Emperan. In Sarawak, however, the feeling was that in 'nine cases out of ten' hostilities arose from onslaughts made by those who were living under Dutch administration.[101]

In January 1875, a group of Iban who crossed into Dutch territory looking for gutta was fired on by local inhabitants.[102] Six months later another group of Rejang Iban crossed into Dutch territory and with hostilities arising, dozens of local 'gutta hunters' were murdered, prompting an outraged letter from the local Rajah.[103] These events were small in themselves, but they could easily have escalated. In 1892, after yet more killings of Kenyah, and the fear that retaliations were about to cause even more hostilities, Charles Brooke involved himself and called a meeting of local leaders: they were warned against retaliations against the 'border tribes'[104] and

> The Dyaks were warned on the other hand against penetrating too far into the interior in small parties after gutta or other jungle produce as they would probably be cut off by the enemy.[105]

Brooke did say in the same address, however, that should such border tribes make an appearance on the Sarawak side in the future then they should be looked upon as 'enemies' who thereby rendered themselves liable to attack. Such qualification seems likely to have given anyone thinking of retaliation hope for an official sanction to do so. A few years later, in 1898, the Sarawak Gazette reported that a group of five Iban from the Sut river (a tributary of the Rejang), who were in the Dutch territory 'working gutta', was attacked by 'inland tribes', probably Kayan. Five were murdered and their heads taken.[106] As overland and cross-border trading routes opened up, such as those outlined in previous sections, great profits could be made and it was difficult for groups not to make the most of these opportunities, notwithstanding the risks that

100 Kater, 1883, p. 4 in Wadley *Conservation and Mobile Indigenous Peoples: Displacement, Forced Settlement, and Sustainable Developments*, p. 319.

101 SG 2 April 1883, p. 30.

102 SG 5 January 1875.

103 SG 1 July 1875, the Rajah of Bunut.

104 SG 1 July 1892, p. 130.

105 SG 1 July 1892, p. 130.

106 SG 1 June 1898, pp. 125, 126.

may have been involved. Underlying disagreements over sharing gutta percha resources bubbled away beneath the surface and in 1901 three Iban in Dutch territory were killed by Kenyah when they were hunting for gutta; as a result, a large group of Iban from the upper Rejang area sought to take revenge. In turn, their attack led to a counter attack resulting in more deaths, of Iban, Kayan and Punan.[107] While these hostilities continued to mount, elsewhere the usual unsettled conditions that had been the hallmark of Brooke Sarawak, continued in the Batang Lupar region. As a result, a large punitive force was mounted in 1902 under Charles Brooke, although when cholera broke out the expedition was called off. The hostilities never settled, either in that year or the following year. The rest of this history need not be relayed here; see Pringle for discussion of this period of what he calls 'virtual anarchy'. Decades later the issue was *still* not entirely settled.[108] The upriver and border areas of Sarawak and the Netherlands Indies were, then, subject to great and seemingly contradictory forces of trade and warfare. That is, uncertain conditions were a part of the overall pattern of life, and within that fractious world, trade and hostilities seemingly went hand in hand.

Further north, Eghenter believes that Iban collectors had been so active in the Apo Kayan area of Dutch territory that, by the end of the nineteenth century, they were causing depletion there as well and that hostilities arose in part due to the depletion of gutta percha resources in Sarawak. The overland linkages facilitated both trade *and* hostilities because the Apo Kayan was within easy overland reach of Sarawak. Eghenter concludes that competition for resources between Sarawak Iban and the Kenyah sometimes broke out into open conflict and that feelings on the Apo Kayan were heightened by anger about the Iban collectors undermining the authority of local leaders by wandering into forests without approval. She also concludes that what further heightened tensions was the alleged unsustainable methods of collecting gutta percha practiced by the Iban (cutting the trees, or else cutting down small rather than larger trees).[109]

In 1921, fifteen Rejang Iban were killed by Kenyah in the Apo Kayan, in what the Sarawak Gazette described as 'exceptionally galling circumstances', following a 'feud' which it said had commenced in 1886.[110] Collectors and traders of gutta percha continued to work the forest for decades – causing hostilities – and did so independently of, even in spite of, formal government attempts to

107 SG 2 January 1902, pp. 6,12.

108 Pringle, pp. 228 ff.

109 Eghenter, *Histories of the Borneo Environment*, p. 97.

110 SG 1 December 1924, pp. 321–324.

control them. Many decades of government intervention halted neither gutta percha collection and trade, nor the violence that arose around it. The issues were complex and emotions strongly felt. As the Iban leader Koh described the late nineteenth century period of warfare, he concluded that while he had himself had taken seven Kayan heads and many captives in revenge for the murder of his uncle, 'I did not just kill, but I took revenge. Vengeance was taken for anyone killed'.[111] Difficulties continued for years, but Koh played a different role when he negotiated with the Third Rajah, Vyner Brooke (son of Charles Brooke) to agree to a formal reconciliation to be held at a Peace Conference that was held in 1924 at Kapit. Thousands of Iban, Kayan, Kenyah and Kajangs from both sides of the border assembled for the event. It aimed to settle the matter once and for all. Although reported in the Sarawak Gazette, no specific treaty or contract was issued from the settlements. The Sarawak Gazette said that the conference was a sequel to preliminary peace-making in the Netherlands East Indies the previous year.[112] According to Eghenter, the Dutch administration, as part of this settlement, asked for 'collaboration of the government of Sarawak to stop unsustainable and unselective collection of gutta percha by Iban in the Apo Kayan ... issuing travel permits to Iban collectors ... limiting access to three main commercial routes across the border and requesting that Iban collectors report to Kenyah customary chiefs before undertaking collecting activities'.[113] The significance of that agreement is still widely known among the Indonesian Apo Kayan Kenyah according to another study, suggesting that the improvement in terms of both peace and better trade was significant enough to be a landmark in common memory.[114] Incursions, hostilities and instability felt on the Netherlands Indies side, or by communities from there looking to travel for trade into Sarawak were well known among Dutch administrators and local inhabitants throughout the late nineteenth and early twentieth century. Nieuwenhuis wrote in 1929 that traders crossing to the Rejang-Baram regions were 'always being ambushed by Hiwans

---

111 Quoted in Sutlive, 1992, p. 30.

112 See SG 1 December 1924, pp. 321–322, also SG 1 May 1924, pp. 168–169; Sutlive pp. 32–33, and Appendix D, *Peace-making at Kapit*, pp. 326–329.

113 Eghenter, p. 97.

114 D. Lumenta, 2003, 'Towards transnational Dayak identities? Changing interconnectedness, identities and nation states – A case study on Iban-Kenyah relations in Sarawak, East Malaysia' in *Economic Prospects, Cultural Encounters and Political Decisions: Scenes in a Moving Asia* (*East and Southeast*), The Work of the 2002/2003 API Fellows, Kuala Lumpur, 2003, p. 9. See also D. Lumenta, 'Moving in a hierarchized landscape. Changing border regimes in Central Kalimantan', *Wacana*, Vol. 13, No. 1, April 2011 (pp. 121–145), p. 131.

[Iban]'.[115] Hostilities continued unabated for decades even after many official expeditions by either the Brooke or Dutch administrations.

In setting out the history of the gutta percha trade, it becomes clear that Sarawak societies embraced the opportunities for engagement with the external commodity trade and did so against considerable odds. This included the need either to negotiate or engage in hostile interactions with others in the trading environment. For Cramb, the transformation that occurred in Iban societies, for example, could not be explained as a passive response to irresistible economic and political forces, but rather that the Iban were active agents in their own transformation who sought to embrace the market and the state on their own terms.[116] To a certain extent Cramb's study builds on earlier studies that have examined aspects of Iban culture, economy and ecological adaptation to change.[117] Freeman considers Iban activities not merely in the context of their agricultural and trading success, but also links those adaptations and achievements with 'invasion and conquest'.[118] He connects trade with the spiritual significance of padi-growing and headhunting in Iban society as important means for valorization and consequent social standing in what was judged as a competitive society. Koepping concluded, in her study of Borneo headhunting in the late nineteenth century, there were many factors in hostilities resulting in headhunting, including disagreements over land assets and forest resources. She found that the economic aspects of feuds and head-taking have been

115 Nieuwenhuis, 1929, p. 32 in Lumenta, p. 129, and see the rest of Lumenta for a further outline of cross-border mobility and hostility related to trade relationships (pp. 121–145).

116 R.A. Cramb, *Land and Longhouse: Agrarian Transformation in the Uplands of Sarawak*, 2007, p. 6.

117 D. Freeman, *Report on the Iban of Sarawak*, Kuching, Government Printing Office, 1955; D. Freeman, *Iban Agriculture: A Report on the Shifting Cultivation of Hill Rice by the Iban of Sarawak*, 1955; D. Freeman, 'Iban Augury', *Bijdragen tot de Taal-, Land- en Volkenkunde*, 117, No.1, 1961, pp. 141–167.; D. Freeman, *Report on the Iban*, University of London, London, The Athlone Press, 1970; D. Freeman, *Some Reflections on the Nature of Iban Society*, Canberra, Australian National University, 1981. C. Sather and J.J. Fox, (eds), *Origins, Ancestry and Alliance: Explorations in Austronesian Ethnography*, Canberra, ANU E Press, 2006; C. Sather, 'Introduction' to *Sources of Iban Traditional History*, B. Sandin, *Sarawak Museum Journal*, Kuching, No.67, Special Monograph No. 7, 1994, pp. 1–78.; C. Sather, ' "All Threads are White" Iban egalitarianism reconsidered' in *Origins, Ancestry and Alliance: Explorations in Austronesian Ethnography*, eds C. Sather and J.J. Fox, Canberra, ANU E Press, 2006, pp. 73–112. V. Sutlive and J. Sutlive (eds), *The Encyclopedia of Iban Studies; Iban History, Society and Culture*, 4 Vols, Sarawak, Tun Jugah Foundation with Borneo Research Council, 2001; V.H. Sutlive, *The Iban of Sarawak*, Arlington Heights, IL AHM Publishing Corporation, 1978; V.H. Sutlive, *Tun Jugah of Sarawak: Colonialism and Iban Response*, 1992; V.H. Sutlive, 'The Iban in historical perspective', *Sarawak Museum Journal*, Vol.40, No.61, 1989, pp. 33–43.

118 Freeman, *Sarawak Museum Journal*, p. 277.

unreasonably down-played.[119] In a study of East Kalimantan forest traders, Sellato has concluded, for example, that disputes over trade resources were far more common than those over subsistence resources,[120] which mirrors the history of the gutta percha trade set out here. Positioning for power and prestige in a changing environment were also influential factors in the Sarawak gutta percha trade because of its linkages with the *pusaka* trade.

It seems clear that while Iban collectors were very much a part of the hostilities set out above, they were not always the instigators, with other collectors also participants in this period of uncertainty as they were responding to the perceived loss of this valuable trade in their own districts. Further, the general hostilities should not be taken to indicate that all collectors were part of some steady stream of headhunters rampaging across Borneo. Most of the reports summarized above come, of course, from European sources which tended to highlight headhunting in newspaper reports for obvious reasons, but the broad trend of hostilities related to the gutta percha trade is clear. It might also be noted we do not hear in the historic record of Iban gutta percha collectors engaging in headhunting while on collecting expeditions in the broader Southeast Asian region. This suggests that because headhunting by gutta collectors occurred in Borneo, rather than elsewhere, it was traditional enmities, or long term collecting activities that were key factors in the hostilities associated with the trade.

Whatever the nuances, the extent to which the trade and hostilities considered in this section fractured collecting societies is illustrated below in Map 6 on which is shown the indicative locations of some of the incidents reported by Sarawak commentators (for the period after 1870). The incidents occur along the length of the Sarawak hinterlands and borderlands. Given the small and sparse population of the time, and the violence of the attacks, such episodes and the consequent exploitation of them by Charles Brooke underline the significance of the changes occurring in the society at this time.

After analysing the trade data and the narrative record of hostilities related to the collection of gutta percha, it becomes apparent that this was a resource that played a role in more than just the economic life of Sarawak (and across its borders). In less than a lifetime, a process of social and economic change occurred that was more revolution than evolution as a society moved into the world-system. Even changes in national borders that remain today hark back

119 E.W. Koepping, *Friendship and Feud in Late 19th Century Borneo*, Centre of Southeast Asian Studies, Melbourne, Monash University, 1988, p. 13.

120 B. Sellato, 'Forests for food, forests for trade – between sustainability and extractivism; The economic pragmatism of traditional peoples and the trade history of northern East Kalimantan' in *Histories of the Borneo Environment*, ed. R.L. Wadley, Leiden, KITLV Press, 2005, p. 67.

to groups of forgotten gutta percha collectors. A short pause here gives occasion to reflect upon the stark differences and motivations that existed between the world of the gutta percha consumers such as cable producers and that of gutta percha collectors in the Borneo forests.

What emerges from the history of the gutta percha trade in Sarawak is a story with contradictory themes. On the one hand is a history of a profitable trade that built upon cultural, topographic and historical knowledge and beliefs. It was a trade that ignored 'national' borders, trekked on overland paths and travelled along river systems that delineated social groupings. In parallel with that was a more harrowing narrative of clashes that inevitably arose among people close enough to share common territories, even common heritage, but also a common desire for the same resources – and a need to protect their rights or seek revenge for past attacks. These parallel histories were shaped on the one hand by the Brooke administrations and Chinese traders; and were forged by the motivations, mores and knowledge of the collectors on whom the trade ultimately depended. It was the manifestation of the local and the external world-systems shaping each other.

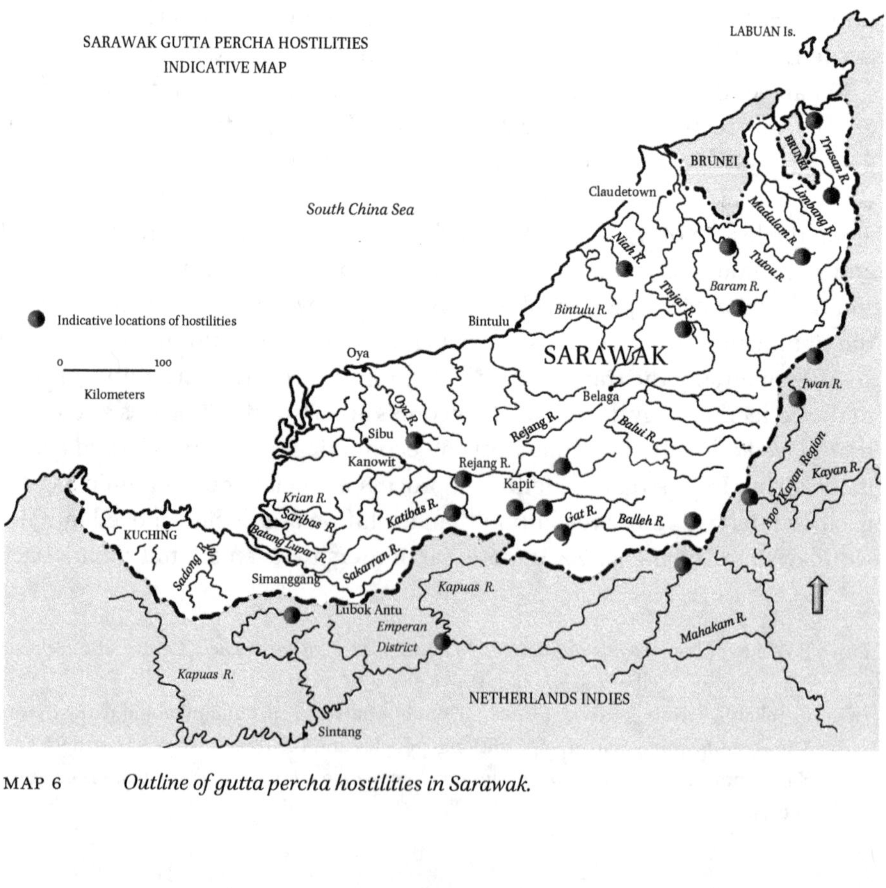

MAP 6 *Outline of gutta percha hostilities in Sarawak.*

CHAPTER 9

Conclusions

> Some Adangs ... have been down here bringing some jungle produce ... They are the first people of this tribe who have visited this station ... They came down with some Dyaks who were gutta hunting in their country and whose assistance they now require for their return journey[1]
>
> SARAWAK GAZETTE, 1 September 1892

Global communications were revolutionized after the somewhat shaky beginnings of submarine telegraphy in 1851. Trade and contact between people moved from the physical and distant to the abstract and the immediate. The world became dependent upon instant communication – and for a half century or so it became dependent upon gutta percha.

Once the use of gutta percha in submarine telegraphy was established, the quest for gutta percha was guaranteed for decades since there were no accepted alternative products to compete with it and the expense of submarine telegraphy discouraged experimentation with alternative products.

There were several factors that shaped the hunt for gutta percha.

The first related to the market demand for gutta percha – the requirements of global telegraphy and the peculiarities of its commercial operations. The relative haste with which this new technology was put in place and the great expense of failed ventures meant that once a reliable technology was identified – based in part upon the use of gutta percha – governments, investors and manufacturers were unwilling to risk trying new ideas by finding alternative products to manufacture cables even if they might be less expensive than gutta percha.

A factor that shaped the European quest for gutta was the need to identify and understand the botanic and scientific features of gutta percha and the trees that supplied it. It also required constant testing in the factory to ensure that the gutta percha being used and applied to cables was fully up to the demanding standards required of this expensive technology. It was a

1 *Sarawak Gazette*, Kuching, 1 September 1892, p. 171.

 | DOI 10.1163/9789004357280_011

quest for scientific knowledge and the application of it in industry, a quest that reflects Kuznets concept of the 'modern economic epoch' as achieved by the application of science to production.[2] While such sophisticated and scientific methods were to be applauded, at the same time expensive failures, and fear of future failure, shaped another feature of the European quest for gutta – the idea that only gutta percha would suit their needs, and that no alternative product could be risked in its place. Fear of failure also meant that a few cable producers dominated the marketplace and influenced demand for gutta percha; these companies had the proven expertise and market connections to produce cables and it was safer to award contracts to such proven contractors. The formation of commercially powerful cable operations also cemented the domination of certain key metropolitan actors in the manufacture of submarine cables. This reinforced the early networks that were established to supply the gutta percha required by the submarine telegraph industry. As a result, the gutta percha commodity chain was, in its metropolitan phase, dominated by a conservative and small group of actors, chain drivers, who used sophisticated manufacturing techniques but relied upon commercial, government or cartel arrangements to win profits. The early entry of the Gutta Percha Co (and the dominance of its partner Telcon) also shaped the market for gutta percha and made it difficult for new entrants.

Another feature that shaped the quest for gutta percha – both for Europeans and of course for the regions that supplied gutta percha – was the very limited global sources of supply. It was one region alone that supplied gutta percha. The very limited geographic source of supply contained the scale of the commodity chain and limited the number of actors in it. Because of these factors, a peculiarity of the gutta percha commodity chain itself was that, despite its global significance, it was relatively limited in the complexity of its operation compared with other commodity chains. British telegraph cable manufacturers had the best access to sources of capital and linkages to a greater commercial and imperial network, so they were well positioned to find customers and establish a reputation in a conservative industry with few competitors. As a result, the gutta percha trade was established around a global economic framework that saw Britain dominate both the demand and supply aspects of the gutta percha trade, focused through the distribution hub of Singapore.

The structural features of the trees supplying gutta, which made them difficult to tap, had an impact on the hunt for gutta percha. As a result, the problem of collecting gutta percha in a sustainable manner shaped the hunt for this

2 S. Kuznets, *Modern Economic Growth: Rate, Structure and Spread*, New Haven, Yale University Press, 1966, p. 9.

product, and caused considerable discussion in European cable and scientific circles for decades. Various European governments, scientists, traders and adventurers attempted to study gutta percha, to look for new forest sources of supply and made varying attempts to establish plantations. These efforts increased botanical knowledge and some made profits by developing and selling patents for the extraction of gutta percha hoping that such extraction methods could be both sustainable and profitable in the longer term. Until the early twentieth century (with the advent of the Tjipetir and Selborne plantations), these ventures proved to be small and were commercial failures other than, perhaps, for those who profited from the sale of the patent rights.

Geography and the peculiarity of the product were then contributing factors in the structure and operation of the trade. While basic issues of demand and supply and the peculiarities of geography and the trees themselves shaped the global hunt for gutta percha, the question was of course more complex than that. Practical and commercial issues and questions that arose from network interaction and the forces of change also shaped the quest for gutta.

Traders in Sarawak and the wider collecting region did not practice collecting methods that would ensure a sustainable future for the supply of gutta percha: but they neither saw the issue as their problem and nor was it within their capability to achieve. Tapping the trees was not practical for indigenous collectors who were attempting to maximise the product they could collect in the shortest time and to minimize risk from either other collectors taking their gutta or attacking them while on expeditions. While the fear was expressed of a threat to global communications due to the problem of sustainable collection, this did not result in the development of plantations as sources of supply. In part, this was because the trees themselves were slow to mature and so they were thought to be too much of a long-term investment for commercial operations. In part, there was also a tendency to blame the problem on collectors' unwillingness to obtain gutta percha by the 'correct' method. Governments in turn did not take up long term investments in plantations; generally speaking, official activities were intermittent and lacked consistent funding or support. When governments did invest resources in scientific research, study visits or small-scale plantings, many of the efforts were subject to personal or policy disagreements, cessation of funding or bungled efforts that led to only half-hearted results at best.

While eventually the Tjipetir and Selborne plantations were developed in the hope of fulfilling the metropolitan need for gutta percha, they were largely too late to answer the needs of submarine telegraphy. Rather, it was trade networks at the periphery that were to fulfil the requirements of cable producers. Cable manufacturers and metropolitan governments relied upon the success of the commercial networks in Singapore and supply areas like Sarawak. This was in part because the functioning of the commodity chain catered very efficiently to their immediate demands. That is, the metropolitan hunt for gutta percha was assigned to others to undertake as a commercial transaction because they could do so efficiently, while manufacturers proceeded with the commerce of cable production. While traders in Sarawak, Singapore and the wider region kept the market supplied, the telegraphic industry kept its attentions focused elsewhere – on winning contracts, government subsidies and producing cables. The quest for quality gutta percha was in large measure set aside for those at the periphery. The periphery was left to do the task assigned to it – supply resources to the metropolitan core. The capability, yet consequent vulnerability of the periphery in its role in the trade was revealed by its relationship to the power and influence of the gutta percha consumer (the metropolitan 'core'). This was revealed in the evidence uncovered from the commercial archives collected here for the first time. The material illustrated the speed with which decisions by 'chain drivers' could ripple along the nineteenth century commodity chain, with evidence traced from the decisions in Directors' Minutes to regional export data in Singapore and onwards to Sarawak. More particularly, the impact on the trade was felt in certain years in Sarawak which was both buffeted by and profited from the powerful operations of the global communications system.

In varying degrees, the roles of actors involved in the hunt for gutta percha were shared along the commodity chain, initially shaped by cable producers and then by prevailing local economic and political conditions. Cable manufacturers could take advantage of the commercial network existing in Singapore. That hub was built upon existing political and economic structures – or the 'ministructures' that were a part of the world-system as it evolved from pre-existing global trade conditions. A bureaucratic, commercial and communications network in Singapore enabled commercial relationships to be established with local intermediaries that linked with local traders in Sarawak. These were the forward and backward linkages in the commodity chain between the metropolitan and the local actors. They were, in effect, elements of the continuing transformation of the ministructures to which Wallerstein attributes successful integration of peripheral economies into the world-system.[3] The peculiarities of the commodity and the industry that used it shaped the gutta percha trade, and thereby the power of actors in the trade and the strategies adopted by local collectors.

---

3 I. Wallerstein, *The Modern World-System III. The Second Era of Great Expansion of the Capitalist World Economy, 1730–1840s*. San Diego, Academic Press, 1989, p. 130.

The success of the hunt for gutta percha in the region that supplied it was based on several factors. Intermediary rulers and traders and the administrative and commercial arrangements – the 'ministructures' that promote the engagement of regional states with the world-system – facilitated the rise of Southeast Asia as a region that supplied commodities to the metropolitan core. In turn, these ministructures ensured that the arrangements for the trade in gutta percha were put in place very quickly in Singapore and the region. Like other commodity trades, the gutta percha trade developed in the context of a beneficial prevailing political regime which had seen relationships built with local political elite. This enabled the demands of submarine telegraphy, and the metropolitan consumer, to be met very quickly. The prevailing economic and political regimes in Singapore and then in Sarawak facilitated the trade and in consequence established the power and role of certain actors in the gutta percha trade. These regimes were essentially conceived by the metropolitan powers of the world-system to enable their commercial (and political) needs to be met at the periphery. In other words, commodity chains need to be envisaged as the product of a wide landscape of features that are not merely commercial. But enabling this to occur both required peripheral responses and in turn created opportunities at the periphery.

In the region that supplied it, the quest for gutta percha was structured around Singapore and centres like Sarawak where Chinese trade networks were built upon pre-existing commercial networks that exploited the demands arising from the new gutta percha trade, and could do so with considerable speed. While they were subject to the vagaries of demand from consumers, these networks established commercial and credit relationships, and had the capacity to supply goods to gutta percha collectors: those needs could not have been met by European agents, and those agents showed no interest in doing so in any case. Chinese traders filled an essential role in the commodity chain: their position in the chain was held by their skill, resources and networks that other others could not provide. The evidence emphasizes the various roles of Chinese entrepreneurs throughout the gutta percha trade – covering trade, shipping, 'manufacture' and credit as well as networking with European agency houses. Those roles in Singapore formed a link with regional collecting areas where existing economies had built upon long-standing produce collecting practices. Because of the trading capacities offered in Singapore, the gutta percha commodity chain was established with great rapidity despite the heterogeneity of actors and considerable geographical distance involved. Sarawak Chinese traders provided the essential local requirements of trade: investment, credit, goods for barter, trade venues such as bazaars and boats, and an interlinking arrangement of trader networks that stretched from hinterland riverine positions in Sarawak through to larger outports, Kuching and thence to Singapore. The impact of the Chinese trade network was a result of specialized skills, local knowledge, competitive commercial practices and fulfilling a unique role in which others could not compete. The leading Teochiu Chinese dominated the gutta percha trade in Singapore and in Sarawak. Other Chinese traders gained from the profits of the trade and so too did local indigenous elites and successful collectors. Sarawak Chinese boat traders and smaller bazaar traders were vulnerable actors in the gutta percha trade since they were often in debt to other traders and were vulnerable to changes in the market.

Another feature of the relationships – indeed the strength of the Sarawak indigenous and Chinese trade networks – was that their specialized knowledge and networks were complementary and inter-dependent but not competitive with one another. In general, this was the strength across the commodity chain: the interlinking nature of the roles and separation of power positions since no group of actors in the local, regional or global setting could afford to alienate the other. As with all actors in the commodity chain, each brought knowledge and connections that provided them a role in the trade and returns on their investment. The significance of local knowledge in the gutta percha trade was in turn shaped by the nature of the environment and the peculiarities of the commodity itself.

Like Singapore, Sarawak possessed the ministructures of political and commercial arrangements required for trade. Brooke policy laid the groundwork that shaped trade, but the gutta percha commodity chain as it functioned in Sarawak was largely left to Chinese traders and indigenous collectors to develop. That process included the need to solve problems of geography, sparse population, disease and feuds between groups – only some of these issues were within the scope of the political and economic grasp of Brooke policies. Instead, local actors devised their own strategies to develop the commodity chain and to incorporate it into their local economies. These strategies relied in large measure upon extensive knowledge of the forest and its resources, and on the use of omens to guide the collecting effort. The Brookes, especially Charles, promoted a Chinese trade presence, used groups of Iban as quasi-military forces and promoted Iban migration as a means of opening up the economic potential of the forest resources of Sarawak. The exploitation of the gutta percha resources of Sarawak played a key role in this process, in part because the Iban were attracted to the gutta percha trade. When hostilities arose – the gutta percha 'wars' – Charles Brooke used them as opportunities to further strengthen his authority over local populations and even as an argument to expand further the political boundaries of Sarawak. Brooke used the attacks on Sarawak gutta percha collectors working in Brunei territories as the pretext to negotiate for the cession of part of that territory. The actions enabled him to exploit events as part of his diplomatic positioning with the Brunei Sultanate and with the British administration. Ultimately, that positioning, that had taken advantage of repeated gutta percha hostilities, changed the shape of the national boundaries of Borneo that continue to this day.

The many actors involved in the gutta percha trade were separated by very considerable differences in space, wealth and political influence – and they lived in very different geographic, social and even spiritual worlds. An interesting conclusion made by Topik, Marichal and Frank is that buyers and growers (or in this case collectors) gave little thought to the origin or destination of commodities; rather, they reacted to one another mainly through price, and actors engaged in a trade are not really in a position to 'see the larger script'. So, metropolitan buyers (of gutta percha) gave no more thought to collectors in Sarawak trading gutta percha for rice, cloth or brassware, than did Kayan or Iban or even Chinese traders to the ultimate consumers of gutta percha. That is, global trade as set out in this study 'created ironies of meaning in which producers and consumers understood the good exchanged in vastly distinct ways'.[4]

Talk of headhunters tends to send a frisson through popular or even scholarly texts, and by no means all of Sarawak engaged in headhunting. The fact remains, however, that hostilities, and at times headhunting, were linked with the gutta percha trade of Sarawak, and of Borneo. In this regard, the links between submarine telegraphy and the consequent hunt for gutta percha represent a unique snapshot of global connections between peoples. The differences in the lives of an Iban or Murut collector, for example, or a wealthy and influential trader such as the Sarawak towkay Ken Wat, or the Chairman of Telcon, Sir John Pender could hardly have been more stark; and each was in large measure unaware of the others in any more than an abstract sense. Actors were interdependent, but found their own systems of trade with the objectives of others being largely forgotten – and left to function based on their own cultural or commercial imperatives. The roles and strategies of the different actors in the hunt for gutta percha functioned through this array of mechanisms. These mechanisms facilitated trade and ensured that the considerable differences that existed among actors were not a weakness but a strength since different groups of actors filled a role in the operation of trade that others could not. This factor affected a shift from potential dissonance

4 S. Topik, C. Marichal and Z. Frank (eds), *From Silver to Cocaine. Latin American Commodity Chains and the Building of the World Economy, 1500–2000*, Durham, Duke University Press, 2006, p. 11.

to a remarkable consonance in the trans-national processes. Some of those mechanisms were controlled by metropolitan agencies, others were developed and resourced by Chinese trade networks. Ultimately, however, it was local indigenous resources and networks upon which the trade depended.

At the practical level, the difficulties of the collecting environment – harsh geographic conditions, lack of transport and transport networks, a hostile environment and spiritual/omen restrictions – made tapping trees a tedious process. For collectors, gutta percha was actively sought for the reasons considered in earlier chapters, but they could well get by without it – the same could not be said for metropolitan interests. Market fluctuations had shown collectors that gutta percha was a significant, but not always reliable part of the indigenous economy. Collectors were well used to switching to trade in other forest commodities – and the focus of much of their attention had to remain on subsistence – managing swidden agriculture or collecting forest resources. These traditional practices had both religious and environmental significance for local indigenous communities – submarine telegraphy and a dependence upon gutta percha belonged to what was for them a world at the periphery.

These true hunters of gutta percha, the collectors, were motivated by varied factors based upon commercial, social and cultural factors quite divorced from the scientific studies that underpinned the European quest. In Sarawak, the importation of prestige items – *pusaka* – and rice and cloth fluctuated with the fluctuations of the trade. Yet those same purchases revealed local actor responses and trade networks in operation, exploiting the opportunities offered by links with the global economy. In effect, the trade data collated here provides a 'paper trail' of how local traders at the periphery were making decisions – just as much as company directors and managers in the metropolitan centres made their decisions. The extent to which Sarawak collectors traversed the region also illustrates that they were not simply passive participants coping with the changes the trade presented, but actively sought to make the most of the opportunity that the gutta percha trade provided them.

The population of Sarawak had been linked to the external economy for many generations and the populace had relied in large measure upon close communal links of bonding social capital. The gutta percha trade, because of its value and geographic dispersal, promoted both hostilities and bonds between groups because of the interaction that trade connections required. As a result, social capital formation within indigenous groups was now changing. Whereas the bonds in the past had been largely within these communities, interaction with intermediaries and thence the external economy changed the nature of the social capital resources that indigenous populations used. The gutta percha trade prompted change in the social capital of indigenous groups from one that was largely group intra-dependent to one that was group inter-dependent. In this sense, the shorter-term material benefits (and costs of hostilities) of the gutta percha trade appear as one feature of the trade, but the independence of intra-group social capital that had been a strength of those communities began a process of disaggregation.

Risk taking, even risk seeking, was a feature of the Sarawak trade network – and a feature of the way in which social capital operated, both between indigenous groups and between those groups and the Chinese trade networks. That is, bonding and bridging social capital in subsistence economies with high levels of hostilities rely on trust *and* risk taking as strategies. While many theories of social capital have been put forward, and some deal with issues of trust, the question of risk taking and risk seeking (as opposed to risk management) deserves further consideration, in particular in subsistence societies. Risk taking enabled local populations to deal with the extreme challenges of the Sarawak trade environment. It enabled them more fully to exploit the new world of opportunity presenting itself because of, amongst other things, the rise of submarine telegraphy. It also provided an opportunity for improved social standing through proving valour and skill by engaging in risky or profitable expeditions to unknown regions. In a theoretical sense, the significance of the trade, newly opened overland routes and trade relationships, is that they serve to emphasize the vigor with which those at the periphery sought engagement with the world-system. The trade figures and evidence such as that from the Sarawak Gazette, show that in the years when gutta percha exports were good, so too were imports of *pusaka* such as brassware or jars, because these were items valued by gutta percha collectors. The gutta percha trade changed the lives of Sarawak inhabitants at this time.

When William Montgomerie first collected the specimens of gutta percha that he sent to the Royal Society in 1843, he found them, he reported, 'at a place much infested by tigers'.[5] Six decades later gutta percha had facilitated a global communications revolution that changed the world economy and social interaction. Its trade had altered the economy and society of the region in which it was discovered and its study had challenged some of the leading scientific minds of the era. The companies that produced this communications revolution, like the scientists that studied it, were all based very far from places infested by tigers or populated with headhunters. The role of

5 W. Montgomerie, in F.J. Mouat, 'Correspondence relative to a remarkable variety of Caoutchouc, termed Gutta Percha', *The Journal of the Agricultural and Horticultural Society of India*, November v, Vol. II, Calcutta, 1843, p. 104.

local-global linkages is a theme throughout this study. If Singapore itself can be at the semi-periphery, then Wallerstein's own words 'The semi-periphery is a necessary structural element in a world-economy' emphasize the significance of the existing pre-conditions of both the Chinese and local economic systems.[6] At the periphery, the significance of local knowledge and the peculiarity of indigenous trade in Sarawak – that evidenced by the regional gutta percha trading of the Iban – further underlines the significance of local system structures. Wallerstein concluded that 'A world-system is a social system, one that has boundaries, structures, member groups, rules of legitimation and coherence'.[7] This provides a perfect summation of these interacting relationships that supported the gutta percha trade. The economic and geo-political significance of telegraphy placed the gutta percha trade in the context of a highly organized, metropolitan system. That trade was built upon commercial and political ministructures in Southeast Asia, and on regional trade hubs like Singapore. Indeed, the telegraph helped to reinforce those structures: they were designed to meet not just the demands of profitable trade networks organized by European outsiders, but to meet the strategic needs of political powers like Britain, France, the Netherlands and the U.S. The influence of the world-system structure is readily evidenced in the region at that time – power and influence from the metropolitan core promoted, amongst other things, a surge in the region's economy. Stern concluded that one of the overriding strengths of Wallerstein's concept of a world-system was to propose how, under colonial conditions – merchant capital, imperial political power and a fluid environment at the frontier – it would exert an aggressive, organizing and transformative impact on local social and technological conditions for production.[8] Such conditions enabled the gutta percha commodity chain to be very quickly organized at the collection, trading and even the political level by the interventions of the Temenggong of Johor, and in Sarawak by the policies of Charles Brooke. Intermediary traders exploited this euro-centric commercial system with their own systems built on long-standing practices and relationships.

Yet in the Sarawak economy, Sverrison's point that 'international and global linkages cannot replace local linkages: on the contrary, they need to go together'

6 I. Wallerstein, *The Modern World-System I: Capitalist Agriculture and the Origins of the European World-Economy in the sixteenth Century, with a New Prologue*, Berkeley, University of California, 2011, p. 347.

7 Ibid.

8 See S.J. Stern, 'Capitalism and the world-system in the perspective of Latin America and the Caribbean', *American Historical Review*, Vol. 93, No.4, October 1988, p. 871.

is especially pertinent.[9] At the local level, indigenous networks likewise had their own practices and relationships. The quest for gutta percha left behind a trail both of violence and social transformation. Such a trail, in world-system thinking, would indicate the ruthless path of incorporation of the periphery into the world-system. Yet the process and its results were also an artefact of local social, cultural, economic, geographic and even spiritual forces. The active engagement of those at the periphery with their own motivations shows that trade – and conceiving the process of globalization – should be unfettered by notions of winners and losers, and commercial transactions alone.

What does this mean for those with a concern for globalization in its broad sweep? Bentley, in considering the nature of globalizing history, noted that 'cross-cultural interaction, communication and exchange ... are defining contexts of human experience'.[10] Aside from revealing the 'experience' of globalization through the prism of the gutta percha trade, the connections discussed here, including in graphical format, add weight to the claim that to understand the periphery and globalization, it is necessary to link the evidence with the social setting in which change occurs.

In both a symbolic and a real sense, it is likely that a small animal figurine, with whatever spiritual or commercial significance, was modelled by a gutta percha collector in the steep hinterland of Sarawak, who carried it with other pieces of gutta percha, overland and downriver to a trading point; perhaps to a Chinese trader in the bazaar at Sibu on the Rejang river in Sarawak. From there the figurine was included in a larger gutta percha shipment, likely to a Chinese towkay's godown in Kuching. By sea the figurine then was shipped with an even larger supply of gutta percha; stored, sorted and bundled with a large shipment that months later ended up at Woolwich in London at one of the cable factories. There, thrown in with other gutta supplies, the small figurine was chopped, boiled and massaged, ending up as a tiny section, unseen in the West India and Panama cable linking the American continents with Britain and beyond. By that time, the figurine had been carried, sold, traded, shipped, commodified, and finally transformed beyond recognition – and tested under a manufacturing regime developed by leading edge scientific and manufacturing practice at the time. This transformation embodied the link that characterized the global hunt for gutta percha – spiritual, cultural, political, commercial and scientific with each in some measure interdependent.

---

9 C. Pietrobelli and A. Sverrisson, (eds), *Linking Local and Global Economies: The Ties That Bind*, London, Routledge, Pietrobelli and Sverrisson, 2003, p. 11.

10 J.H. Bentley, 'Globalizing History and historicizing globalization', in *Globalization and Global History*, B.K. Gills and W.R. Thompson (eds), London, Routledge, 2006, p. 26.

APPENDIX 1

Gutta Percha

1.1 Characteristics of Gutta Percha – Implications for Trade

There were two features of 'gutta percha' that presented difficulties for both the trade and for submarine telegraphy:

> *Low Yield* – approximately 250 gm (around ½ lb) per tree was assessed as being collected from a 30 year old tree[1] – with up to three times that on a 50 year old tree with a 39 inch circumference, using the indigenous collection method.[2] This latter amount would equate to around 89 trees per pikul of 133.3 lb. James Brooke said the trees grew as large as six feet in diameter in Sarawak, but he of course would have been writing about early years of trade when old trees were still being accessed.[3] Burn-Murdoch quoted Sherman as claiming to have obtained 8½ lb from a tree of 42 inch circumference,[4] but Sherman's amount is considerably higher than any figures given in other accounts. Obach concluded that even an average of 2½ lb was too high; French and Dutch analysts also report collections of around 1 lb.[5] Clearly, there was considerable variation according to size and health of the tree, season of collection and so on. In 1900, Curtis experimented with likely yields, and even on a large, healthy, 50-year-old specimen could collect 1.5 lb (just over half a kilogram) of gutta; and on 52 foot high tree in 1901 extracted a similar amount – Williams cites similar findings of up to around just over a kilogram, with more being collected from inferior species [but these would be less likely to be suitable for telegraphic applications].[6]

1 W.R. Tromp de Haas, *The Netherlands India Government Gutta-Percha Plantations in Java*, Tjipetir, Batavia Government Gutta-Percha Plantations, 1908, p. 11.

2 A.M. Burn-Murdoch, 'Some facts about gutta percha', *Agricultural Bulletin of the Straits and Federated Malay States*, Vol. IV, No. 2 (February), 1905, p. 38.

3 C.B. Buckley, *An Anecdotal History of Old Times In Singapore* 1819–1867, Singapore, Fraser & Neave, Reprinted 1965, University of Malaya Press, 1902, p. 404.

4 Burn-Murdoch, p. 38.

5 See E.F.A. Obach, 'Cantor lectures on gutta percha', *Journal of the Society of the Arts*, Vol. 46, 1897, p. 120, who lists various amounts from Oxley, Logan, Wray and Burck; see T. Seeligmann, L. Torrilhon and H. Falconnet, *Indiarubber and Gutta Percha. A Complete Practical Treatise*, 2nd English edn, London, Scott, Greenwood & Son, 1910, p. 329; A. Combanaire, *Au Pays des Coupeurs de Têtes: À travers Bornéo*, 1902, Singapore, Editions Pagode Reprint 1993, p. 202.

6 C. Curtis, 'Notes on Gutta Percha Trees', *Agricultural Bulletin of the Straits and Federated Malay States*, H.N. Ridley ed., Vol. 1, No. 6, March 1902, p. 222 and L. Williams, 'Lactiferous

 | DOI 10.1163/9789004357280_012

Difficulties Establishing Plantations – trees were slow growing. It could take up to thirty years before they could be fully productive if gutta percha was harvested by collecting the latex from the tree trunks. This fact and low yield made trees unappealing for cultivation by European developers notwithstanding the demand for cable communications. It was for similar reasons unsuitable for cultivation by local small holders and so gutta percha cultivation was not adopted into the economic life of the region as occurred with para rubber in the twentieth century. Only after a means of collecting gutta percha from leaves and branches was developed were two plantations developed, but this occurred largely after gutta percha was in demand for submarine cables – see next section.

By 1910, one other calculation concluded that 400 gms (around 0.9 lb) per tree was a more reliable yield and using a conservative estimation of the average useage per nautical mile of cable they concluded that should the global submarine cable network need replacing this would require 27 million gutta percha trees to be felled.[7] Not included in that amount was any calculation for the tens of thousands of miles of underground cabling insulated with gutta percha. In addition to the amounts required for global communications, industrial and commercial applications of lower grade gutta percha continued to add to demand. Even in 1900, around the height of the submarine telegraph trade, Telcon was producing 100,000 gutta percha golf balls per week.[8]

The words *gutta percha* were not scientific or botanical terms, but popular and commercially used words that covered the product of many types of trees. As the trade in gutta percha developed and segmented, aside from various indigenous or botanical names, even more 'trade names' (for example, *Sarawak mixed* or *Pahang white*) were developed. These names were used to help identify the preferred types of gutta percha which were sought by the market. The prices achieved for gutta percha varied significantly: the poor quality product used for industrial purposes achieved a fraction of the price of the best qualities sought by cable producers. For example, 3½ d per lb. for 'inferior' and five times that, or 1s 5d per lb. for 'good white' in London in 1873 (poorer qualities were also more difficult to sell).[9] In 1901, Singapore figures reported in the *India Rubber World* show that premium (cable) grade gutta percha obtained

plants of economic importance v. Resources of gutta-percha-*Palaquium* species (Sapotaceae)', *Economic Botany*, Vol.18, No.1 (January-March), 1964, p. 17.

7 Seeligmann et al., p. 322, follows Obach, 1897.

8 G.L. Lawford and L.R Nicholson, *The Telcon Story*, London, The Telegraph Construction & Maintenance Co. Ltd, 1950, p. 88.

9 *London and China Telegraph*, 3 June 1873, p. 364.

more than ten times that of the lower quality product.[10] Annual trade figures, such as those for Britain, were present as total import value for all varieties combined. From a research perspective, this makes the use of annual 'average' prices for gutta percha largely meaningless in relation to studying gutta percha used in submarine telegraphy, since the true price paid by cable manufacturers could not be disambiguated from total import values for all types of gutta.

The various commercial qualities, names and prices that were accorded varieties of better quality gutta percha on sale in the Singapore market reflected the testing shown in Obach's tables; that is, variations in content of dirt, water, resin and gutta were matched against the (real or perceived) origin of the various types of gutta percha. So, for example, the following prices and varieties are seen from the 1902 market prices[11]

Varieties, qualities and prices of Singapore Gutta Percha

	Dirt	Water	Resins	Gutta	$ Price per Picul
Bila of Red Soondi..........	33.6	7.0	31.4	28.0	150
Sarawak Soondi No.2........	37.1	6.8	25.5	29.6	135
Penang Gutta Palelo No. 1......	2.1	5.8	53.8	38.3	180
Sarawak Red Soondi No.1......	19.0	3.9	35.5	41.6	350
Bagan White Soondi No.1......	0.7	8.6	36.5	54.2	350
Koatei Gutah Merah No.2......	21.7	5.1	28.5	44.7	360
Indragiri White Soondi.......	2.0	4.1	46.2	47.7	370
Sambas White Soondie........	1.0	4.4	53.6	41.0	380
Koatei Guta Merah No.1.......	14.8	3.8	34.2	46.6	500
Pahang White Soondie No.1....	4.2	0.5	12.8	82.5	500

The figures are interesting for several reasons. Firstly, they reveal the considerable difference in market prices that different varieties could achieve. That is, the varying characteristics of gutta percha itself helped to shape the operation of the trade. Secondly, the better quality Sarawak (*Red Soondi No. 1*) achieved a reasonably good price. Also of note is that the different types were all accorded different names in the market place and there were many more varieties marketed – see Seeligmann et al. who devote a complete chapter to the topic.[12] Some of the permutations in name and characteristics may have reflected the extent to which the gutta percha of commerce

10 India Rubber World, New York, 1 July 1901, p. 312.

11 Seeligmann et al., p. 350.

12 Seeligmann et al., Ch. IV.

was a 'manufactured' product developed by Chinese traders as well as a commodity with naturally occurring variation. Also of interest is the considerable difference in the measurements that different varieties achieved: particularly the high levels of dirt and resin in the poorer qualities, and the high degree of gutta (and very low amount of resin) in the most sought after variety – *Pahang white* (although for some reason the *Koatei Guta Merah No. 1* also achieved an equivalent price). The figures reveal the degree to which testing and measurement was a part of marketplace practice (and marketing) of gutta percha – unusual for a forest product, and indeed any commodity at that time. Plate 15 below shows examples of trees supplying gutta percha.

PLATE 15 *Trees* (Palaquium borneense) *at the plantation at Tjipetir.*
Note: These plantation trees at Tjipetir were '23½ years old'. The trees seen here were likely grown to test varieties, growing conditions etc, and to obtain seeds if possible. They were established before a technique to obtain gutta percha from leaf material was developed. The trees grown in later years at Tjipetir and at Telcon's Selborne plantation were pruned to grow considerably lower, making it possible to collect leaves from ground level or with the use of small ladders.
Source: W.R. Tromp de Haas, *The Netherlands India Government Gutta-Percha Plantations in Java*, Tjipetir, Government Gutta-Percha Plantations, Batavia, 1908, p. 1.

1.2 What Happened to the Trade in Gutta Percha?

As outlined in earlier chapters, there was a fall in the production of submarine telegraph cables and therefore in the demand for gutta percha after the Pacific cables were manufactured in 1901–2. That is, it was a reduction in demand, rather than a lack of supply that diminished the trade in gutta percha. The *Straits Settlements Annual Report* for 1906 concluded of gutta percha:

> This once great trade shews little signs of revival[13]

In the early years of the twentieth century, the needs of telegraphy were slowly being met in part by wireless telegraphy and in later years, some cables were insulated with paper, or with the addition of *balata*, another type of latex which had been adopted by Telcon, for example, in 1910.[14] During the four years of World War I, Telcon's cable production was around 16,000 nautical miles.[15] There were sales as part of post-war cable re-construction and expansion, but the period and the few years that followed saw mainly government cables being produced – a time described by a Telcon executive as one of 'great difficulty' for all the cable companies.[16] This period coincided with the phase during which the Tjipetir plantation was achieving export levels of production and when Telcon's Selborne plantation had not long been established (in 1915). That is, after years of study and disagreement, then finally agreement being reached upon which variety ought to be grown – and the desired extraction methods determined – the market and the telegraphic world was slowly changing. Fears about the impact of wireless telegraphy, and the fall in cable production led the Dutch government to restrict investment at Tjipetir until the early 1920s when it decided to build a new factory to replace the older, 'primitive' one that had been in operation.[17] The slump in cable production also led to the conclusion that the Tjipetir product would be more likely of use for industrial applications.[18]

---

13 Straits Settlements Report for 1906, No.540, in R.L. Jarman, *Annual Reports of the Straits Settlements 1855–1941*, 12 Vol. set, Slough, Archive Editions, 1998, Volume 5: 1901–1907, p. 57.

14 Lawford and Nicholson, p. 101; used when mixed with gutta percha rather than as a universal replacement for it.

15 Lawford and Nicholson, p. 92.

16 E. Wilshaw, *Cable Communications and how to use them to the Best Advantage*, paper presented to Chartered Institute of Secretaries (Country Conference), Manchester, 22 May 1924, p. 8.

17 H. van Lennep, *The Netherland Indies Government Gutta Percha Estate Tjipetir (Java)*, Buitenzorg, 1922, pp. 7, 20–21.

18 van Lennep, p. 21.

Nonetheless, gutta percha insulated cables continued in production and gutta percha continued to be collected and traded, including for industrial and golfing applications. By the early 1930s the general trade depression had lowered the demand for gutta percha to 'negligible proportions' and much of the Selborne output was sold to the Dutch for production of white gutta percha used to manufacture golf balls.[19] The balance was used for cable production, although the cable industry was shifting towards using gutta percha mainly in repairs of existing cables.[20] During the war the Japanese Government took over the Telcon plantation in Malaya – it was returned to Telcon ownership after the war, but there were no exports from the plantation to Japan during the wartime period.[21] The Selborne gutta percha plantation was closed in the 1970s.[22]

Telcon and Siemens developed synthetic compounds under proprietary names such as 'K-gutta' or 'Para-gutta'.[23] Eventually polythene insulation was developed by Telcon (1937) and was used for wartime telephone cables[24] and for submarine telegraph cables in the 1950s and became widely used thereafter.[25] It is unclear when the last gutta percha insulated cable was laid – among the last appear to have been Japanese military cables that were laid in World War II.[26]

The view expressed, including in scholarly writing, that the gutta percha trade diminished because of lack of demand because it had been replaced by *para* rubber as the insulation for submarine telegraph cables is incorrect. While it was a latex of trade significance, rubber remained as limited in its use for cable insulation as it had been in the nineteenth century.

Dove notes that for indigenous communities in the region, the contemporary patterns of collecting latex from the forest for trade to the external market were established by the gutta percha trade.[27] In Borneo for example, *jelutong*, was a wild rubber

19 Lawford and Nicholson, p. 125.

20 J.N. Dean, 'Gutta percha and balata, with particular reference to their use in submarine cable manufacture', *Journal of the Royal Society of Arts*, Vol. 92, No. 4668, June 23rd, 1944, p. 381.

21 Lawford and Nicholson, p. 124.

22 L. Edwards, *A Ramble in the World of Gutta Percha*, revision of speech originally delivered 30 November 1989. Unpublished paper, copy provided by author, 1993, p. 11.

23 Lawford and Nicholson, p. 132, see also Dean 1944.

24 Lawford and Nicholson, p. 138.

25 W.G.R. Jacob, *Notes for the Managing Director – Cable and Wireless. Polythene*, Non-Accessioned Document, Page File Ref. 1532/61, 1961, Cable and Wireless Archives, Porthcurno.

26 See W. Glover, *Cable Timeline 1901–1950*. http://www.atlantic-cable.com/Cables/CableTimeLine/index1901.htm, 2017.

27 M.R. Dove, 'Transition from native forest rubbers to *Hevea brasiliensis* (Euphorbiaceae) among tribal smallholders in Borneo', *Economic Botany*, Vol. 48, 1994, pp. 382–396 and

that was collected and traded. It first appears in the Sarawak trade returns in 1900.[28] The jelutong trade continued to grow and for some years helped to fill the shortfall traders found with the lessening in demand for gutta percha. The Telcon orders extant at the Maritime Museum Greenwich, UK, show that Jewesbury supplied jelutong to the company, which suggests it may have been combined with gutta percha – whether this was to reduce costs is unclear. Jelutong was mainly used in industrial applications and in chewing gum rather than having widespread application in the telegraphic industry.

In the 1840s, it was a dentist, Edwin Truman (later to become dentist to the British Royal Household) who first purified gutta percha to use to fill the spaces in the roots left of a tooth following root canal therapy. He patented this purification technique and sold the rights to the Gutta Percha Co which adopted it. That technique, enhanced by the research of Willoughby Smith, underpinned the quality controls that established the Company's pre-eminent position. These were the techniques endorsed by the British Parliamentary Committee that set out the need for quality standards for the telegraph industry and which other cable producers emulated. Truman worked for thirty years on various techniques to further refine gutta percha, and patented another process used for underground cables. It was adopted by the British Post Office which paid him an annual royalty of £500.[29] The whole question of gutta percha, be it purification, manufacture, and its commercial application became a large part of Truman's life, or, as Bright puts it, he 'exercised himself much in such matters' and a large piece of machinery designed to slice raw gutta percha in the factory setting was designed by him – known as the 'Truman Masticator'.[30] Now, more than a century and a half later, it is almost entirely in the dental field – still in relation to root canal treatment – that gutta percha is used. Plus ça change.

M.R. Dove, *The Banana Tree at the Gate. A History of Marginal Peoples and Global Markets in Borneo*, New Haven, Yale University Press, 2011.

28 See B.E Smythies, 'History of forestry in Sarawak', *The Malayan Forester*, Vol. 26, No. 4, (October) 1963, p. 235.

29 H.W. Bruton, 'Edwin Thomas Truman', *Supplement to Dictionary of National Biography*, S. Lee, ed., London, Smith, Elder & Co, 1912, n.p.

30 C. Bright, *Submarine Telegraphs Their History, Construction and Working*, London, Crosby Lockwood and Son, 1898, p. 289.

APPENDIX 2

Statistical Material

TABLE 1 *Estimated nautical miles submarine cables and gutta percha exports: Singapore (Pikuls) and Sarawak (Sarawak $ and Pikuls) 1850–1903*

	Cables	GP Sing	GP S'k $	GP S'k Pikuls
1850	25	2,274		
1851	94	5,748		
1852	195	6,886		
1853	1,096	11,644		3,500
1854	860	10,554		
1855	524	5,041		
1856	97	10,389		
1857	3,079	12,394		
1858	4,066	14,197		
1859	6,531	18,083		
1860	1,139	20,903		
1861	1,578	10,271		
1862	733	14,162		
1863	211	17,707	181,918	
1864	1,437	30,592	299,566	13,169
1865	2,858	18,832	114,409	
1866	2,767	12,935	27,843	
1867	662	11,901	55,590	
1868	1,679	9,044		
1869	10,155	14,078		
1870	9,808	33,275	826,903	10,209
1871	2,969	26,038	132,694	3,159
1872	3,251	34,401	235,275	
1873	4,678	36,854	162,087	3,241
1874	9,121	21,672	72,207	
1875	1,462	13,774	59,202	
1876	4,398	17,082	75,588	
1877	873	25,336	137,352	2,647
1878	466	27,573	147,401	
1879	10,641	55,910	320,507	
1880	4,957	49,729	244,855	
1881	5,743	62,051	181,430	3,709

 | DOI 10.1163/9789004357280_013

	Cables	GP Sing	GP S'k $	GP S'k Pikuls
1882	5,861	64,729	169,015	3,576
1883	7,646	59,724	141,721	3,397
1884	12,023	50,036	161,144	3,165
1885	280	45,234	121,503	2,542
1886	2,546	33,946	75,724	1,553
1887	2,790	25,539	74,752	1,625
1888	2,272	25,713	72,897	1,509
1889	1,996	59,493	167,705	1,997
1890	7,535	78,930	241,595	2,907
1891	8,080	54,026	246,473	2,666
1892	3,929	41,990	174,537	2,169
1893	3,502	38,045	158,186	2,432
1894	6,300	42,841	162,233	1,937
1895	3,504	43,910	194,120	2,782
1896	3,637	43,769	190,939	2,820
1897	5,991	45,417	185,532	2,867
1898	4,578	93,398	275,735	3,745
1899	3,141	78,343	565,621	8,980
1900	14,362	97,399	788,298	7,964
1901	15,928	73,815	769,749	5,660
1902	9,576	63,559	450,250	2,915
1903	7,793	35,661	179,441	1,283
TOTAL	**231,423**	**1,856,847**	**8,841,997**	**110,125**

Note: Blank spaces indicate data was not available for that year. A small number of cables insulated with India rubber were laid during the period; these are not included in this calculation. *Sources*: For Cables: Estimates calculated from data in *Pall Mall Gazette*, 'The West Indian Telegraph Expedition', 1870, July: n.p.; United States Hydrographic Office, *Submarine Cables*, Washington, Dept. of Navy, Bureau of Navigation, 1892, pp. 41–59; *India Rubber World*, New York, 1 February 1902, p. 136; Telcon, *The Telegraph Construction and Maintenance Co. Ltd on the Occasion of the Visit of the Delegates to the International Telegraph Conference*, London, Telegraph Construction and Maintenance Co. Ltd, 1903, Appended; C. Bright, *The Life Story of Sir Charles Tilston Bright Civil Engineer, with Which is Incorporated the Story of the Atlantic Cable and the First Telegraph to India and the Colonies*, London, Constable & Company, 1908, p. 313 and W. Glover, *Cable Timeline 1845–1900* and, *Cable Timeline 1901–1950*; 2017/Continuing, http://www.atlantic-cable.com. Domestic cables, such as along rivers, lakes and so on are not included. See also 'Nomenclature des Câbles formant le Réseau Sous-marin du Globe Dressée d'Après des Documents Officiels par le Bureau International des Administrations Télégraphiques', published in *Journal Telegraphique*, various issues, listed in Bibliography.

For Gutta Percha: Singapore: H.W. Jewesbury, *A Short History of the Introduction of Gutta Percha into Europe*, London, H.W. Jewesbury, 1883, p. 12; E.F.A. Obach, 'Cantor Lectures on Gutta Percha', London, *Journal of the Royal Society of Arts*, 1898, p. 96; A.M. Burn-Murdoch, 'Some facts about gutta percha', *Agricultural Bulletin of the Straits and Federated Malay States*, Vol. IV, No.2 (February) 1905, p. 40.

Sarawak: *Sydney Morning Herald*, 3 October 1854, p. 4., Treasurer's Reports reprinted in Sarawak Gazettes various issues 1870–1904; and Consular Reports in British Sessional Papers (BSP): 1865 [3478], p. 28; 1866 [3582], p. 805; and 1866 [3669], p. 266.

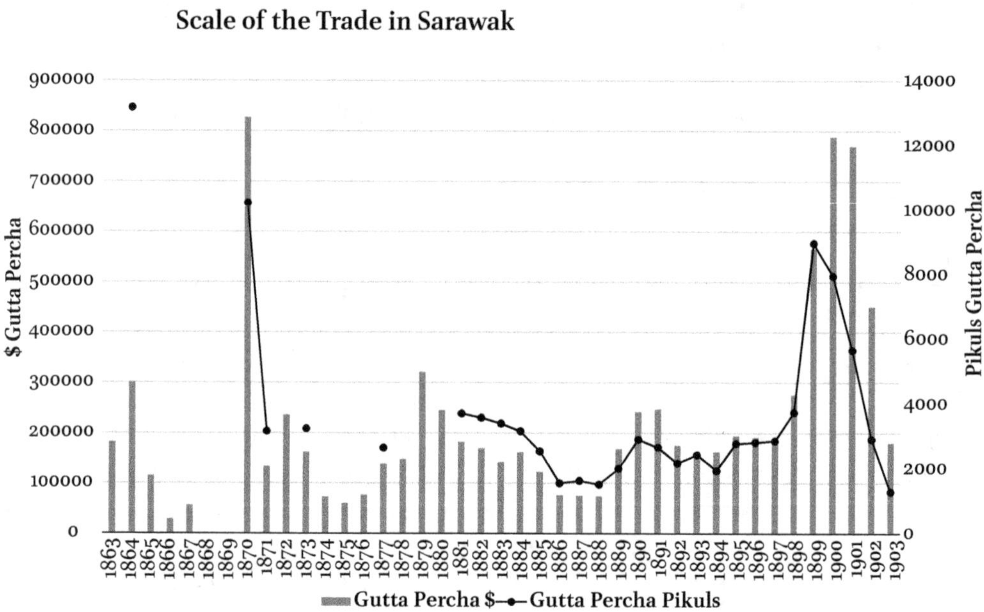

FIGURE 23 *Sarawak gutta percha exports – Sarawak dollars and pikuls 1863–1903*
Source: Sarawak Gazette various issues 1870–1904; and Consular Reports in British Sessional Papers (BSP): 1866 [3582], p. 805; and 1866 [3669], p. 266 for the years 1863–67. Figures for 1868 and 1869 were not available. See Table 1 above for statistics.

Sarawak exports were not infrequently close to 3,000 pikuls per year as can be seen in Figure 23 – in theory this could require around 267,000 trees to be felled and the gutta collected from them based on the Burn-Murdoch (1905) calculation noted earlier in Appendix 1. That is, 1.5 lb per tree (excluding any allowance for adulteration). Of the years for which statistics are extant, it was in 1864 that the largest amount of gutta percha exports was recorded (by pikuls weight) – at 13,169 pikuls. On the Burn-Murdoch figure – calculated for collection from decades old very large trees – that would require more than one million trees to have been felled (1,170,285). The total amount by weight exported for the period recorded in the statistics presented here for the years 1863–1903 (with 13 years of figures not available), would have required 9,786,441 trees on that calculation. While it is impossible to quantify what allowance should be made for adulteration, collection from other varieties or the likelihood of much collecting being from less mature trees, the scale of the labor involved in felling, processing and transporting the gutta percha even as far as Kuching must surely be regarded with some awe.

TABLE 2 *Sarawak commodity export sectors 1870–1903 (Sarawak dollars)*

	Forest Prod.	Minerals	Cultivations	Sea Produce	Gutta Percha	Forest Excl. GP
1870	924,195	90,963	147,238	16,054	826,903	97,292
1871	300,193	82,719	182,711	14,728	132,694	167,499
1872	489,307	191,545	195,141	8,444	235,275	254,032
1873	369,907	242,186	225,102	69,888	162,087	207,820
1874	229,597	165,944	304,390	18,818	72,207	157,390
1875	180,579	152,754	308,372	25,609	59,202	121,377
1876	254,958	161,937	344,463	34,498	75,588	179,370
1877	267,480	139,266	358,959	33,125	137,252	130,228
1878	254,538	83,086	451,348	20,353	147,401	107,137
1879	457,661	137,063	495,187	40,333	320,507	137,154
1880	380,979	144,149	458,906	23,620	244,855	136,124
1881	379,204	138,683	460,731	55,100	181,430	197,774
1882	406,470	163,010	463,383	43,540	169,015	237,455
1883	385,121	120,349	562,431	24,336	141,721	243,400
1884	376,755	84,845	468,909	27,005	161,144	215,611
1885	374,844	96,178	478,259	35,549	121,503	253,341
1886	359,068	110,087	567,579	15,277	75,724	283,344
1887	323,133	155,110	621,217	29,250	74,752	248,381
1888	382,851	118,915	738,185	40,784	72,897	309,954
1889	500,642	107,444	938,031	9,065	167,705	332,937
1890	577,629	133,562	759,283	11,843	241,595	336,034
1891	659,721	133,057	921,798	15,922	246,473	413,248
1892	582,620	174,324	1,172,879	13,554	174,537	408,083
1893	578,104	172,474	1,157,043	27,060	158,186	419,918
1894	576,944	200,676	1,088,458	11,992	162,233	414,711
1895	582,423	196,154	1,184,077	8,464	194,120	388,303
1896	744,664	184,714	1,180,066	11,971	190,939	553,725
1897	839,532	273,000	1,380,635	15,403	185,532	654,000

TABLE 2 *Sarawak commodity export sectors 1870–1903 (Sarawak dollars)* (cont.)

	Forest Prod.	Minerals	Cultivations	Sea Produce	Gutta Percha	Forest Excl. GP
1898	1,409,243	326,230	1,740,205	19,637	275,735	1,133,508
1899	1,426,579	783,630	1,888,729	22,799	565,621	860,958
1900	1,627,209	71,008	2,247,679	27,238	788,298	838,911
1901	1,858,677	1,305,835	2,355,023	35,559	769,749	1,088,928
1902	1,634,616	1,410,038	3,211,004	55,578	450,250	1,184,366
1903	1,334,007	1,832,515	3,892,966	35,886	179,441	1,154,566
TOTAL	22,029,450	9,883,450	32,950,387	898,282	8,162,571	13,866,879

Notes: The Treasurer's Report set out Sarawak's key exports into the major export sectors as follows:

Forest Produce	Termed 'Jungle Produce' in the original data, this sector usually included Beeswax Birds' Nests Camphor Canes Dammar Gutta Percha India-rubber Rattans Timber Vegetable Tallow
Minerals	Antimony Coal Diamonds Gold Quicksilver Silver
Cultivations/ Manufactures	Coprah Gambier Padi Pepper Sago (flour) Sago (pearl) Sago (raw)
Sea Produce	Fish Tortoiseshell Trepang

Commodities included in these categories varied in later years:
Betel-nuts were added and Vegetable Tallow was dropped from the 'Jungle Produce'
Diamonds and Silver were dropped from the 'Minerals'
Coffee and Cutch were added to 'Cultivations' and Sago in Flour form only was included in that category, and Sea Produce in later years simply listed 'Fish (all kinds)'.

Sarawak Gutta Percha Leaf Exports

	Dollars	Pikuls
1894	200	132.68
1895	2520	1285.5
1896	0	0
1897	0	0
1898	1000	270
1899	1036	248.5
1900	700	186.25

	Dollars	Pikuls
1901	3	3.75
1902	476	316.25
TOTAL	5,935	2,442.93

Source: Sarawak Gazette (SG), various issues 1870–1904, and Consular Report in British Sessional Paper (BSP) 1871 [C.429], p. 809.

TABLE 3 *Sarawak six major commodity exports including gutta percha 1863–1903 (Sarawak dollars)*

	Sago Fl'r	Pepper	GP	Gold	Rattan	Gambier
1863	64,634		181,918	5,220	–	
1864	111,509		299,566	9,482	–	
1865	67,207		114,409	5,394	12,683	
1866	56,647		27,843	2,250	19,679	
1867	97,399		55,590	6,998	2,340	
1868						
1869						
1870	128,025	1,681	826,903		151	607
1871	164,935	2,022	132,694	4,952	769	432
1872	190,815	1,570	235,275	7,485	7,412	26
1873	218,826	1,919	162,087	6,262	7,797	–
1874	301,988	1,008	72,207	745	28,838	–
1875	306,464	515	59,202	1,383	23,332	–
1876	313,559	358	75,588	2,749	47,301	–
1877	337,530	380	137,352	1,280	16,064	686
1878	409,212	226	147,401	1,730	12,684	20,875
1879	363,895	362	320,507	2,258	28,162	88,148
1880	343,134	3,850	244,855	3,663	29,183	97,852
1881	345,023	19,125	181,430	2,918	91,796	87,266
1882	285,274	48,610	169,015	1,380	99,607	118,351
1883	310,617	87,078	141,721	234	67,281	153,507
1884	205,445	108,952	161,144	3,621	111,454	135,844
1885	222,118	121,997	121,503	21,754	145,093	121,369
1886	263,079	138,171	75,724	28,281	152,650	145,511

TABLE 3 *Sarawak six major commodity exports including gutta percha 1863–1903* (cont.)

	Sago Fl'r	Pepper	GP	Gold	Rattan	Gambier
1887	314,536	138,581	74,752	8,810	180,454	142,376
1888	317,878	217,655	72,897	6,521	234,456	158,715
1889	401,290	325,519	167,705	10,759	19,223	168,345
1890	343,635	237,476	241,595	16,599	179,983	133,235
1891	553,485	231,813	246,473	20,602	297,490	101,109
1892	751,149	281,161	174,537	36,882	271,909	113,256
1893	766,099	223,889	158,186	21,871	272,825	135,918
1894	641,939	240,211	162,233	52,740	236,421	152,346
1895	687,262	240,248	194,120	29,240	163,163	210,438
1896	704,667	230,693	190,939	26,111	279,615	179,539
1897	689,702	400,689	185,532	29,975	345,424	168,642
1898	756,275	736,760	275,735	23,685	189,291	171,346
1899	790,583	859,188	565,621	541,711	169,531	168,050
1900	750,266	1,254,422	788,298	843,709	279,993	200,609
1901	559,723	1,477,499	769,749	1,246,114	354,713	258,704
1902	644,020	2,205,762	450,250	1,366,794	361,653	297,291
1903	766,463	2,733,301	179,441	1,784,600	186,819	311,568
TOTAL	**15,546,307**	**12,572,691**	**8,841,997**	**6,186,762**	**4,927,239**	**4,041,961**

Source: Treasurer's Reports reprinted in Sarawak Gazettes (SG), various issues 1870–1904; and Consular Reports in British Sessional Papers (BSP): 1865 [3478], p. 28; 1866 [3582], p. 805; 1866 [3669], p. 266 for the years 1863–65 respectively. Figures for 1868–69 are not available. For other years, blank sections reflect no exports recorded: for example, Gambier, Pepper, Coal and Quicksilver do not appear in Sarawak exports in all years, such as in the 1860's. Commodities selected to give the major products that appear most consistently, and for reasonable levels, for each of the three key export sectors ('Jungle [Forest] Produce, Minerals, and Cultivations') over the period.

TABLE 4 *Sarawak gutta percha exports – imports of treasure, rice and cloth 1863–1903 (Sarawak dollars)*

	GP Exp	Treasure	Rice	Cloth
1863	181,918	56,578	15,631	146,252
1864	299,566	109,361	11,925	176,857
1865	114,409	39,891	13,907	151,155
1866	27,843	28,343	3,882	84,638
1867	55,590	41,824	3,713	91,773
1868				
1869				
1870	826,903	530,028	99,430	293,825
1871	132,694	113,337	89,699	212,359
1872	235,275	117,293	126,968	240,124
1873	162,087	91,278	176,250	200,510
1874	72,207	165,522	114,344	261,444
1875	59,202	94,056	116,898	186,345
1876	75,588	94,985	129,601	250,630
1877	137,352	138,318	85,674	255,630
1878	147,401	151,048	130,607	250,489
1879	320,507	174,462	227,505	264,521
1880	244,855	131,008	165,931	259,059
1881	181,430	104,437	145,224	283,769
1882	169,015	170,787	138,521	204,323
1883	141,721	158,147	155,577	254,761
1884	161,144	85,328	155,094	240,236
1885	121,503	186,379	154,426	292,741
1886	75,724	139,638	161,207	274,942
1887	74,752	164,180	187,780	266,533
1888	72,897	139,011	186,334	237,244
1889	167,705	216,530	208,918	295,314
1890	241,595	168,063	240,426	237,737
1891	246,473	260,407	232,257	285,402
1892	174,537	334,903	257,456	239,279
1893	158,186	221,853	243,115	206,208
1894	162,233	262,503	315,967	235,869
1895	194,120	210,222	333,076	277,127
1896	190,939	281,624	422,113	342,294
1897	185,532	313,509	347,266	388,247

TABLE 4 *Sarawak gutta percha exports – imports of treasure* (cont.)

	GP Exp	Treasure	Rice	Cloth
1898	275,735	434,265	391,469	416,812
1899	565,621	292,779	573,445	449,464
1900	788,298	453,513	598,217	497,394
1901	769,749	485,612	745,896	537,895
1902	450,250	682,420	816,029	499,278
1903	179,441	625,943	1,041,065	746,104
1904	96,271	867,739	974,056	581,626
TOTAL	**8,938,268**	**9,337,124**	**10,536,899**	**11,616,210**

Source: Treasurer's Reports reprinted in Sarawak Gazettes, various issues 1870–1904; Consular Reports in British Sessional Papers (BSP): 1865 [3478], p. 28; 1866 [3582], p. 805; 1866 [3669], p. 266 for the years 1863–65 respectively. Figures for 1868–69 are not available.

TABLE 5 *Sarawak gutta percha exports – imports of jars, brassware, ironware, sugar and opium 1863–1904* (*Sarawak dollars*)

	GP Exp $	Jars	Brassware	Ironware	Sugar	Opium
1863	181,918		27,729			14,500
1864	299,566		70,454	6,702	2,969	4,665
1865	114,409		37,468	4,713	9,597	8,059
1866	27,843	11,680	15,624	5,981	5,452	8,868
1867	55,590	30,901	2,178	3,164	1,279	10,810
1868						
1869						
1870	826,903	6,714	40,162	15,654	11,889	39,140
1871	132,694	15,571	23,564	20,263	12,305	15,358
1872	235,275	3,058	29,405	25,988	11,387	35,220
1873	162,087	7,006	24,926	14,598	15,244	36,800
1874	72,207	746	23,088	14,523	8,652	35,220
1875	59,202	3,354	15,794	17,563	12,993	35,372
1876	75,588	8,417	18,108	32,082	11,491	34,820
1877	137,352	4,570	17,001	16,230	13,391	29,970
1878	147,401	3,952	20,576	15,517	16,492	55,190
1879	320,507	6,301	35,845	26,976	22,704	61,225

	GP Exp $	Jars	Brassware	Ironware	Sugar	Opium
1880	244,855	5,789	41,390	21,591	23,463	57,108
1881	181,430	11,974	51,601	23,114	18,524	48,320
1882	169,015	13,518	46,807	24,011	20,957	67,350
1883	141,721	12,806	38,125	17,558	18,437	57,156
1884	161,144	14,717	32,842	18,354	22,290	68,375
1885	121,503	9,420	26,264	23,140	19,649	73,257
1886	75,724	4,123	33,431	23,076	25,293	55,031
1887	74,752	10,817	41,996	20,882	22,229	48,362
1888	72,897	13,459	41,153	22,979	25,050	76,916
1889	167,705	16,248	49,779	22,915	19,649	73,257
1890	241,595	14,642	53,514	22,832	30,553	76,068
1891	246,473	11,046	68,479	33,257	27,221	59,746
1892	174,537	7,261	56,239	23,483	37,138	60,525
1893	158,186	8,275	64,806	26,430	33,132	73,928
1894	162,233	10,437	30,518	32,471	46,236	72,120
1895	194,120	16,179	24,109	32,471	40,484	78,348
1896	190,939	14,400	35,159	42,008	49,189	87,292
1897	185,532	9,105	26,235	83,705	52,741	66,218
1898	275,735	8,929	45,693	47,916	61,050	78,938
1899	565,621	5,157	40,604	74,201	75,272	93,875
1900	788,298	9,087	43,739	79,338	99,826	134,283
1901	769,749	14,440	47,020	86,416	105,567	173,090
1902	450,250	25,546	44,924	124,085	101,403	184,852
1903	179,441	43,055	44,923	145,753	134,643	284,130
1904	96,271	3,824	21,961	106,411	128,736	206,250
TOTAL	8,938,268	416,524	1,453,233	1,398,351	1,394,577	2,780,012

Source: Treasurer's Reports reprinted in Sarawak Gazettes, various issues 1870–1904; Consular Reports in British Sessional Papers (BSP): 1865 [3478], p. 28; 1866 [3582], p. 805; 1866 [3669], p. 266 for the years 1863–65 respectively. Figures for 1868–69 are not available.

TABLE 6 *Singapore gutta percha exports (Pikuls) 1844–1903*

Year	Pikuls
1844	168
1845	169
1846	5,364
1847	9,296
1848	12,326
1849	9,734
1850	2,274
1851	5,748
1852	6,886
1853	11,644
1854	10,554
1855	5,041
1856	10,389
1857	12,394
1858	14,197
1859	18,083
1860	20,903
1861	10,271
1862	14,162
1863	17,707
1864	30,592
1865	18,832
1866	12,935
1867	11,901
1868	9,044
1869	14,078
1870	33,275
1871	26,038
1872	34,401
1873	36,854
1874	21,672
1875	13,774
1876	17,082
1877	25,336
1878	27,573
1879	55,910

Year	Pikuls
1880	49,729
1881	62,051
1882	64,729
1883	59,724
1884	50,036
1885	45,234
1886	33,946
1887	25,539
1888	25,713
1889	59,493
1890	78,930
1891	54,026
1892	41,990
1893	38,045
1894	42,841
1895	43,910
1896	43,769
1897	45,417
1898	93,398
1899	78,343
1900	97,399
1901	73,815
1902	63,559
1903	35,661
TOTAL	**1,856,847**

Source: H.W. Jewesbury, *A Short History of the Introduction of Gutta Percha into Europe*, London, Jewesbury, 1883, p. 12.
E.F.A. Obach, 'Cantor Lectures on Gutta Percha', London, *Journal of the Royal Society of Arts*, 1898, p. 96.
A.M. Burn-Murdoch, 'Some facts about gutta percha', *Agricultural Bulletin of the Straits and Federated Malay States*, Vol IV, No. 2 (February), 1905, p. 40.

References

Primary Sources

Parliamentary Papers and Official Records

British Parliamentary Papers:

Consular Report on the Trade of Sarawak for 1863	Paper 1865 [3478]
Consular Report on the Trade of Sarawak for 1864	Paper 1866 [3582]
Consular Report on the Trade of Sarawak for 1865	Paper 1866 [C.3669]
Consular Report on the Trade of Sarawak for 1866	Paper: 1867–68 [3953-VII]
Consular Report on the Trade of Sarawak for 1867	Paper: 1867–68 [3953-VIII]
Consular Report on the Trade of Sarawak for 1869	Paper: 1871 [C.343-LXV]
Consular Report on the Trade of Sarawak for 1870	Paper: 1871 [C.429-LXVI]
Consular Report on the Trade of Sarawak for 1885	Paper: 1887 [C.4915 C.4923]
Consular Report on the Trade of Sarawak for 1899	Paper: 1900 [Cd.1] [Cd.352]
Statistical Abstract for the United Kingdom 1863–77	Paper: 1878 [C.2073]
Statistical Abstract for the United Kingdom 1869–83	Paper: 1884 [C.4117]
Statistical Abstract for the United Kingdom 1885–99	Paper: 1900 [Cd. 306]

Report of the Joint Committee Appointed by the Lords of the Committee of the Privy Council for Trade and the Atlantic Telegraph Company to Inquire into Construction of Submarine Telegraph Cables, Together with the Minutes of Evidence and Appendix, 1861, Command Papers, [2744] LXII.591, London, HMSO.

Select Committee on Newspaper Stamps: Report, together with Proceedings of the Committee, Minutes of Evidence, Appendix, and Index, 1851; Reports of Committees (558) Vol. XVII.1.

Bureau International des Administrations Télégraphiques, *Nomenclature des câbles formant le réseau sous-marin du globe dressée d'après des documents officiels par le Bureau International des Administrations Télégraphiques, Journal Télégraphique*, Supplements to

1877 Vol 3 No. 29
1883 Vol 7 No. 5
1887 Vol 11 No. 4
1889 Vol 13 No. 9

1892 Vol 16 No. 4
1894 Vol 18 No. 10
1897 Vol 21 No. 11
1901 Vol 25 No. 5
1903 Vol 27 (no number)

Congrés International des Électriciens 1881, *Comptes Rendus des Travaux*, Paris, Ministère des Postes et des Télégraphes, 1882.

Sherman, P.L. *Report of the Special Agent of the Forestry Bureau Sent to Investigate Gutta-Percha and Rubber in the Straits Settlements, Java and Sumatra*, Manila, Forestry Bureau, 1901.

Sherman, P.L., *The Gutta Percha and Rubber of the Philippine Islands*, Manila, Bureau of Public Printing, 1903.

Tromp de Haas, W.R., *The Netherlands India Government Gutta-Percha Plantations in Java*, Tjipetir, Batavia Government Gutta-Percha Plantations, 1908.

United States Hydrographic Office, *Submarine Cables*, Dept of Navy, Washington, Bureau of Navigation, 1892.

United States Department of Commerce and Labor, Bureau of the Census, *Manufactures 1905, Part IV, Special Reports on Selected Industries*, Washington Government Printing Office, 1908.

Vander Laan, J.W., *Production of Gutta-Percha, Balata, Chicle and Allied Gums*, Trade Promotion Series No. 42, Crude Rubber Series, Washington, Department of Commerce, Bureau of Foreign and Domestic Commerce, Government Printing Office, 1927.

Van Lennep, H., *The Netherland Indies Government Gutta Percha Estate Tjipetir (Java)*, Buitenzorg, 1922.

Unpublished Official Records

Colonial Office Records: CO 273 Straits Settlements
CO 874 British North Borneo Company series.

Foreign Office Records: FO 12/52 Correspondence

Kew Director's Correspondence, Royal Botanic Gardens Kew, Archives 165; 166; 168.

Kew Royal Botanic Gardens, Archives, *Malaya. Gutta Percha, 1876–1904*, Kew.

Ramsay, W. 'Report on the Extraction of gutta-percha from the leaves of the Isonandra gutta-percha tree', *Malaya, Gutta Percha, 1876–1904*, Royal Botanic Gardens Kew, Archives, ff 160–162, 1897.

Thiselton-Dyer, W.T., Correspondence to C.P. Lucas (Colonial Office), in *Kew: Malaya, Gutta Percha, 1876–1904*, Royal Botanic Gardens Kew, Archives.

Wray, L., Report to Resident Hugh Low, Perak, September 25, 1883. 'On the great loss of gutta resulting from the wasteful means of extraction employed by the Malays', *Kew: Malaya, Gutta Percha, 1876–1904*, Royal Botanic Gardens Kew, Archives, ff 160–162, 1883.

Manuscripts and Archives

Borneo Company Ltd (BCL), *Board Minutes*, London Metropolitan Archives, Ms.27 178, Reference Code CLC/B.123-11.

Cable & Wireless Ltd. Public Relations Office, (n.d.), *Copper and Cables*, Non-Accessioned Document, Page File Ref. A8/48, Cable and Wireless Archives, Porthcurno.

Gutta Percha Company, 1864–1903, *Directors Minutes*, British Insulated Callenders Cables (BICC) Archives, BICC Item I 18/103–110, Archives Department, Business Archives, UK National Museums, Liverpool.

India Rubber, Gutta-Percha and Telegraph Works Company Limited (Silvertown), *Directors Minutes*, 1864–1965, B/BTR/IRGP, London Metropolitan Archives.

Jacob, W.G.R., 1961 *Notes for the Managing Director – Cable and Wireless. Polythene*, Non-Accessioned Document, Page File Ref. 1532/61, Cable and Wireless Archives, Porthcurno.

Joint Committee on Chemistry, Colonies and Trade, *Minutes*, 1843–1852, Royal Society for the Encouragement of Arts, Manufacture and Commerce, London, Royal Society Arts Archives, London.

Norddeutsche Seekabelwerke (NSW) GmbH, Archives, Nordenham, Germany.

Sarawak Chamber of Commerce, *Minutes of Meetings 1897–98*, Borneo Company Ltd Archives, London Metropolitan Archives (Guildhall), Ms.27293A, Reference Code CLC/B.123-11.

Selborne Plantation Co. 1916–1928, *Minute Book*, British Insulated Callenders Cables (BICC) Archives, BICC Item I/18, Archives Department, Business Archives, UK National Museums, Liverpool Maritime Museum.

Telcon (TCM) *Telcon Archives*, TCM 21/9; TCM 21/26/13; TCM 21/50-53, various years, National Maritime Museum, Greenwich, London.

Newspapers, Periodicals and Directories

Directory and Chronicle for China, Japan, Corea [sic], Indo-China, Straits Settlements, Malay States, Sian, Netherlands India, Borneo, the Philippines &c, Hong Kong, Hong Kong Daily Press, 1909.

Freeman's Journal, Dublin.

Illustrated Times, London, William Dover, (1855–1872).

India Rubber and Gutta Percha and Electrical Trades Journal, 1884–1910 (later the *India-Rubber Journal* 1911–1915), London.
India Rubber World, New York.
Le Caoutchouc & La Gutta-Percha Organe Mensuel, Paris.
London and China Telegraph, London, James West Publishers.
Manufacturer and Builder, New York.
Money Market Review, London, Cornhill.
New York Times, New York, New York Times Co.
Pall Mall Gazette, London, J.K. Sharpe.
Sarawak Gazette, 1870–1908, Kuching, Kuching Government Printer.
Singapore and Straits Directory, Singapore, various publishers, 1879–1910.
Singapore Directory for the Straits Settlements 1876–79, Singapore Straits Times Office.
Singapore Free Press and Mercantile Advertiser (also known as *Singapore Free Press and Daily News*), Singapore, 1887–1893.
Straits Calendar and Directory (Including Sarawak and Labuan), Singapore, various publishers, 1870–1873.
Straits Times and Singapore Journal of Commerce, Singapore, 1845–1852.
Straits Times, Singapore.
Sydney Morning Herald, Sydney, Charles Kemp and John Fairfax.
Telegraphic Journal and Electrical Review, London, Henry Gillman.
The Argus, Melbourne, Australia.
The Phoenix, A Monthly Magazine for China, Japan & Eastern Asia, London.
The Spectator, London.
The Sydney Chronicle, Sydney, Edward John Hawkesely.
The Times, London, various issues 1840–1915.
Tijdschrift voor Nijverheid in Nederlandsch-Indië, Batavia, Nederlandsch-Indische Maatschappij van Nijverheid.

Published Contemporary Accounts

Abbott, L, 'Nature's Common Carrier', *Harper's New Monthly Magazine*, No. CCXXXVIII, Vol. XL, March 1870, pp. 507–522.
Agricultural Bulletin of the Straits and Federated Malay States, 'Gutta Percha From Leaves', New Series, 1902, Vol. I, No. 4, p. 153.
Agricultural Bulletin of the Straits and Federated Malay States, 'The Dissemination of Seeds by Natural Means – Gutta Percha', New Series, 1902, Vol. I, No. 4, p. 154.
American Journal of International Law, 'Censorship of telegrams transmitted by Cable and Wireless', Vol. 9, No. 3, Supplement: Diplomatic Correspondence Between the United States and Belligerent Governments Relating to Neutral Rights and Commerce, 1915, pp. 270–313.

Anderson, J., *The Colonial Office List Comprising Historical and Statistical Information Respecting the Colonial Dependencies of Great Britain*, London, Harrison and Sons, 1892.

Armstrong, T.B., 'Gutta Percha', *Transactions of the American Institute, Sixth Annual Report of the American Institute of the City of New York*, 1848, pp. 242–245.

Ayrton, W.E., 'Sixty years of submarine telegraphy: Address to the Imperial Institute', *The Electrician*, 42, (February), 1897, pp. 546–548.

Baring-Gould, S. and C.A. Bampfylde, *A History of Sarawak under its Two White Rajahs, 1839–1908*, London, Henry Southeran, 1909.

Beccari, O., *Wanderings in the Great Forests of Borneo. Travels and Researches of a Naturalist in Sarawak*, London, Archibald Constable & Co, 1904, Reprinted Singapore, Oxford University Press, 1989.

Beech, M.W.H., 'Punans of Borneo', *Man*, 1911, Vol. 11, pp. 17–18.

Blackwoods Edinburgh Magazine, 'The All-British Trans-Pacific Cable', (February), 1897, pp. 269–279.

Bock, C., *The Head-hunters of Borneo. A Narrative of Travel up the Mahakkam and Down the Barito*, 2nd ed., London, Sampson, Low, Marston, Searle & Rivington, 1881.

Boyle, F., *Adventures Among the Dyaks of Borneo*, London, Hurst and Blackett, 1865.

Braddell, T. 'The Sultan of Johore', *The Journal of the Indian Archipelago and Eastern Asia*, New Series, Vol. II, 1858, pp. 46–67.

Bright, C., *Submarine Telegraphs Their History, Construction and Working*, London, Crosby Lockwood and Son, 1898.

Bright, C., *The Story of the Atlantic Cable*, London, George Newnes, 1903.

Bright, C., *The Life Story of Sir Charles Tilston Bright Civil Engineer, with Which is Incorporated the Story of the Atlantic Cable and the First Telegraph to India and the Colonies*, Revised & Abridged edn, London, Constable & Company, 1908.

Bright, C., *Imperial Telegraphic Communication*, London, P.S. King & Son, 1911.

Bright, C., *Telegraphy, Aeronautics and War*, London, Constable & Company, 1918.

Bright, C.T., *The Telegraph to India and its Extension to Australia and China*, London, William Clowes and Sons, 1866.

Brooke, C., *Ten Years in Sarawak*, 2 Vols, London, Tinsley Brothers, 1866.

Brooke, J., 'Speech of Sir James Brooke on Borneo', *The Leader*, issue 2, No. 88, 1851, pp. 1130–1131.

Brown, F.J., *The Cable and Wireless Communications of the World*, 2nd edn, London, Sir Isaac Pitman & Sons, 1930.

Bruton, H.W., 'Edwin Thomas Truman', *Supplement to Dictionary of National Biography*, S. Lee ed., London, Smith, Elder & Co, 1912,

Buckley, C.B., *An Anecdotal History of Old Times In Singapore* 1819–1867, Singapore, Fraser & Neave, Reprinted 1965, University of Malaya Press, 1902.

Buenting, T, *Buenting's International Cotton Code*, Second Edition, Bremen, Theodor Buenting & Co., 1929.

Burbidge, F.W., 'Notes on gutta percha and caoutchouc-yielding trees', *Journal of the Straits Branch Royal Asiatic Society*, 1879, No. 3, (July), pp. 52–59.

Burbidge, F.W., *The Gardens of the Sun or A Naturalists Journal on the Mountains and in the Forests and Swamps of Borneo and the Sulu Archipelago*, London, John Murray, 1880.

Burkill, I.H. 'Botanical Collectors, Collections and Collecting Places in the Malay Peninsula', *The Gardens Bulletin, Straits Settlements,* Vol. IV, Nos 4 and 5, August. 1927, pp. 113–202.

Burkill, I.H. (ed.), with contributions by W. Birtwistle, F.W. Foxworthy and J.B. Scrivenor, *A Dictionary of the Economic Products of the Malay Peninsula*, 2 Vols, London, Crown Agents, 1935.

Burn-Murdoch, A.M., 'Some facts about gutta percha', *Agricultural Bulletin of the Straits and Federated Malay States*, Vol. IV. No. 2 (February), 1905, pp. 33–42.

Burns, R., 'The Kayans of the Northwest of Borneo', *Journal of the Indian Archipelago and East Asia*, 3, 1849, pp. 140–152.

Cameron, J., *Our Tropical Possessions in Malayan India*, Smith, Elder and Co., London, Oxford University Press, 1865, Reprinted 1965.

Chalmers, T., *100 Years of Gutta Percha*, Glasgow, R & J. Dick Ltd, 1946.

Clauson-Thue, W., *ABC Universal Commercial Electrical Telegraphic Code. Specially adapted for the use of Financiers, Merchants, Shipowners, Brokers, Agents Etc.* London, Published by the Author, 1873.

Clauson-Thue, W., *The Seventh Edition Five-Letter ABC Universal Commercial Telegraphic Code*, William Droege ed., London, Sir Isaac Pitman & Sons, 1936.

Cleghorn, H., J.F. Royle, R.B. Smith and R. Strachey, 'Report of the Committee appointed by the British Association to consider the probable effects in an economical and physical point of view of the destruction of tropical forests', *Report of the British Association for the Advancement of Science,* London, John Murray, 1852, pp. 78–102.

Clouth, F., *Rubber, Gutta-Percha and Balata*, London, Maclaren & Sons, 1903.

Collins, J., *Report on the Gutta Percha of Commerce*, London, W.H. Allen, 1878.

Combanaire, A. and J. de la Fresnaye, Patent Application 'Improved process for the purification of Gutta Percha', English Patent Record No. 22,758, November 11, 1902, in India Rubber World, 1 February 1902, p. 149.

Combanaire, A., *Au Pays des Coupeurs de Têtes: À travers Bornéo*, 1902, Singapore, Editions Pagode, Reprint 1993.

Conrad, J., *Almayer's Folly. A Story of an Eastern River*, London, T. Fisher Unwin, 1896.

Crocker, W.M., 'Notes on Sarawak and Northern Borneo', *Proceedings of the Royal Geographical Society and Monthly Record of Geography*, New Monthly Series, Vol. 3. No. 4 (April.), 1881, pp. 193–208.

Curtis, C., 'Notes on Gutta Percha Trees', *Agricultural Bulletin of the Straits and Federated Malay States*, H.N. Ridley ed., Vol. 1, No. 6, March 1902, pp. 220–223.

d'Abbans, J., La Gutta-Percha et Le Caoutchouc en Malaisie. Conference de M. Le Comte de Jouffroy d'Abbans, L'Agriculture Pratique des Pays Chauds, *Journal du Jardin Colonial et des Jardins d'Essai Des Colonies Françaises*, Juillet 1902-Juin 1903, pp. 586–603.

Daly, D.D., 'Explorations in British North Borneo, 1883–87', *Proceedings of the Royal Geographical Society and Monthly Record of Geography*, New Monthly Series, Vol. 10. No. 1 (Jan.) 1888, pp. 1–24.

De Haan, C., Report in *Tijdschrift voor Nijverheid in Nederlandsch-Indië*, Volume XII, 1866, pp. 66–68.

De Vriese, W.M., *De Handel in Getah-Pertja* (Gutta-Percha), Leyden, A.W. Sythoff, 1856.

De Windt, H., *On the Equator,* London, Cassell, Petter, Galpin & Co, 1882.

Dean, J.N., 'A visit to Selborne Plantation', *Zodiac, House Magazine*, Telcon Staff Association Social and Athletic Club and Shop Stewards Committee, (Mar.), 1949, pp. 1–3.

Denison, N., 'Journal when on a trip from Sarawak to Meri on the North-West Coast of Borneo in the Brunei Territory', *Journal of the Straits Branch of the Royal Asiatic Society,* Vol. 10, (December), 1883, pp. 173–188.

Dodd, G., *The Curiosities of Industry and the Applied Sciences*, London, George Routledge & Co, 1852.

Faraday, M., *Faraday's Diary, Volume 5, September 6 1847 – October 17 1851*, London, G Bell and Sons Ltd, Rep. 1934, 1848.

Faraday, M., 'On the use of gutta percha in electrical insulation', *The Philosophical Magazine* xxxii, 1848, pp. 165–167.

Foxworthy, F.W., 'Minor forest products of the Malay Peninsula', *Malayan Forest Records* No. 2, 1922, pp. 151–217.

Furness, W.H., *The Home-Life of Borneo Head-Hunters: Its Festivals and Folk-Lore*, Philadelphia, J.B. Lippincott Company, 1902.

Gomes, E.H., *Seventeen Years Among the Sea Dyaks of Borneo*, London, Seeley & Co. Limited, 1911.

Grabowsky, F.S., 'Ueber die „djawet's" oder heiligen Töpfe der Oloh ngadju (Dajaken) von Süd-Ost-Borneo', *Zeitschrift für Ethnologie*, 17, 1885, pp. 121–128.

Haddon, A.C., *Head-Hunters Black, White, and Brown*, London, Methuen & Co., 1901.

Ham, S.P., 'De proefaanplantingen van caoutchouc-en getah-gertja leverende hourtsoorten te Tjipetir. Onder de leiding van het Boschwezen', *Tijdschrift voor Nijverheid in Nederlandsch-Indië*, 1900, Volume LXI, pp. 64–77.

Hemmeon, J.C., *The History of the British Post Office*, Cambridge, Mass., Harvard University, 1912.

Herbertson, A.J. and O.J.R. Howarth (eds), *The Oxford Survey of the British Empire*, London, Oxford University Press, 1914.

Hiller, H.M., 'A brief report of a journey up the Rejang river in Borneo', *Proceedings of the American Philosophical Society*, Vol. 35, No. 153 (December), 1896, pp. 321–328.

Hobson, J.A., Imperialism. A Study, New York, James Pott & Company, 1902.

Hornaday, W.T., *Two Years in the Jungle. The Experiences of a Hunter and Naturalist in India, Ceylon, the Malay Peninsula and Borneo*, 7th edn, New York, Charles Scribner's Sons, 1904.

Hose, C. and McDougall, W., *The Pagan Tribes of Borneo*, 2 Vols, London, Macmillan and Co, 1912.

Hose, C., 'A journey up the Baram river to Mount Dulit and the highlands of Borneo', *The Geographical Journal*, Vol. 1, No. 3 (Mar.), 1893, pp. 193–206, with map at p. 288.

Hose, C., 'The natives of Borneo', *The Journal of the Anthropological Institute of Great Britain and Ireland*, Vol. 23, 1894, pp. 156–172.

Hose, C., 'In the heart of Borneo', *Geographical Journal*, Vol. 16, No. 1, (July), 1900, pp. 39–59.

Hose, C., *Natural Man. A Record from Borneo*, London, Macmillan & Co, 1926, Reprinted Singapore, Oxford University Press 1988.

Indisch Genootschap, 'De Gouvernements getah pertja-kultuur', *Indisch Genootschap Algemeene Vergadering* van 20 November 1900, pp. 203–235.

Jackson, J.R., *Commercial Botany of the Nineteenth Century*, London, Cassell & Co., 1890.

Jewesbury, H.W., *A Short History of the Introduction of Gutta Percha into Europe*, London, H.W. Jewesbury, 1883.

Johnson, G., *The All Red Line. The Annals and Aims of the Pacific Cable Project*, Ottawa, James Hope & Sons, 1903.

Jumelle, H., *Les Cultures colonials. Plantes a Latex et a Résines*, Paris, Librairie J.B. Baillière et Fils, 1915.

Kappey, F.E., 'Electric Telegraph Service' in *The British Empire Series*, Vol. V, London, Keagan Paul, Trench Trubner & Co Ltd, 1902.

Kater, C., 'Iets over de bij de Dajaks in de Wester-afdeeling van Borneo zoo gezochte tempajans of Tadjau's, door den Assistent-Resident,' *Tijdschrift voor Indische Taal-, Land-, en Volkenkunde*, TBG, 1867, 16, pp. 438–449.

Kirby, S.W., 'Johore in 1926', *The Geographical Journal*, Vol. 71, No. 3 (Mar.) 1928, pp. 240–257.

Lardner, D., *The Electric Telegraph Popularised*, London, Walton and Maberly, 1855.

Lecomte, H., *Les Arbres a Gutta-Percha. Leur Culture. Mission Relative A L'Acclimatation de ces Arbres Aux Antilles et a La Guyane*, Paris, Georges Carré et C. Naud, 1899.

Lourme, M. 'Rapport sur les travaux de la commission de la gutta-percha', *Bulletin Economique de l'Indochine*, Vol. 2, No. 10, 1 April 1899, pp. 320–328.

Leon, L., 'L'Industrie de la gutta-percha en Indo-Chine', *Bulletin de la Société Académique de Brest*, Deuxième Série, Tome XXV, 1899–1900, pp. 51–109.

Logan, J.R., 'The Orang Binua of Johore', *The Journal of the Indian Archipelago and East Asia*, Vol. I, 1847, pp. 242–293.

Logan, J.R., 'The Orang Sabimba of the extremity of the Malay Peninsula', *The Journal of the Indian Archipelago and East Asia*, Vol. I, 1847, pp. 295–298.

Logan, J.R., 'Range of the gutta taban collectors, and present amount of imports into Singapore', *Journal of the Indian Archipelago and Eastern Asia*, Vol. II, 1848, pp. 529–533.

Low, H., *Sarawak. Notes During a Residence in That Country with H.H. The Rajah Brooke*, Richard Bentley, London, 1848, Reprinted Singapore, Oxford University Press, 1988.

Lumholtz, C., *Through Central Borneo An Account of Two Years' Travel in the Land of the Head-Hunters Between the Years 1913 and 1917*, 2 Vols, New York, Charles Scribner's Sons, 1920.

MacDonald, G., *Gutta Percha Willie. The Working Genius*, London, Henry S. King & Co., 1873.

Marquis of Salisbury [Prime Minister], 'Address to Dinner of the Institution of Electrical Engineers', *The Electrician*, Vol. 24, (November 8), 1889, p. 13.

Marx, K. and F. Engels, *Manifesto of the Communist Party,* New York, New York Labor New Co, 1908.

McDougall, H., *Letters from Sarawak: Addressed to a Child*, London, Thomas Priest, 1854.

Moering, C.A. and T. Neal, *The New General and Mining Telegraph Code*, London, William Clowes and Sons, 1891.

Money Market Review, 1871, 4 March.

Mouat, F.J., 'Correspondence relative to a remarkable variety of Caoutchouc, termed Gutta Percha', *The Journal of the Agricultural and Horticultural Society of India*, November V, Vol. II, Calcutta, 1843, p. 101–106.

Nederlandsch-Indische Maatschappij van Nijverheid en Landbouw, 'Getah Pertjah van den Nederlandsch Indischen Arcipel', *Tijdschrift voor Nijverheid in Nederlandsch-Indië*, Volume IV, 1858, pp. 396–410.

Obach, E.F.A., 'Cantor lectures on gutta percha', *Journal of the Society of the Arts*, Vol. 46, 1897, pp. 97–114, 117–133; 1898, pp. 137–164; 1898, pp. 169–197.

Obach, E.F.A., *Cantor Lectures on Gutta Percha* [Lecture Reprint with addition of Appendices], London, Society for the Encouragement of Arts, Manufactures & Commerce, 1898.

Oxley, T., 'Gutta percha', *Journal of the Indian Archipelago and Eastern Asia*, Vol. I, 1847, pp. 22–29.

Pall Mall Gazette, 'The West Indian Telegraph Expedition', *Pall Mall Gazette*, July 1870.

Pinardière, de la, L., 'L'industrie de la Gutta-Percha', *Revue des Eaux et Forêts*, 33, (Deuxième série 8), pp. 71–72.

Prescott, G.B., History, Theory, and Practice of the Electric Telegraph, Boston, Ticknor and Fields, 1866.

Reith, G.M., *Handbook to Singapore*, Singapore, The Singapore and Straits Printing Office, 1907.

Ross, J.D., *Sixty Years: Life and Adventure in the Far East*, 2 Vols, London, Hutchinson & Co, 1911.

Roth, H.L., *The Natives of Sarawak and British North Borneo*, 2 Vols, London, Truslove & Hanson, 1896.

Russell, W.H., *The Atlantic Telegraph*, London, Day & Son, 1865.

Sala, G., *The Great Exhibition "Wot is to Be"*, London, The Committee of the Society for Keeping Things in their Places, 1850.

Schlechter, R., *Westafrikanische Kautschuk-Expedition 1899/1900*, Berlin, Kolonial-Wirtschaftlichen Komitee, 1900.

Schlechter R., Über die neue Guttapercha von Neuguinea, *Der Tropenpflanzer Zeitschrift für Tropische Landwirtschaft*, Nr 10, Oktober 1903, pp. 467–471.

Schmitt, B.E., 'The Peace Conference of 1919', *The Journal of Modern History*, Vol. 16, No. 1 (Mar.), 1944, pp. 49–59.

Scientific American, 'Gutta Percha', Vol. 65, No. 7, New York, Munn & Co, 1891, p. 96.

Scott, C.P.G., 'The Malayan words in English', *Journal of the American Oriental Society*, Vol. 17, 1896, pp. 93–144.

Seeligmann, T., L. Torrilhon, and H. Falconnet, *Indiarubber and Gutta Percha. A Complete Practical Treatise*, 2nd English edn, London, Scott, Greenwood & Son, 1910.

Séligmann-Lui, G., *Rapport sur les Origines de la Gutta-Percha*, Paris, Dunod, Libraire des Corps Nationaux des Ponts et Chaussées, des Mines et des Télégraphes, 1883.

Shepperson, A.B., *Telegraphic Cipher Code Especially Adapted to the Cotton Trade*, New York, Shepperson Publishing Company, 1878.

Siemens, W., 'President's Address', *Journal of the Society of Telegraph Engineers*, Vol. VII, No. 21, Jan. 23, 1878, pp. 3–19.

Siemens von W., *Scientific and Technical Papers*, 2 Vols, London, John Murray, 1892.

Siemens von W. (Trans, Coupland W.C.), *Personal Recollections*, London, Asher and Co, 1893.

Smith, W., in H.C. Forde, E.B. Bright, T. Russell Crampton, W.H. Preece, W. Smith, S. Trott, and F.A Hamilton, "Discussion on Submarine telegraph cables: their decay and renewal," *Journal of the Society of Telegraph Engineers and Electricians*, Vol. 12, No. 50, 1884, pp. 504–532.

Smith, W., *The Rise and Extension of Submarine Telegraphy*, London, J.S. Virtue & Co, 1891.

Society for Promoting Christian Knowledge, *Caoutchouc and Gutta Percha*, London, 1852.

Song, O.S., *One Hundred Years History of the Chinese in Singapore*, London, J. Murray, 1923.

Squier, G.O., 'An American Pacific Cable', Paper Presented at Thirteenth Meeting of the *American Institute of Electrical Engineers*, New York and Chicago, December 27, 1899, Vol. XVI Transactions, pp. 605–626.

St. John, S., *Life in the Forests of the Far East*, Two Volumes, London, Smith Elder and Co., 1862.

Swart, A.G.N, *Rubber Companies in the Netherland East Indies*, Amsterdam, The Netherland Commission for the International Rubber Exhibition, London, 1911.

Telcon, *The Telegraph Construction and Maintenance Co. Ltd. On the Occasion of the Visit of the Delegates to the International Telegraph Conference*, London, Telegraph Construction and Maintenance Co. Ltd, 1903.

Telegraphic Journal and Electrical Review, Vol. XI, No. 241, July 8, London, Henry Gillman, 1882.

Templer, J.C. (ed.), *Private Letters of Sir James Brooke K.C.B Rajah of Sarawak, Narrating the Events of His Life*, 3 Vols, London, Richard Bentley, 1853.

Terry, H.L., *India-Rubber and Its Manufacture. With Chapters on Gutta-Percha and Balata*, London, Archibald Constable & Co. Ltd., 1907.

Thompson, S.P., *The Life of Lord Kelvin*, 2 Vols, Chelsea Publishing Company, New York, 1910, as 2nd edn Reprinted New York, Chelsea Publishing Company, 1976.

Thomson, J., *The Straits of Malacca Indo-China and China*, London, Sampson Low, Marston, Low & Searle, 1875.

Thomson, W., *Atlantic Telegraph Cable. Address Delivered before the Royal Society of Edinburgh December 18, 1865*, with other Documents, London, William Brown & Co, 1866.

Todd, C., 'Telegraphic enterprise in Australia', *Proceedings of the Royal Colonial Institute*, Seventh Vol., 1886, pp. 144–165.

Train, G.F., *An American Merchant in Europe, Asia and Australia: A series of Letters from Java, Singapor, China... etc*, Broadway, G.P Putnam & Co, 1857.

Tucker, A.L.P., *Sir Robert G. Sandeman K.C.S.I. Peaceful Conqueror of Baluchistan*, London, Society for Promoting Christian Knowledge, 1921.

Von Gaffron, W, G, 'Nota Omtrent den getah-pertja-boom', *Nutuurkundig Tijdschrift voor Nederlandsch-Indië*, 16, 1858, pp. 224–228.

Wallace, A.R., *The Malay Archipelago*, 10th edn, 1869, London, MacMillan & Co, UK, Reprinted Singapore, Graham Brash Pte Ltd, 1987.

Western Journal and Civilian, 'Manufacture of Gutta Percha', Vol. XII, No. 2, 1854, pp. 85–93, St. Louis M. Tarver and H. Cobb.

Wilkinson, H.D., *Submarine Cable Laying and Repairing*, 2nd edn, London, "The Electrician" Printing and Publishing Company, 1908.

Wilshaw, E., *Cable Communications and how to use them to the Best Advantage*, paper presented to Chartered Institute of Secretaries (Country Conference), Manchester, May 22, 1924.

Wray, L., 'Gutta-producing trees', *Journal of the Straits Branch of the Royal Asiatic Society*, No. 12. (December), 1883, pp. 207–221.

Wright, A. and H.A. Cartwright, *Twentieth Century Impressions of British Malaya*, London, Lloyds Greater Britain Publishing Company Ltd., 1908.

Secondary Sources

Adas, M., *Machines as the Measure of Men: Science, Technology, and Ideologies of Western Dominance*, Ithaca, NY, Cornell University Press, 1990.

Addo, H., *Imperialism: The Permanent Stage of Capitalism*, Tokyo, United Nations University, 1986.

Ahvenainen, J., *The Far Eastern Telegraphs, The History of Telegraphic Communications between the Far East, Europe and America before the First World War*, Helsinki, Suomalainen Tiedeakatemia, 1981.

Alexander, J. and P. Alexander, 'Commodification and consumption in a central Borneo community', *Bijdragen tot de Taal-, Land- en Volkenkunde*, 151, No. 2, 1995, pp. 179–193.

Anderson, R.O'G., *Imagined Communities: Reflections on the Origins and Spread of Nationalism*, London, Verso, 1991.

Andaya, B.W. and L.Y. Andaya, *A History of Malaysia*, Basingstoke, Macmillan, 2001.

Antunes, C., and K. Fatah-Black, (eds), *Explorations in History and Globalization*, Abingdon, Routledge, 2016.

Baines, J., *Joseph Conrad A Critical Biography*, Harmondsworth, Penguin, 1971.

Bair, J., 'Global Capitalism and Commodity Chains: Looking Back, Going Forward', *Competition and Change*, Vol. 9, No. 2, June 2005, pp. 153–180.

Bakar, M.A., A. Kaur, and A.Z. Ghazali, *Historia. Essays in Commemoration of the 25th Anniversary of Dept. of History, University of Malaya*, Kuala Lumpur, Malaysian Historical Society, 1984.

Banerjee, A. and E. Duflo, 'The economic lives of the poor', *Journal of Economic Perspectives*, Vol. 21(1), (Winter), 2007, pp. 141–167.

Barlow, C., 'The rubber smallholder economy', *Bulletin of Indonesian Economic Studies*, 18, No. 2, 1982, pp. 86–119.

Barlow, C., 'Growth, structural change and plantation tree crops: the case of rubber', *World Development*, Vol. 25, (10), 1997, pp. 1589–1607.

Baron, S., Field, J. and T. Schuller (eds), *Social Capital: Critical Perspectives*, Oxford, Oxford University Press, 2000.

Barty-King, H., *Girdle Round the Earth: The Story of Cable and Wireless and its Predecessors*, London, Heinemann, 1979.

Batabyal, A.A. and H. Beladi, 'Swidden agriculture in developing countries', *Review of Development Economics*, Vol. 8, No. 2, May 2004, pp. 255–265.

Batu Bala, S., *Kelabits Story. The Great Transition* (ebook), Singapore, Partridge Publishing, 2014.

Beauchamp, K, *History of Telegraphy*, Stevenage, Institution of Electrical Engineers, 2001.

Bentley, J.H., 'Globalizing History and historicizing globalization', in *Globalization and Global History,* B.K. Gills and W.R. Thompson eds., London, Routledge, 2006.

Bergesen, A.J., 'Review Symposium on Andre Gunder Frank's "Reorienting the Nineteenth Century: Global Economy in the World-system After Andre Gunder Frank"', *Journal of World-System Research* Vol. 21, Number 1, (Winter/Spring), 2015, pp. 147–161.

Boomgaard, P., 'Oriental nature, its friends and its enemies: conservation of nature in late- colonial Indonesia 1889–1949', *Environment and History*, Vol. 5, No. 3, October 1999, pp. 257–292.

Boomgaard, P., 'The making and unmaking of tropical science: Dutch research on Indonesia, 1600–2000', *Bijdragen tot de Taal-, Land- en Volkenkunde,* Vol. 162, No. 2/3 (2006), 2006, pp. 191–217.

Boomgaard, P., F. Combijn, and D. Henley (eds), *Paper Landscapes: Explorations in the Environmental History of Indonesia*, Leiden, KITLV Press, 1997.

Booth, A., 'The economic development of Southeast Asia in the colonial era: c.1870–1942', *History Compass*, 6/1, 2008, pp. 25–53.

Boyce, R.W.D., 'Imperial dreams and national realities: Britain, Canada and the struggle for a Pacific telegraph cable, 1879, 1902', *The English Historical Review*, (February), Vol. 115, Iss. 460, 2000, pp. 39–70.

Britton, J.A. and J. Ahvenainen, 'Showdown in South America: James Scrymser, John Pender, and United States-British cable competition', *Business History Review*, Vol. 78, 1, (Spring), 2004, pp. 1–27.

Brookfield, H., L.M. Potter, and L. Byron (eds), *In Place of the Forest: Environmental Socio-Economic Transformation in Borneo and the Eastern Malay Peninsula*, Tokyo, United Nations University Press, 1995.

Cain, P.J. and A.G. Hopkins, *British Imperialism 1688–2015*, Abingdon, Routledge, 2016.

Carlson, J.D. 'The 'Otter-Man' empires: The Pacific fur trade, Incorporation and the zone of ignorance', *Journal of World-system Research*, VIII, III, Fall, 2002, pp. 390–442.

Chambert-Loir, H and Dupoizat, M-F, 'Les jarres de Bornéo d'après un texte malais de 1839', *Archipel*, Volume 66, 2003. pp. 113–160.

Chandler, A.D. with T. Hikino, *Scale and Scope. The Dynamics of Industrial Capitalism*, Belknap Press, Cambridge, Mass., Harvard University Press, 2004.

Chapman, S., *The Rise of Merchant Banking*, London, Routledge, 2010.

Chee, B.T., *Chinese Overseas Comparative Cultural Issues*, Hong Kong, Hong Kong University Press, 2004.

Chew, D., *Chinese Pioneers on the Sarawak Frontier*, Singapore, Oxford University Press, 1990.

Chiang, H.D., *A History of Straits Settlements Foreign Trade 1879–1915*, Singapore National Museum, 1978.

Chin, J.M., *The Sarawak Chinese*, Kuala Lumpur, Oxford University Press, 1981.

Clarke, A.C., *Voice Across the Sea*, New York, Harper & Brother, 1958.

Cleary, M.C., 'Indigenous trade and European economic intervention in North-West Borneo c. 1860–1930', *Modern Asian Studies*, Vol. 30, No. 2, 1996, pp. 301–324.

Cleary, M.C., 'From hornbills to oil? Patterns of indigenous and European trade in colonial Borneo', *Journal of Historical Geography*, Vol. 23, No. 1, 1997, pp. 29–45.

Coates, V.T. and B. Finn, *A Retrospective Technology Assessment: Submarine Telegraphy. The Transatlantic Cable of* 1866, San Francisco, San Francisco Press Inc., 1979.

Cookson, G., *The Cable: The Wire that Changed the World*, Brimscombe Port, Tempus, 2003.

Cooper, F., 'What is the Concept of Globalization Good for? An African Historian's Perspective', *African Affairs*, Vol. 100, No. 399, April 2001, pp. 189–213.

Cowan, C.D., *Nineteenth Century Malaya, The Origins of British Political Control*, London, Oxford University Press, 1961.

Cramb, R.A. and R.H.W Reece, *Development in Sarawak: Historical and Contemporary Perspectives*, Centre of Southeast Asian Studies, Melbourne, Monash University, 1988.

Cramb, R.A., 'The commercialization of Iban agriculture', in *Development in Sarawak: Historical and Contemporary Perspectives*, eds R.A. Cramb and R.H.W. Reece, Melbourne, Centre of Southeast Asian Studies, Monash University, 1988, pp. 105–134.

Cramb, R.A., *Land and Longhouse: Agrarian Transformation in the Uplands of Sarawak*, Copenhagen, Nias Press, 2007.

Curtin, P.D., *Cross-Cultural Trade in World History*, Cambridge, Cambridge University Press, 1984.

Davis, S.H. and A. Wali, 'Indigenous land tenure and tropical forest management in Latin America', *Ambio*, Vol. 23, No. 8 (December), 1994, pp. 485–490.

Dean, J.N., 'Gutta percha and balata, with particular reference to their use in submarine cable manufacture', *Journal of the Royal Society of Arts*, Vol. 92, No. 4668, June 23rd, 1944, pp. 367–386.

Dean, J.N., 'A visit to Selborne Plantation', *Zodiac House Magazine*, Telcon Incorporating Staff Association Social and Athletic Club and Shop Stewards Committee, (Mar.) 1949, pp. 1–3.

De Vries, J., 'The limits of globalization in the early modern world', *The Economic History Review*, New Series, Vol. 63, No. 3, August 2010, pp. 710–733.

Dixon C., *South East Asia in the World-economy*, Cambridge, Cambridge University Press, 1991.

Dougherty, M.L., 'Theorizing Theory: Origins and Orientations of Commodity Chain Analysis', *The Global Studies Journal*, Volume 1, Number 3, 2008, pp. 29–38.

Dove, M.R., *The Real and Imagined Role of Culture in Development, Case Studies from Indonesia*, Honolulu, University of Hawai'i Press, 1990.

Dove, M.R., 'Transition from native forest rubbers to *Hevea brasiliensis* (Euphorbiaceae) among tribal smallholders in Borneo', *Economic Botany*, Vol. 48, 1994, pp. 382–396.

Dove, M.R., 'The life-cycle of indigenous knowledge, and the case of natural rubber production' in *Indigenous environmental knowledge and its Transformation. Critical Anthropological Perspectives. Studies in Anthropology*. A. Bicker, R. Ellen and P. Parkes eds, Amsterdam, Harwood Academic Publishers, 2000, pp. 209–248.

Dove, M.R., *The Banana Tree at the Gate. A History of Marginal Peoples and Global Markets in Borneo*, New Haven, Yale University Press, 2011.

Drabble, J.H. and P.J. Drake, 'The British agency houses in Malaysia: Survival in a changing world', *Journal of Southeast Asian Studies*, 12(2), (Sep.) 1981, pp. 297–328.

Drabble, J.H., *An Economic History of Malaysia, c. 1800–1990 The Transition to Modern Economic Growth*, Basingstoke, Macmillan, 2000.

Dransfield, J. and N. Manokaran, (eds.), *PROSEA: Plant Resources of South-East Asia 6, Rattans*, Bogor Indonesia, PROSEA, Yayasan Obor, 1994.

Drayton, R.H., *Nature's Government. Science, Imperial Britain, and the Improvement of the World*, New Haven, Yale University Press, 2000.

Du Boff, R.B., 'Business demand and the development of the telegraph in the United States, 1844–1860', *Business History Review*, Vol. LIV, No. 4 (Winter), 1980, pp. 459–479.

Dunn, F.L., *Rain-forest Collectors and Traders: A Study of Resource Utlilization in Modern and Ancient Malaya*, Kuala Lumpur, Monographs of the Malaysian Branch, Royal Asiatic Society, No. 5, 1982.

Dwyer, R.E.P., Rubber Production in New Guinea and Papua (continued), *New Guinea Agricultural Gazette*, Vol. 7, No. 4, November 1941, pp. 249–279.

Edwards, L. 'Gutta percha: The plastic of its time', *Plastics Historical Society* (n.d.).

Edwards, L., *A Ramble in the World of Gutta Percha*, revision of speech originally delivered 30 November 1989, Unpublished paper, copy provided by author, revised version 1993.

Edwards, L., Manager of the Telcon Balata [and Gutta Percha] Processing Plant, Greenwich; *Telcon and Gutta Percha, Discussion with Author*, 19 July 2007.

Edwards, R.W., 'The North American Fur Trade in the World System', *Field Notes: A Journal of Collegiate Anthropology*, 1 (1), 2009, pp. 46–64.

Eghenter, C., 'Towards a causal history of a trade scenario in the interior of East Kalimantan, Indonesia, 1900–1999', *Bijdragen tot de Taal-, Land- en Volkenkunde,* 157, (2001), No. 4, 2001, pp. 739–769.

Eghenter, C., 'Histories of conservation or exploitation? Case studies from the interior of Indonesian Borneo' in *Histories of the Borneo Environment,* ed. R.L. Wadley, Leiden, KITLV Press, 2005, pp. 87–107.

Ellis, F., *Peasant Economics Farm Households and Agrarian Development,* 2nd edn, Cambridge, Cambridge University Press, 2000.

Elson, R.E., 'International commerce, the state and society, economic and social change', in *The Cambridge History of Southeast Asia, Volume Three – from c. 1800–1930s,* ed. N. Tarling, Cambridge, Cambridge University Press, 1999, pp. 127–191.

Engel, A., 'Buying time: futures trading and telegraphy in nineteenth-century global commodity markets', *Journal of Global History,* Vol. 10, Iss. 2, July 2015, pp. 284–306.

Evans, H.J., 'The path to freedom? Transocean and German wireless telegraphy, 1914–1922', *Historical Social Research,* 35 (1), 2010, pp. 209–233.

Fasseur, C., 'The Cultivation System and its Impact on the Dutch Colonial Economy and the Indigenous Society in Nineteenth-Century Java', in *Two Colonial Empires: Comparative Essays on the History of India and Indonesia in the Nineteenth Century,* Comparative Studies in Overseas History 6, C.A. Bayly and D.H. Kolff, (Eds), Dordrecht, Martinus Nijhoff Publishers, 1986, pp. 137–154.

Feenstra, R.C and G.H. Hanson, 'Intermediaries in entrepot trade: re-exports of Chinese goods', *Journal of Economics and Management Strategy,* Vol. 13, Iss. 1, March 2004, pp. 3–35.

Field, A.J., 'The Magnetic Telegraph, Price and Quantity Data, and the New Management of Capital', *The Journal of Economic History,* Vol. 52, No. 2 (June. 1992), pp. 401–441.

Flynn, D.O. and A. Giráldez, 'Path dependence, time lags and the birth of globalisation: A critique of O'Rourke and Williamson', *European Review of Economic History,* Vol. 8, No. 1 (April), 2004, pp. 81–108.

Francis, A. *Culture and Commerce in Conrad's Asian Fiction,* New York, Cambridge University Press, 2015.

Frank, A.G. *ReOrienting the 19th Century. Global Economy in the Continuing Asian Age,* Boulder, Paradigm Publishers, 2014.

Frank, A.G., 'Immanuel and Me with-out Hyphen', *Journal of World-system Research,* VI, 2, Summer/Fall, 2000, pp. 216–231.

Frank, Z. and A. Musacchio, 'Brazil in the international rubber trade, 1870–1930', in *From Silver to Cocaine. Latin American Commodity Chains and the Building of the World Economy, 1500–2000,* eds S. Topik, C. Marichal and Z. Frank, Durham, Duke University Press, 2006, pp. 271–299.

Freeman, D., *Report on the Iban of Sarawak*, Kuching, Government Printing Office, 1955.

Freeman, D., *Iban Agriculture: A Report on the Shifting Cultivation of Hill Rice by the Iban of Sarawak*, London, HMSO, 1955.

Freeman, D., 'Iban Augury', *Bijdragen tot de Taal-, Land- en Volkenkunde*, 117, No. 1, 1961, pp. 141–167.

Freeman, D., *Report on the Iban*, London, University of London, The Athlone Press, 1970.

Freeman, D., 'The Iban of Sarawak and their religion: a review article', *Sarawak Museum Journal*, Vol. 23, 1975, pp. 275–288.

Freeman, D., *Some Reflections on the Nature of Iban Society*, Canberra, Australian National University, 1981.

Friedman, W.F., *Report on the Use of Codes and Code Language, The International Telegraph Regulations Pertaining thereto, and the Bearing of this History on the Cortina Report, International Radiotelegraph Conference of Washington 1927*, Washington, U.S. Government Printing Office, 1928.

Frodin, D.G., 'The Biological Exploration of New Guinea' in *The Ecology of Papua Part 1*, Marshal A.J. and Beehler, B.M. eds., Singapore, Periplus Publishing, 2007.

Frost, M.R., 'Transcultural diaspora: The Straits Chinese in Singapore 1819–1918', *Asian Research Institute Working Paper Series* No. 10, National University of Singapore, 2003.

Garbade, K.D. and W.L. Silber, 'Technology, communication and the performance of financial markets: 1840–1975', *The Journal of Finance*, Vol. 33, No. 3, (June.), 1978, pp. 819–832.

Gereffi, G. and M. Korzeniewicz (eds), *Commodity Chains and Global Capitalism*, Westport, Praeger, 1994.

Giddens A., *The Consequences of Modernity*, Stanford, CT, Stanford University Press, 1990.

Gills, B.K. and W.R. Thompson, *Globalization and Global History*, London, Routledge, 2006.

Glover, W. *The Safety Insulated Wire & Cable Company*, New York, http://atlantic-cable.com/CableCos/SafetyCable/index.htm, 2013.

Glover, W., (Continuously Updated), *Cable Timeline 1845–1900*. http://www.atlantic-cable.com/Cables/CableTimeLine/index1850.htm, 2017a.

Glover, W., (Continuously Updated), *Cable Timeline 1901–1950*. http://www.atlantic-cable.com/Cables/CableTimeLine/index1901.htm, 2017b.

Godfrey, H.L., *The Forgotten Trade. Submarine Telegraphy and the Gutta Percha Trade. The Response in Nineteenth Century Sarawak*, PhD Thesis, Armidale, University of New England, 2011.

Godfrey, H.L., 'Trade and Transformation on the Borneo Frontier: towards a study of the Baram region of Sarawak in the late nineteenth century', *Borneo Research Bulletin*, 2014, Vol 45, pp. 87–135.

Gooday, G.J.N., *The Morals of Measurement: Accuracy, Irony, and Trust in Late Victorian Electrical Practice*, Cambridge, Cambridge University Press, 2004.

Gordon, R.J., 'Does the "New Economy" measure up to the great inventions of the past?', *Journal of Economic Perspectives*, Vol. 14, No. 4, (Fall), 2000, pp. 49–74.

Gordon, R.J., *The Rise and Fall of American Growth: The U.S. Standard of Living since the Civil War (The Princeton Economic History of the Western World)*, Princeton, Princeton University Press, 2016.

Goss, A., *The Floracrats. State-sponsored Science and the Failure of the Enlightenment in Indonesia*, Madison, University of Wisconsin Press, 2011.

Greaves, R.L., *Persia and the Defence of India, 1884–1892 A Study in the Foreign Policy of the Third Marquis of Salisbury*, London, Athlone Press, 1959.

Hall, P. and P. Preston, *The Carrier Wave. New Information Technology and the Geography of Innovation*, London, Unwin Hyman, 1988.

Hall, T.D., 'Incorporation in the world-system: Towards a critique', *American Sociological Review*, Vol. 51, (June), 1986, pp. 390–402.

Hanson, J.R., *Trade in Transition. Exports from the Third World, 1840–1900*, New York, Academic Press, 1980.

Harrisson, B., *Pusaka. Heirloom Jars of Borneo*, Oxford, Oxford University Press, 1986.

Headrick, D.R. and P. Griset, 'Submarine telegraph cables: business and politics, 1828–1939', *Business History Review*, (Autumn) Vol. 75, No. 3, 2001, pp. 543–578.

Headrick, D.R., *The Tools of Empire: Technology and European Imperialism in the Nineteenth Century*, New York, Oxford University Press, 1981.

Headrick, D.R., *The Invisible Weapon: Telecommunications and International Politics, 1851–1945*, New York, Oxford University Press, 1991.

Headrick, D.R., 'Review: "Under the Wire: How the Telegraph Changed Diplomacy"', *Victorian Studies*, (Spring), Vol. 47, No. 3, 2005, pp. 491–493.

Healey, C., 'Tribes and states in "pre-colonial" Borneo: Structural contradiction and the generation of piracy', *Social Analysis*, No. 18, 1985, pp. 3–39.

Held, D., A. McGrew, D. Golblatt, and J. Perratton, *Global Transformations: Politics, Economics and Culture*. Stanford, Stanford University Press, 1999.

Heng, D.T.S and S.M.K Aljunied, *Reframing Singapore. Memory, Identity, Trans-regionalism*, Amsterdam, Amsterdam University Press, 2009.

Hoffman, C.L., *The Punan: Hunters and Gatherers of Borneo*, Ann Arbor, UMI Research Press, 1986.

Hoffman, C.L., 'The "Wild Punan" of Borneo: A matter of economics' in *The real and Imagined Role of Culture in Development: Case Studies from Indonesia*, ed. M. Dove, Honolulu, University of Hawai'i Press, 1990, pp. 89–118.

Hopkins, T.K. and I. Wallerstein, 'Commodity chains in the capitalist world-economy prior to 1800. Commodity chains: Construct and research', in *Commodity Chains and Global Capitalism*, eds G. Gereffi and M. Korzeniewicz, Westport, Praeger, 1994, pp. 17–19.

Howe, S., 'David Fieldhouse and 'Imperialism': Some historiographical revisions', in *Managing the Business of Empire: Essays in Honour of David Fieldhouse*, eds P. Burroughs and A.J. Stockwell, London, Frank Cass, 1998, pp. 213–232.

Huff, W.G., 'Bookeeping barter: Money, credit, and Singapore's international rice trade, 1870–1939', *Explorations in Economic History*, Vol. 26, 1989, pp. 161–189.

Huff, W.G., *The Economic Growth of Singapore: Trade and Development in the Twentieth Century*, Cambridge, Cambridge University Press, 1995.

Huff, W.G., 'Monetization and financial development in Southeast Asia before the Second World War', *Economic History Review*, New Series, Vol. 56, No. 2 (May), 2003, pp. 300–345.

Hugill, P.J., *Global Communications since 1844. Geopolitics and Technology*, Baltimore, Johns Hopkins University Press, 1999.

Hugill, P.J., *World Trade since 1431: Geography, Technology and Capitalism*, Baltimore, Johns Hopkins University, 1995.

Hunt, B.J., 'Doing science in a global empire: cable telegraphy and electrical physics in Victorian Britain', pp. 312–333, in *Victorian Science in Context* edited by B. Lightman, Chicago, University of Chicago Press, 1997.

Hunt, B.J., 'Insulation for an empire. Gutta percha and the development of electrical measurement in Victorian Britain', in *Semaphores to Short Waves*, ed. F.A.J.L. James, London, Royal Society of the Arts, 1998, pp. 85–104.

Hunt, B.J., *The Maxwellians*, Ithaca, Cornell University Press, 2005.

Hwang, K., 'International Collaboration in Multilayered Center-Periphery in the Globalization of Science and Technology', *Science Technology and Human Values*, 33, 2008, pp. 101–133

International Council for Science, ICSU *Series on Science for Sustainable Development No. 4: Science. Traditional Knowledge and Sustainable Development*, Paris, 2002.

International Telecommunication Union, *CCITT 50 Years of Excellence 1956–2006*. www.itu.int/ITU-T/50, 2006.

Irwin, G.W., *Nineteenth Century Borneo: A Study in Diplomatic Rivalry*, Singapore, Donald Moore, 1967.

Jarman, R.L., *Annual Reports of the Straits Settlements 1855–1941*, 12 Vol. set, Slough, Archive Editions, 1998.

Jessup, D.E., 'Connecting Alaska: The Washington-Alaska Military Cable and Telegraph System', *The Journal of the Gilded Age and Progressive Era*, Vol. 6, No. 4 (October) 2007, pp. 384–408.

Jevons, W.S., *Money and the Mechanism of Exchange*, New York, D. Appleton & Company, 1919.

John, R.R., 'Recasting the information infrastructure for the industrial age' in *A Nation Transformed by Information: How Information Has Shaped the United States from Colonial Times to the Present*, eds A.D. Chandler and J.W. Cortada, New York, Oxford University Press, 2000, pp. 55–105.

Kathirithamby-Wells, J., 'The Politics of Commerce in Southeast Asia: An historical perspective', *Inaugural Lecture, University of Malaya, Kuala Lumpur,* December 15 1992.

Kathirithamby-Wells, J., *Nature and Nation. Forests and Development in Peninsular Malaysia,* Singapore, Nias Press, 2005.

Kaur, A., 'The babbling Brookes: Economic change in Sarawak 1841–1941', *Modern Asian Studies,* Vol. 29, No. 1, 1995, pp. 65–109.

Kaur, A., *Economic Change in East Malaysia. Sabah and Sarawak since 1850,* Basingstoke, Macmillan Press Ltd, 1998.

Kahn, D., *The Codebreakers: The Comprehensive History of Secret Communication from Ancient Times to the Internet,* New York, Scribner, 1996.

Kedit, P.M., *Iban Bejalai,* Kuala Lumpur, Sarawak Literary Society, 1993.

Kemp, A.R., 'Paragutta, A New Insulating Material for Submarine Cables', *Bell Systems Technical Journal,* Vol. 10, Iss. 1, 1931, pp. 132–148.

Kennedy, P.M., 'Imperial cable communication and strategy', *The English Historical Review,* Vol. 86, No. 341 (October), 1971, pp. 728–752.

Kessler, C.S., 'Review Article. Colonial Science and the Creation of a Postcolonial Scientific Tradition in Indonesia', *Akademia* 37, Disember 1990, pp. 91–105.

Kieve, J., *The Electric Telegraph: A Social and Economic History,* Newton Abbot, David & Charles, 1973.

Knapen, H., *Forest of Fortune? The Environmental History of Southeast Borneo, 1600–1880,* Leiden, KITLV, 2001.

Koepping, E.W., *Friendship and Feud in Late 19th Century Borneo,* Melbourne, Centre for Southeast Asian Studies, Monash University, 1988.

Kopytoff, I., 'The cultural biography of things: Commoditization as process' in *The Social Life of Things: Commodities in Cultural Perspectives,* ed. A. Appadurai, Cambridge, Cambridge University Press, 1986, pp. 64–91.

Kreps, C., 'Non-Western models of museums and curation in cross-cultural perspective' in *A Companion to Museum Studies,* ed. S. MacDonald, Malden, Blackwell Publishing, 2006, pp. 457–472.

Kuznets, S., 'Modern economic growth: Findings and reflections', *The American Economic Review,* Vol. 63, No. 3, (June.), 1973, pp. 247–258.

Kuznets, S., *Modern Economic Growth: Rate, Structure and Spread,* New Haven, Yale University Press, 1966.

Langdale, J., 'Review [The Carrier Wave: New Information Technology and the Geography of Innovation]', *Annals of the Association of American Geographers,* Vol. 80, No. 4 (December), 1990, pp. 634–635.

Latham, A.J.H. and N. Neal, 'The international market in rice and wheat, 1868–1914', *The Economic History Review,* New Series, Vol. 36, No. 2 (May), 1983, pp. 260–280.

Lawford, G.L. and L.R. Nicholson, *The Telcon Story,* London, The Telegraph Construction & Maintenance Co. Ltd, 1950.

Lawrence, P., *Road Belong Cargo: A Study of the Cargo Movement in the Southern Madang District, New Guinea*. Manchester, Manchester University Press, 1964.

Leach, D.L., 'The Survey of Sarawak', *The Geographical Journal*, Vol. 100, No. 3, Sep. 1942, pp. 98–105.

Lee, Y.L. 'The Population of British Borneo', *Population Studies*, Vol. 15, No. 3, Mar. 1962, pp. 226–243.

Lee, Y.L. 'The Population of Sarawak', *The Geographical Journal*, Vol. 131, No. 3, Sep. 1965, pp. 344–356.

Lew, B. and B. Cater, 'The telegraph, co-ordination of tramp shipping, and growth in world trade, 1870–1910', *European Review of Economic History*, Vol. 10, 2006, pp. 147–173.

Lian, K.F. and K.W. Koh, 'Chinese enterprise in colonial Malaya: the case of Eu Tong Sen', *Journal of Southeast Asian Studies*, 35 (3), 2004, pp. 415–432.

Lindblad, J.T., 'Review [Of Free Trade and Native Interests: The Brookes and the Economic Development of Sarawak, 1841–1941 by Ooi Keat Gin]', *Bulletin of the School of Oriental and African Studies, University of London*, Vol. 62, No. 1, 1999, pp. 195–196.

Lipton, M., 'The theory of the optimising peasant', in *Development Studies Revisited: Twenty Five Years of the Journal of Development Studies*, eds. C. Cooper and E.V.K Fitzgerald, London, Frank Cass and Co Ltd, 1989, pp. 411–435.

Lockard, C.A., *From Kampong to City: A Social History of Kuching, Malaysia 1820–1970*, Athens, Ohio, Ohio University Monographs in International Studies, Southeast Asia Series, No. 75, 1987.

Lockard, C.A., *Chinese Society and Politics in Sarawak: Historical Essays*, Jalan Salim, Sarawak Chinese Association, 2009.

Lumenta, D., 'Towards transnational Dayak identities? Changing interconnectedness, identities and nation states – A case study on Iban-Kenyah relations in Sarawak, East Malaysia' in *Economic Prospects, Cultural Encounters and Political Decisions: Scenes in a Moving Asia (East and Southeast)*, The Work of the 2002/2003, API Fellows, Kuala Lumpur, 2003, pp. 1–18.

Lumenta, D., 'Moving in a hierarchized landscape. Changing border regimes in Central Kalimantan', *Wacana*, Vol. 13, No. 1, April 2011, pp. 121–145.

Mackenzie, C., *Realms of Silver. One Hundred Years of Banking in the East,* London, Routledge & Kegan Paul, 1954.

Mahony, M.A., 'The local and the global: Internal and external factors in the development of Bahia's cacao sector', in *From Silver to Cocaine. Latin American Commodity Chains and the Building of the World Economy, 1500–2000*, eds. S. Topik, C. Marichal and Z. Frank, Durham, Duke University Press, 2006, pp. 174–203.

Marichal, C., S. Topik and Z. Frank, 'Conclusion: Commodity chains and globalization in historical perspective' in *From Silver to Cocaine. Latin American Commodity Chains and the Building of the World Economy, 1500–2000*, eds S. Topik, C. Marichal and Z. Frank, Durham, Duke University Press, 2006, pp. 352–360.

McGowan, P.J. and B. Kordan, 'Imperialism in world-system perspective: Britain 1870–1914', *International Studies Quarterly*, Vol. 25, No. 1, World-System Debates (Mar.) 1981, pp. 43–68.

McSweeney, K., 'The dugout canoe trade in Central America's Mosquitia', *Annals of the Association of American Geographers*, Vol. 94, issue 3, (September.) 2004, pp. 638–661.

Mering Ngo, T.H.G., 'A new perspective on property rights: Examples from the Kayan of Kalimantan' in *Borneo in Transition. People, Forests, Conservation, and Development*, eds C. Padoch and N.L. Peluso, Kuala Lumpur, Oxford University Press, 1996, pp. 137–149.

Metcalf, P., 'Birds and Deities in Borneo', *Bijdragen tot de Taal-, Land- en Volkenkunde*, 132, No. 1, Leiden, 1976, pp. 99–123.

Michie, R.C., *The London Stock Exchange: A History*, Oxford, Oxford University Press, 1999.

Michie, R.C., *The Global Securities Market: A History*, New York, Oxford University Press, 2006.

Miller, R. and R. Greenhill, 'The fertilizer commodity chains: Guano and nitrate, 1840–1930' in *From Silver to Cocaine. Latin American Commodity Chains and the Building of the World Economy, 1500–2000*, eds S. Topik, C. Marichal and Z. Frank, Durham, Duke University Press, 2006, pp. 228–270.

Morus, I.R., *Frankenstein's Children: Electricity and Experiment in early Nineteenth-century London*, Princeton, Princeton University Press, 1998.

Moustakerski, P. and Shang-Jin Wei *Outsourcing Tariff Evasion: A New Explanation for Entrepot Trade, IMF Working Paper* WP/05/102, Washington, International Monetary Fund, 2005.

Muller, S.M. and H.J.S. Tworek, 'The telegraph and the bank: on the interdependence of global communications and capitalism, 1866–1914', *Journal of Global History*, 10, 2015, pp. 259–283.

Munan, H., *Beads of Borneo*, Singapore, Editions Didier Millet, 2005.

Nickles, D.P., 'Telegraph diplomats. The United States' relations with France in 1848 and 1870', *Technology and Culture*, Vol. 40, No. 1, (Jan.) 1999, pp. 1–25.

Nickles, D.P., *Under the Wire: How the Telegraph Changed Diplomacy*, Cambridge, Mass, Harvard University Press, 2003.

Nyiri, B., *Chasing Dragons through Time and Space: Martabani dragon jars in the Kelabit highlands, Sarawak, East Malaysia*, Ph. D thesis, Leicester, University of Leicester, School of Archaeology and Ancient History, 2016.

Ohff, H-J., *Empires of Enterprise: German and English commercial interests in East New Guinea 1884 to 1914*, Ph. D thesis, Adelaide, University of Adelaide, School of History and Politics, 2008.

O'Rourke K.H. and J.G. Williamson, 'When did globalisation begin?', *European Review of Economic History*, Vol. 6, No. 1, April 2002, pp. 23–50.

O'Rourke, K.H. and J.G. Williamson, 'Once more: When did globalisation begin?', *European Review of Economic History*, Vol. 8, No. 1, April 2004, pp. 109–117.

Ooi K.G., *Of Free Trade and Native Interests, The Brookes and the Economic Development of Sarawak, 1841–1941*, Kuala Lumpur, Oxford University Press, 1997.

Ooi, K.G., 'For want of rice: Sarawak's attempts at rice self-sufficiency during the period of Brooke Rule 1841–1941', *Journal of South East Asian Studies*, Vol. 29, No. 1, 1998, pp. 8–23.

Orozco, M., 'Worker remittances in an international scope', *Inter-American Dialogue, Research Series*, Remittances Project, (March) 2003, pp. 1–12.

Osborne, M., *Southeast Asia. An Introductory History,* Tenth Ed., Crows Nest, Allen & Unwin, 2010.

Oslin, G.P., *The Story of Telecommunications*, Macon, Mercer University Press, 1992.

Padoch, C., *Migration and Its Alternatives Among the Iban of Sarawak*, PhD Thesis, Columbia University, 1978.

Padoch C. and N.L. Peluso, (eds), *Borneo in Transition. People, Forests, Conservation, and Development*, Kuala Lumpur, Oxford University Press, 1996.

Palladino P. and M. Worboys, 'Science and Imperialism', *Isis*, 84, 1993, pp. 84–102.

Palumbo-Liu, D., B.W. Robbins and N. Tanoukhi, eds, *Immanuel Wallerstein and the Problem of the World: System, Scale, Culture*, Durham, Duke University Press, NC, 2011.

Payne, R., *The White Rajahs of Sarawak*, Robert Hale, 1960, Reprinted Singapore, Oxford University Press, 1992.

Péhaut, Y., *Les plantations allemandes des Mers du Sud avant 1914*, CRET, Bordeaux, Presses Univ de Bordeaux, 1990.

Phalen, W.J., *How the Telegraph Changed the World,* Jefferson, North Carolina, McFarland & Company Ltd, 2015.

Pietrobelli, C. and A. Sverrisson, (eds.), *Linking Local and Global Economies: The Ties That Bind,* London, Routledge, 2003.

Popkin, S.L., *The Rational Peasant: The Political Economy of Rural Society in Vietnam*, Berkeley, University of California Press, 1979.

Potter, L.M., 'A forest product out of control. Gutta percha in Indonesia and the wider Malay World 1845–1915', in *Paper Landscapes: Explorations in the Environmental History of Indonesia,* eds P. Boomgaard, F. Combijn and D. Henley, Leiden, KITLV Press, 1997, pp. 281–308.

Pringle, R.M. and O.C. Doering, Preface to *Rajah's Servant,* A.B. Ward, Cornell University Southeast Asia Program, Data Paper No. 61, Ithaca, Cornell University Press, 1966, p. vi–vx.

Pringle, R.M., *Rajahs and Rebels: The Ibans of Sarawak under Brooke Rule, 1841–1941*, Ithaca, Cornell University Press, 1970.

Putnam, R.D., *Bowling Alone*, New York, Simon & Schuster, 2000.

Pyenson, L., *Empire of Reason: Exact Sciences in Indonesia, 1840–1940,* Leiden, Brill, 1989.

Ranjit Singh, D.S., 'The structure of the indigenous economy in Sabah in the 1860s and 1870s' in *Historia. Essays in Commemoration of the 25th Anniversary of the Department of History, University of Malaya,* eds M.A. Bakar, A. Kaur and A.Z. Ghazali, Kuala Lumpur, Malaysian Historical Society, 1984, pp. 382–402.

Reece, R.H.W., 'Economic development under the Brookes' in *Development in Sarawak,* eds R.A. Cramb and R.H.W. Reece, Centre of Southeast Asian Studies, Melbourne, Monash University, 1988, pp. 21–34.

Reinhardt, J.M. 'Administrative Policy and Practice in Sarawak: Continuity and Change Under the Brookes', *The Journal of Asian Studies,* Vol. 29, No. 4 (August.) 1970, pp. 851–862.

Reid, A., *A History of Southeast Asia: Critical Crossroads. Blackwell History of the World,* Chichester, Wiley-Blackwell, 2015.

Richards, A.J.N., 'Review [Rajahs and Rebels: The Ibans of Sarawak under Brooke Rule, 1841–1941]', *Modern Asian Studies,* Vol. 8, No. 4, 1974, pp. 555–557.

Richards, A.J.N., *An Iban-English Dictionary,* Oxford, Clarendon Press, 1981.

Roberts, S., *Distant Writing: A History of the Telegraph Companies in Britain between 1838 and 1868,* http://www.atlantic-cable.com/Article/DistantWriting/index.htm. 2007.

Runciman, S., *The White Rajahs: A History of Sarawak from 1841–1946.* Cambridge, Cambridge University Press, 1960.

Sack, P. and D. Clark, (eds), *German New Guinea. The Annual Reports, Edited and Translated,* Canberra, Australian National University Press, 1979.

Sack, P. and D. Clark, (eds), *German New Guinea. The Draft Annual Report 1913–14, Edited and Translated.* Canberra, Australian National University Press, 1980.

Sandin, B., 'Sources of Iban traditional history', *Sarawak Museum Journal,* Kuching, No. 67, Special Monograph No. 7, 1994.

Sather, C. and Fox, J.J. (eds), *Origins, Ancestry and Alliance: Explorations in Austronesian Ethnography,* Canberra, ANU E Press, 2006.

Sather, C., 'Introduction' to *Sources of Iban Traditional History,* B. Sandin, *Sarawak Museum Journal,* Kuching, No. 67, Special Monograph No. 7, 1994, pp. 1–78.

Sather, C., '"All Threads are White" Iban egalitarianism reconsidered' in *Origins, Ancestry and Alliance: Explorations in Austronesian Ethnography,* eds C. Sather and J.J. Fox, Canberra, ANU E Press, 2006, pp. 73–112.

Saunders, G., 'Epilogue. In the eye of the beholder – development or exploitation? Changing perceptions of the Borneo environment' in *Histories of the Borneo Environment,* ed. R.L. Wadley, Leiden, KITLV Press, 2005, pp. 271–293.

Schutze, C., *Kabel, unser Leben: Narrative Identitätskonstruktionen eines Unternehmens und seiner Belegschaft,* Oldenburg, Isensee Verlag, PhD Dissertation, 2012.

Scott, J.C., *The Moral Economy of the Peasant, Rebellion and Subsistence in Southeast Asia,* New Haven, Yale University Press, 1976.

Sellato, B., *Preface*, to Adolphe Combanaire, *Au pays des Coupeurs de Têtes. A travers Bornéo*, Singapore, Editions Pagodes, 2003.

Sellato, B., 'Forests for food, forests for trade – between sustainability and extractivism; The economic pragmatism of traditional peoples and the trade history of northern East Kalimantan' in *Histories of the Borneo Environment*, ed. R.L. Wadley, Leiden, KITLV Press, 2005, pp. 61–86.

Seth, S., 'Putting knowledge in its place: science, colonialism, and the postcolonial', *Postcolonial Studies*, Vol. 12, No. 4, 2009, pp. 373–388.

Siemens, G., *History of the House of Siemens Vol. I: The Era of Free Enterprise*, (Trans. A.F. Rodger), Freiburg/Munich, Karl Alber, 1957.

Smith, C. and M.N. Wise, *Energy and Empire. A Bibliographical Study of Lord Kelvin*, Cambridge, Cambridge University Press, 1989.

Smythies, B.E., 'History of forestry in Sarawak', *The Malayan Forester*, Vol. 26, No. 4, (October), 1963, pp. 232–253.

Somers Heidhues, M.F., *Golddiggers, Farmers, and Traders in the "Chinese Districts" of West Kalimantan*, Indonesia SEAP Publications, Ithaca, Cornell University, 2003.

Standage, T., *The Victorian Internet*, London, Weidenfeld Nicholson, 1998.

Stanley, B.E., '"Going Global": Wannabe world cities in the Middle East' in *Emerging Issues in the 21st Century World-system: New Theoretical Directions for the 21st Century World-System*, ed. W.A. Dunaway, Westport, Praeger, 2003, pp. 151–170.

Stern, S.J. 'Capitalism and the world-system in the perspective of Latin America and the Caribbean', *American Historical Review*, Vol. 93, No. 4, October 1988, pp. 829–872.

Storey, G., *Reuters Century 1851–1951*, London, Max Parrish, 1951.

Sutlive, V. and J. Sutlive (eds), *The Encyclopedia of Iban Studies: Iban History, Society and Culture*, 4 Vols, Sarawak, Tun Jugah Foundation with Borneo Research Council, 2001.

Sutlive, V.H., *The Iban of Sarawak*, Arlington Heights, IL AHM Publishing Corporation, 1978.

Sutlive, V.H., *Tun Jugah of Sarawak: Colonialism and Iban Response*, Kuala Lumpur, Sarawak Literary Society, 1992.

Sutlive, V.H., 'The Iban in historical perspective', *Sarawak Museum Journal*, Vol. 40, No. 61, 1989, pp. 33–43.

Sverrisson, A., 'Local and global commodity chains' in *Linking Local and Global Economies: The Ties That Bind*, eds C. Pietrobelli and A. Sverrisson, London, Routledge, 2003, pp. 17–35.

Swettenham, F., *British Malaya An Account of the Origin and Progress of British Influence in Malaya*, 2nd edn. London, George Allen and Unwin, 1955.

Tagliacozzo, E., 'Onto the coasts and into the forests; Ramifications of the China trade on the ecological history of northwest Borneo, 900–1900 CE', in *Histories of the Borneo Environment*, ed. R.L. Wadley, pp. 25–60, Leiden, KITLV Press, 2005.

Talib, N.S., *Administrators and Their Service. The Sarawak Administrative Service under the Brooke Rajahs and British Colonial Rule*, Selangor Darul Ehsan, Oxford University Press, 1999.

Tarling, N., *Britain, the Brookes and Brunei*, Kuala Lumpur, Oxford University Press, 1971.

Tarling, N., *The Cambridge History of Southeast Asia, Vol. 2 Part 1, From c. 1800 to the 1930s*, Cambridge, Cambridge University Press, 1999.

Tarling, N., *Imperialism in Southeast Asia: 'A fleeting, passing phase'*, London, Routledge, 2001.

Taylor J.G., *Global Indonesia*, London, Routledge, 2013.

Thambiah, S., 'The emergence of the ethnic category Bhuket: Diversity and the collective hunter-gatherer identity in Borneo' in *Beyond the Green Myth: Borneo's Hunter-gatherers in the Twenty-First Century*, eds. P.G. Sercombe and B. Sellato, Copenhagen, Nias Press, 2007, pp. 91–109.

The Economist, 'Economic Focus, Another day, Another $1.08. Even those in Absolute Poverty have Choices', April. 28, 2007, p. 98.

Thiessen, T., *Borneo: Sabah, Sarawak, Brunei (Bradt Travel Guide)*, Chalfont St Peter, Bradt Travel Guides, 2016.

Topik, S. and M. Samper, 'The Latin American commodity chain' in *From Silver to Cocaine. Latin American Commodity Chains and the Building of the World Economy, 1500–2000*, eds S. Topik, C. Marichal and Z. Frank, Durham, Duke University Press, 2006, pp. 118–146.

Topik, S., C. Marichal and Z. Frank (eds), *From Silver to Cocaine. Latin American Commodity Chains and the Building of the World Economy, 1500–2000*, Durham, Duke University Press, 2006.

Trocki, C.A., *Opium, Empire and the Global Political Economy: A Study of the Asian Opium Trade, 1750–1950*, London, Routledge, 1999.

Trocki, C.A., 'Opium and the beginnings of Chinese capitalism in Southeast Asia', *Journal of Southeast Asian Studies*, 33, (2), 2002, pp. 297–314.

Tucker, K.H., *Anthony Giddens and Modern Social Theory*, London, Sage Publications, 1998.

Turnbull, C.M., 'The Johore gambier and pepper trade in the mid-nineteenth century', *Journal of the South Seas Society*, Vol. 15, No. 1, 1959, pp. 43–55.

Turnbull, C.M., *The Straits Settlements, 1826–67: Indian Presidency to Crown Colony*, London, Athlone Press, 1972.

Turnbull, C.M., *A History of Singapore 1819–1975*, Kuala Lumpur, Oxford University Press, 1977.

Viswanathan, M., S. Sridharan and R. Ritchie, 'Understanding consumption and entrepreneurship in subsistence marketplaces', *Journal of Business Research*, 63, 2010, pp. 570–581.

Wadley, R.L., 'Environmental Challenges in South-East Asia. (Book Reviews)', *Borneo Research Bulletin* (Annual), Vol. 30, 1999, pp. 168–171.

Wadley, R.L., 'The history of displacement and forced settlement in West Kalimantan, Indonesia', in *Conservation and Mobile Indigenous Peoples: Displacement, Forced Settlement, and Sustainable Developments*, eds D. Chatty and M. Colchester, New York, Berghahn Books, 2002, pp. 313–328.

Wadley, R.L. (ed.), *Histories of the Borneo Environment,* Leiden, KITLV Press, 2005.

Wadley, R.L., 'Boundaries, territory, and resource access in West Kalimantan, Indonesia, 1800–2000' in R.L. Wadley, ed., *Histories of the Borneo Environment,* pp. 137–158, Leiden, KITLV Press, 2005.

Wadley, R.L., 'Slashed and burned: war, environment, and resource insecurity in West Borneo during the late nineteenth and early twentieth centuries', *Journal of the Royal Anthropological Institute* (New Series), Vol. 13, 2007, pp. 109–128.

Walker, J.H., *Power and Prowess. The Origins of Brooke Kingship in Sarawak*, Hawai'i, Allen & Unwin, University of Hawai'i Press, 2002.

Walker, J.H., 'Hikayat Panglima Nikosa and the Sarawak Gazette: Transforming texts in nineteenth century Sarawak', *Modern Asian Studies*, Vol. 39 No. 2, 2005, pp. 427–460.

Wallerstein, I., *The Modern World-System I. Capitalist Agriculture and the Origins of the European World-Economy in the Sixteenth Century*, New York, Academic Press, 1974.

Wallerstein, I., *The Modern World-System III. The Second Era of Great Expansion of the Capitalist World Economy, 1730–1840s*, San Diego, Academic Press, 1989.

Wallerstein, I., 'The actor and world-systems analysis: comments on Blau and Wierviorka', *Contemporary Sociology*, Vol. 34, No. 1 (Jan.), 2005, pp. 9–10.

Wallerstein, I., *The Modern World-System I: Capitalist Agriculture and the Origins of the European World-Economy in the sixteenth Century, with a New Prologue*, Berkeley, University of California, 2011.

Ward, A.B., *Rajah's Servant*, Cornell University Southeast Asia Program Data Paper No. 61, Ithaca, Cornell University Press, 1966.

Warner, K., *Shifting Cultivators. Local Technical Knowledge and Natural Resource Management in the Humid Tropics*, Rome, Food and Agriculture Organization of the United Nations, 1991.

Warren, J.F., 'Joseph Conrad's fiction as Southeast Asian History: Trade and politics in East Borneo in the late nineteenth century', *Brunei Museum Journal*, 1977, pp. 21–34.

Warren, J.F., *Ah Ku and Karayuki-san: Prostitution in Singapore 1870–1940*, Singapore, Singapore University Press, 2003.

Warren, J.F., *The Sulu Zone 1768–1898: The Dynamics of External Trade, Slavery and Ethnicity in the Transformation of a Southeast Asian Maritime State*, Singapore, NUS, 2007.

Wenzlhuemer, R, *Connecting the Nineteenth-Century World: The Telegraph and Globalization*, Cambridge, Cambridge University Press, 2012.

Williams, L., 'Lactiferous plants of economic importance v. Resources of gutta-percha-*Palaquium* species (Sapotaceae)', *Economic Botany*, Vol. 18, No. 1 (Jan.-Mar.), 1964, pp. 5–26.

Williamson, J.G., 'Land, labor, and globalization in the third world, 1870–1914', *The Journal of Economic History*, Vol. 62, No. 1 (Mar.), 2002, pp. 55–85.

Winkler, J.R., *Nexus. Strategic Communications and American Security in World War I*, Cambridge, Mass., Harvard University Press, 2008.

Winseck, D.R. and R.M. Pike, *Communication and Empire: Media, Markets and Globalization 1860–1930*, Durham, Duke University Press, 2007.

Winston, B., *Media, Technology and Society: A History: From the Telegraph to the Internet*, London, Routledge, 2000.

Wong L.K., 'The Trade of Singapore 1819–1869.', *Journal of the Malayan Branch Royal Asiatic Society*, Vol. 33, No. 4, 1960, pp. 5–308.

Wong L.K., 'Singapore: Its growth as an entrepot port, 1819–1941' *Journal of Southeast Asian Studies*, 9, (1), 1978, pp. 50–84.

Woodman, H.D., *King Cotton and His Retainers: Financing and Marketing the Cotton Crop of the South, 1800–1925*, Washington, Beard Books, 2000.

Woolcock, M. and D. Narayan, 'Social capital: Implications for development theory, research, and policy', *The World Bank Research Observer*, (August.), Vol. 15, No. 2, 2000, pp. 225–249.

Woshner, M., *India-rubber and Gutta Percha in the Civil War Era*, Alexandria, O'Donnell Publications, 1999.

Wright, L.R., *The Origins of British Borneo*, Hong Kong, Hong Kong University Press, 1970.

Zhou, Y., *Historicizing Online Politics: Telegraphy, the Internet and Political Participation in China*, Stanford, Stanford University Press, 2006.

Index

www.ingramcontent.com/pod-product-compliance
Lightning Source LLC
LaVergne TN
LVHW010640300926
849161LV00005B/997
* 9 7 8 9 0 0 4 3 4 4 3 3 4 *